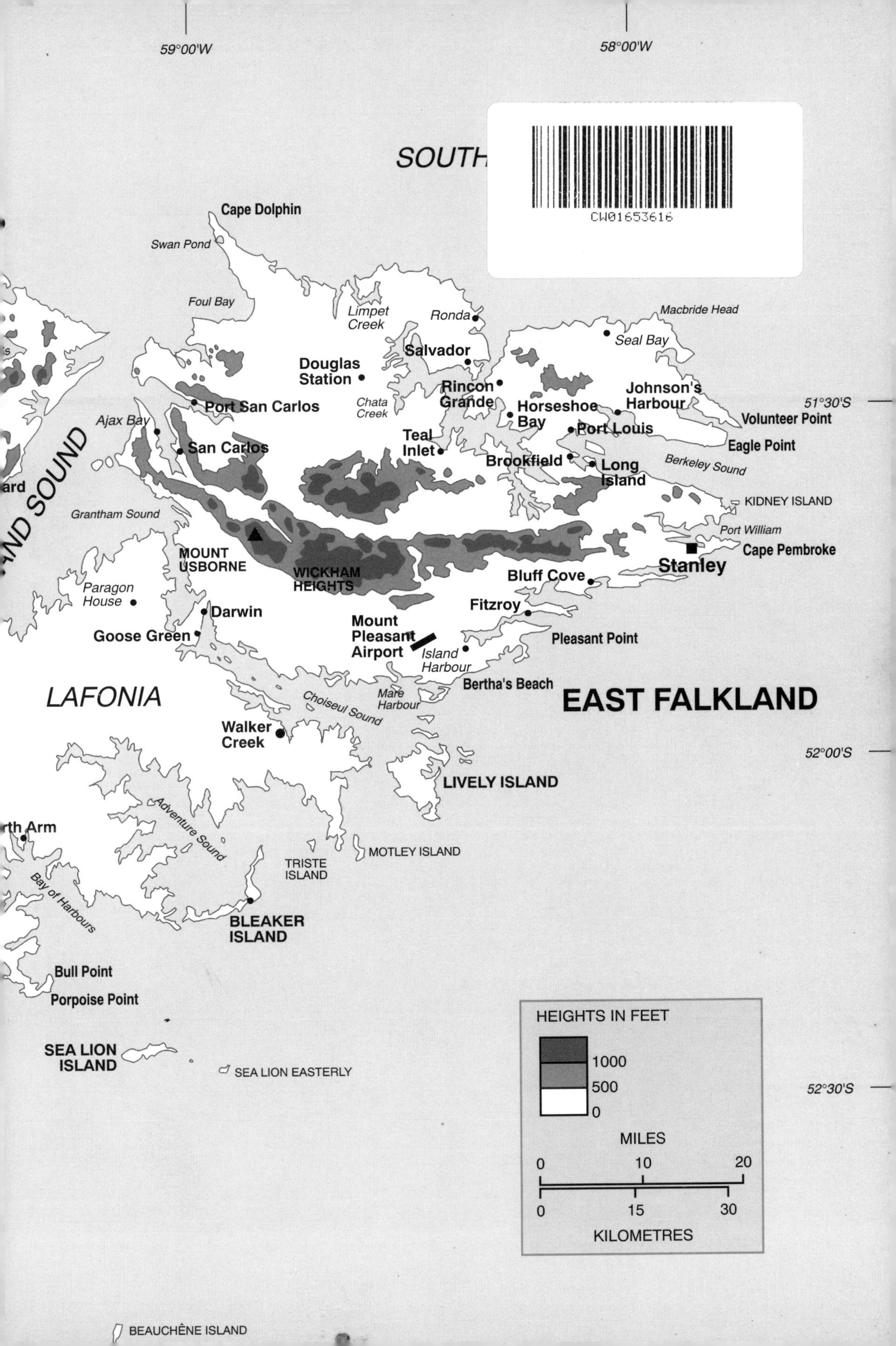

59°00'W
58°00'W
SOUTH
Cape Dolphin
Swan Pond
Foul Bay
Limpet Creek
Ronda
Macbride Head
Seal Bay
Salvador
Douglas Station
Rincon Grande
Johnson's Harbour
51°30'S
Port San Carlos
Chata Creek
Horseshoe Bay
Volunteer Point
Ajax Bay
Port Louis
Eagle Point
San Carlos
Teal Inlet
Brookfield
Long Island
Berkeley Sound
ard
ND SOUND
KIDNEY ISLAND
Grantham Sound
Port William
Cape Pembroke
MOUNT USBORNE
Stanley
WICKHAM HEIGHTS
Bluff Cove
Paragon House
Fitzroy
Darwin
Mount Pleasant Airport
Goose Green
Pleasant Point
Island Harbour
Bertha's Beach
LAFONIA
Mare Harbour
Choiseul Sound
EAST FALKLAND
Walker Creek
52°00'S
LIVELY ISLAND
Adventure Sound
rth Arm
MOTLEY ISLAND
TRISTE ISLAND
Bay of Harbours
BLEAKER ISLAND
Bull Point
Porpoise Point
HEIGHTS IN FEET
1000
500
0
MILES
0
10
20
0
15
30
KILOMETRES
SEA LION ISLAND
SEA LION EASTERLY
52°30'S
BEAUCHÊNE ISLAND

Atlas of Breeding Birds of the Falkland Islands

Robin W and Anne Woods

Illustrated by Geoffrey McMullan

Anthony Nelson

First published in 1997 by
Anthony Nelson, P.O. Box 9, Oswestry, Shropshire SY11 1BY, England

ISBN 0904614 60 3

Published in association with Falklands Conservation with support from WWF-UK

Designed by
Quetzal Communications, Basingstoke, Hampshire

Printed in Great Britain by
Redwood Books, Trowbridge, Wiltshire

Contents

Preface

In 1983 I visited the Falkland Islands to study the songbirds in tussac grassland. I thought that a list of the breeding birds would be useful when talking with local people about their birds and made up a single page record sheet while travelling southwards from Ascension Island on SS *Uganda*. I used the list with Roddy Napier for West Point, Dunbar and Low Islands and also collected records for Kidney Island, Grand and Steeple Jason, Carcass and Outer Northwest Island. Later I realised that this record sheet could be used as the basis for a survey leading to an Atlas of Falkland breeding birds.

I suggested the Atlas project to Tom Davies, then Chairman of the Falkland Islands Trust in Stanley and he agreed that it would be a good way to collect the knowledge of local people and to strengthen the activities of the Trust. He discussed the project with the Trust committee and received their support. In England, I broached the idea of a Breeding Birds Survey with Simon Lyster, then Secretary of the Falkland Islands Foundation. Their committee also supported the project and it was agreed that the Foundation would pay the costs of printing and distributing the forms. The survey record form was redrafted, checked by Tom Davies and Andy Douse in Stanley and by Kevin Standring of the Royal Society for the Protection of Birds and finally appeared as an A4 sheet of thin card, with the record form on one side and information on completion and a return address on the reverse. The Trust committee suggested that the form should be creased so that it could be folded, sealed with tape and sent free to the Trust in Stanley, who would pay postage. Survey forms reached the Falklands in November 1984 and the first returns were received by February 1985.

The information presented in the maps and tables of this Atlas is based on records from about 160 observers who sent reports equivalent to over 550 Breeding Birds Survey forms over a period of ten breeding seasons between 1983/84 and 1992/93. About 80 observers (51%) were Falkland Islanders or contract workers, 47 (30%) were military personnel and 29 (19%) were visitors to the Falklands.

This is the first countrywide biological survey in the Falklands for which the local knowledge of residents was sought, valued and considered to be essential. The results are substantial, considering that this country of 12,200 square kilometres (4,700 square miles), with land in 255 10-km grid squares, has a small permanent human population of only about 2,100. Of these, over 1,500 people live in the only town of Stanley and the others live on about 90 farms and settlements in the camp*. Most of the squares were visited but there were difficulties in travelling around East and West Falkland and to the 700 or so uninhabited islands in the archipelago. There is also a military exclusion zone in the Wickham Heights.

The records for the 234 10-km squares surveyed are valuable in many ways. They show that some species are found almost everywhere, while others have a restricted distribution and their habitats may need special protection. The settlements are shown to have an important effect on distribution of some species but not on others, and records of the common birds show the general uniformity of bird habitats in the Falklands when considered on the basis of 10-km squares. Analyses of the distribution of records in relation to the topographical features have revealed the importance of the coasts and surrounding seas to more than half of the breeding species and shown that some species are dependent on the wetland areas for survival.

This Atlas provides basic information on the distribution, populations and behaviour of Falkland breeding birds and is a good starting point for further studies. It can be seen that information on some of the breeding petrels is minimal and knowledge of the breeding behaviour for some species, the rarer waterfowl for example, is sparse. The Atlas shows how valuable all observations can be and I shall be very interested to receive further records of presence or breeding, particularly for those bird species which appear to be rare, and for the 10-km grid squares which could not be fully surveyed. I also welcome comments about the accuracy of the information in the maps or text.

I hope this Atlas will be of interest to all Falkland Islanders and visitors to the Islands, that it will show the importance of protecting the marine environment from pollution, draw attention to some uncommon species and their habitats which could be vulnerable, and that it will be useful when plans for development are being considered in the Falklands.

Robin W Woods, 68 Aller Park Road, Newton Abbot, Devon, TQ12 4NQ, England.

Footnote *Camp is derived from the Spanish 'campo' meaning countryside, and is the local name for all land outside Stanley.

Acknowledgements

I am grateful for the support of the Falkland Islands Trust and Falkland Islands Foundation, now merged as Falklands Conservation. I thank Jane Cameron, Government Archivist, for her help with additional historical material, for access to records made by JE Hamilton and for putting me on the track of AF Cobb's manuscript notebook. Through the kindness of Joan and Terry Spruce, I was thrilled to read and tape Cobb's valuable records, made from 1906 to 1923 when he lived on Lively and Bleaker Islands, and I should particularly like to thank Harold and Grace Bennett for the loan of unpublished papers and access to some very interesting records of birds made by his father, AG Bennett.

In common with most other Atlas projects, almost all of the work has been voluntary and I am particularly grateful to all those who sent records during the survey. Without their interest and effort this Atlas would not exist. However, any errors in content or in presentation of the material are my responsibility. Contributions varied from single records of interesting or uncommon bird species nesting near a settlement to several forms with many species recorded. Several people drew maps showing locations of colonies while others wrote letters which included some historical records. Survey forms did not always show the observer's name in full and some could not be checked. My apologies to anyone whose name has been omitted from the list.

Contributors to the Breeding Birds Survey 1983-93

Mike Alazia
Brian Aldridge
Corporal Ashlin (REME)
C Bailey (HMS *Achilles*)
J Bennison
Arina Berntsen
Leon Berntsen
Major Nicholas W Beswick
Yeoman Brian Bewsher
Tony Blake
Junior Technician Blake
Graham Bound
WRP (Bill) Bourne
Petty Officer Simon Bray
Diana Bradley
Lisa Brothers
Flight Lieutenant E Buckley
Corporal Butt
Andrew Carter
Tony Chater
Gerry S Clark
Andrew Clarke
Squadron Leader MCA Clarke
Lieutenant Colonel CN Clayden
Jennifer Clement
Squadron Leader John A Cliffe
Marie Clifton
Lieutenant Colonel M Conroy
Brigadier Robert JS Corbett
Sergeant Tim P Cowley
Squadron Leader Craggs
John P Croxall
Radio Officer WF (Bill) Curtis
Tom H Davies
Reg Davis
Roy Dennis
Tui de Roy
Danny Donnelly
Andy Douse
Lieutenant Colonel E Duggan
Captain Tim R Duggleby
Paul Dukes
Leading Weapons Engineer Mechanic Edwards (HMS *Broadsword*)
Norma Edwards
Cyril Ellis
Val Ellis
Raymond Evans
Faith Felton
P Felton
Sonia Felton
Sergeant Steve Felton
Tony Felton
Ethel Ferguson
David Fisher
Petty Officer Paul K Fitzpatrick
Staff Sergeant Forrest
French photographers - Marion C, Marion L & Jacob L
Neil Goodwin
Kurt Grant
Phil Gregory
Dann Hale
Lieutenant Colonel Helen B Hannah
Brook Hardcastle
Eileen Hardcastle
Gavin Hardcastle
Staff Sergeant NP Harris
Lieutenant Philip Heath
Nicholas Hill
Richard Hill
Janet & Rodney Hutching
Philip Hutton
Lance Corporal Jackson
Captain JW Johnsen
Mark Jones
Amy E Jonson
David Lampard
Josie Larsen
Kate Layman
David Lee
Major IME Lloyd
Major JY McCallum
Robin McGill
Captain Colin MacGregor (HMS *Endurance*)
Clara & Bill MacKay
Ian J McLeod
Staff Sergeant Geoffrey McMullan
Mandy McRae
Marlane Marsh
Rodney Martins
Petty Officer Ken Maule
Bruce R May
Michael (Port Stephens school)
Dennis Middleton
Phil Middleton & 1st Falkland Scouts
Simon R Miller
Carol Miller
Michael J Morrison
Harro H Müller
Lily & Roddy Napier
North Arm schoolchildren
Flight Sergeant J Nugent
Donna O'Daniel
Captain Carol O'Nians
Flight Sergeant David Osborn
Kathy Patric
Graham Paul
Personnel from HMS *Penelope*
Albert Phillips
Robin Pitaluga
Lisa Pole-Evans
Sally Poncet
Port Stephens schoolchildren
Stan da Prato
Chief Petty Officer Pringle (HMS *Bristol*)
Keith Reddick
Mike Reeves
A John Reid
Mike Riddy
Ann Robertson
Alison Robson
Major Rusby
Flight Lieutenant Andrew Shaw
Captain Sherring
Sergeant K Shield
Neil Simpson and Cumbria & Lancashire Falklands Expedition
CFN Smith (REME)
George Smith
Roger K Spink
Ian J Strange
Michael Strickland
Major Simon WL Strickland
Kate Thompson
Warrant Officer J Thwaite
Major Towersley
Corporal P Triggs
The late Bobby Tulloch
Christophe Verheyden
Neil Watson
Steve Whitley
J Roger Wilson
Commander Michael Winter
Dick Wolsey
Shane Wolsey.

Thanks are due to Dr John P Croxall, chairman of Falklands Conservation (UK) for his support and critical reading of text and to Geoffrey McMullan for offering to illustrate the Atlas at just the right time. I am very grateful to Tony Nelson for his encouragement and enthusiasm for publishing books about the Falklands.

Introduction

Before 1983, very little information on the distribution of Falkland breeding birds had been recorded. In 1932/33, AG Bennett as Acting Government Naturalist, had organised a survey of penguin and albatross colonies and reported to the Governor (Bennett 1933). In the early 1960s, many of the highly visible colonies of several large seabirds, including Gentoo and Rockhopper Penguins, Black-browed Albatross, Southern Giant Petrel, King Shag, Falkland Skua and Dolphin Gull were partly mapped by me from information supplied by Roddy Napier. Steve Whitley recorded many seabird colonies from 1976 to 1984 and marked them on a copy of the Directorate of Overseas Surveys 1:250,000 map (DOS 653, 1965). Croxall *et al.* (1984) summarised available records and commented: 'There is an urgent need for a comprehensive review of Falkland Islands requirements for the conservation of wildlife, of which seabirds, although important and spectacular, are only a part'. The smaller or less conspicuous birds had received hardly any attention and their distribution in the Falkland archipelago could only be discussed in very general terms.

Development of the Atlas project

The Breeding Birds Survey record form (Figures 1a and 1b), was based on the form used for *The Atlas of Breeding Birds in Britain and Ireland* (Sharrock 1976). Fifty-nine possible breeding species were listed and a few rare species, unlikely to be recorded, were omitted. This decision was only partly correct because several records of breeding Barn Owls were received and six other species bred or possibly bred during the survey. The names for species were those used in *Falkland Islands Birds* (Woods 1982) with local names for 45 species and a code number to help in sorting the records.

Three levels of breeding evidence were requested; possible, probable and confirmed. From 1986/87 onwards, the category 'possible' was broadened to 'present in square', which allowed the inclusion of birds seen offshore in coastal squares. Observers were asked to tick column 1 and use the right code letters in columns 2 or 3 if they had found evidence of breeding. A fourth column (N) was added so that some information on the number of each species could be collected. Code letters A (1-10), B (11-100), C (100-1000) or D (1000+) were given to be used for estimates of pairs seen, if they could not be counted. These ranges were adapted from those used by the Devon Bird Watching and Preservation Society between 1977 and 1985, for their *Tetrad Atlas of the Breeding Birds of Devon* (Sitters 1988). Observers were also asked for information on presence or absence and estimated numbers of the introduced mammals (cats, rats, mice, sheep, cattle, horses). A paragraph on the back of the form explained that this information was requested because these animals could do considerable damage to native birds or their habitats. Space was allowed for adding the names of other mammals, such as rabbits, hares, silver foxes and guanacos if they were present.

Publicity, data collection and processing the information

In 1984/85, Tom Davies publicised the survey through personal contacts and talks from the Falkland Islands Broadcasting Station. He collected some records by visiting settlements and completing the forms with residents. Survey forms were also distributed to military personnel through successive Conservation Officers of British Forces Falkland Islands (BFFI). In the first full season, 19 observers completed 45 survey forms and, as the forms were returned, it became clear that the description 'island or farm' for a locality was unsatisfactory because data could not be placed sufficiently accurately on maps. Sharrock (1976) had used a recording system based on 10-km grid squares to give a broad picture of bird distribution. Therefore, we decided to use 10-km squares for recording and mapping, but unfortunately, the 29 sheets of the 1:50,000 maps (DOS 453, 1961-62) were not overprinted with this grid though coordinates appeared in the margins. A sketch map of the Falklands was drawn and the Universal Transverse Mercator Grid was added, from the grid on the 1:250,000 two-sheet map.

The survey forms were checked on receipt and almost all records were accepted, because most Falkland birds can be identified easily. Observers appeared to have some difficulty with a few species, particularly the two smallest landbirds, the Grass Wren and Cobb's Wren. Both occur on some islands with good stands of tussac grass, but Cobb's Wren is very dependent on tussac habitat whereas the Grass Wren is widely distributed on the main islands. Two records of Cobb's Wren in mainland squares were not accepted as the observers did not record the presence of the Grass Wren. An identification leaflet, 'Which Wren is that ?', was made available through the FI Trust. There was also some difficulty over the identification of the Crested Caracara (Carancho) and Striated Caracara (Johnny Rook) and differences between

Name of Observer: ....
....

Place: ....
Island or farm

Year: ....

Period of Observation: ....

	Code No.	1	2	3	N
Common name of species (Local name in brackets)					
King Penguin	01				
Gentoo Penguin	02				
Rockhopper Penguin (Rocky)	03				
Macaroni Penguin	04				
Magellanic Penguin (Jackass)	05				
White-tufted Grebe (Black/Golden Grebe)	06				
Silvery Grebe (White Grebe)	07				
Black-browed Albatross (Mollymawk)	08				
Giant Petrel (Stinker)	09				
Slender-billed Prion (Firebird)	10				
White-chinned Petrel (Cobbler)	11				
Great Shearwater	12				
Sooty Shearwater (Shyer bird)	13				
Wilson's Storm- Petrel	14				
Grey-backed Storm-Petrel	15				
Falkland Diving Petrel (Diver)	16				
Rock Cormorant (Black Shag)	17				
King Cormorant (King Shag)	18				
Black-crowned Night Heron (Quark)	19				
Black-necked Swan	20				
Ashy-headed Goose	21				
Ruddy-headed Goose (Brent, Brant)	22				
Upland Goose	23				
Kelp Goose	24				
Domestic Goose – living wild	25				
Patagonian Crested Duck (Grey Duck)	26				
Flightless Steamer Duck (Logger)	27				
Flying Steamer Duck (Canvasback)	28				
Speckled Teal (Teal)	29				
Chiloe Wigeon	30				
Yellow-billed Pintail (Grey Teal)	31				
Silver Teal (Pampa Teal)	32				
Cinnamon Teal (Red Teal)	33				
Turkey Vulture (Turkey Buzzard)	34				
Red-backed Hawk (Hawk, Blue Hawk)	35				
Striated Caracara (Johnny Rook)	36				
Crested Caracara (Carancho)	37				
Peregrine Falcon (Sparrow Hawk)	38				
Magellanic Oystercatcher (Pied Oystercatcher, Curlew)	39				
Black Oystercatcher (Black Curlew)	40				
Two-banded Plover (Plover)	41				
Rufous-chested Dotterel (Dotterel)	42				
Common Snipe (Snipe)	43				
Great Skua (Sea-hen)	44				
Dolphin Gull	45				
Kelp Gull (Big Gull, White Gull)	46				
Brown-hooded Gull (Pink-breasted Gull)	47				
South American Tern (Split-tail)	48				
Short-eared Owl (Owl)	49				
Tussock Bird (Black Bird)	50				
Dark-faced Ground-tyrant (Blue Bird, News Bird)	51				
Short-billed Marsh Wren (Grass Wren, Tomtit)	52				
Southern House Wren (Rock Wren)	53				
Falkland Thrush (Thrush)	54				
Falkland Pipit (Skylark, Lark)	55				
Long-tailed Meadowlark (Robin, Starling)	56				
Black-throated Finch (Sparrow)	57				
Black-chinned Siskin (Siskin)	58				
House Sparrow (English Sparrow)	59				

GRID SQUARE

COLUMN 1: POSSIBLE BREEDING

Bird in possible nesting area or present in square.

COLUMN 2: PROBABLE BREEDING

S Singing male
T Bird holding territory
D Display/agitated behaviour
N Visiting probable nest site
B Nest-building

COLUMN 3: CONFIRMED BREEDING

DD Distraction display
UN Used nest found
FL Recently fledged young
FD Adult carrying droppings
NE Nest and eggs or bird sitting tight
NY Nest and young or downy young
FY Adult with food for young

NOTE:Column 3 entries are more important than numbers in Column 'N'

COLUMN N: HOW MANY BIRDS?

Enter estimated number of PAIRS if known OR use the following letters

A 1 – 10 pairs
B 11 – 100 pairs
C 100 – 1000 pairs
D 1000+ pairs

RECORD PRESENCE/ABSENCE AND ESTIMATED NUMBER OF:–

Cats
Rats
Mice
Sheep
Cattle
Horses
Rabbits
Hares

Figure 1a

NO STAMP NEEDED

THE SECRETARY
FALKLAND ISLANDS TRUST
PORT STANLEY
FALKLAND ISLANDS

EXPLANATORY NOTES

Notes on the Survey Form
The names of 59 different birds are printed, and there are extra spaces for any others that you find. Since this is a breeding birds survey, the names of common non-breeding birds such as Sheathbill and White-rumped Sandpiper have not been included.

The names used are those accepted in the literature. Local names are in brackets.

How to use the Form
Year: Please write the year for the breeding season when you use the form – e.g. 1984-5.

Period of observation: Please put the length of time over which you have made the observation – e.g. one day, two months, five years.

Collecting the records
If you see a bird in a place where it could nest, put a ✓ in Column 1 opposite its name. If you see evidence of probable breeding such as a bird singing regularly in one area or nest building, write S, T, D, N or B, as appropriate, in Column 2. A nest with eggs or young, an adult carrying droppings or small non-flying young birds means confirmed breeding. In this case, write one of the pairs of letters, NE, UN etc. in Column 3.If you are able to count or estimate numbers of breeding pairs, either write the number in Column N or use the code letters on the form to show the approximate number.

Introduced Species
It would be useful to have some idea of presence and numbers of non-native animals, particularly cats, rats and mice which can sometimes do a great deal of damage to native birds. If there are none in the area you survey, please write NONE by the animal's name. Add the names of other introduced species such as hares, rabbits, foxes, guanacos if you know they are present.

Why do this Survey?
Very little of people's knowledge of Falkland birds has been recorded. By filling in this form, no matter how few entries are on it, you will be adding your bit to the picture we are trying to build up. Please send this form to the Secretary, Falkland Islands Trust, Port Stanley.

THANK YOU FOR YOUR HELP

FALKLAND ISLANDS TRUST

BREEDING BIRDS SURVEY

Take this form with you when you are next in Camp and help add to our knowledge of Falkland Islands breeding birds. Complete the form and return it to the Secretary, Falkland Islands Trust by folding it so that the address is showing and stick down with a fragment of sellotape, stamp edging or similar.

Postage will be paid by the Trust.

Printed with the aid of a grant from the Falkland Islands Foundation.

Figure 1b

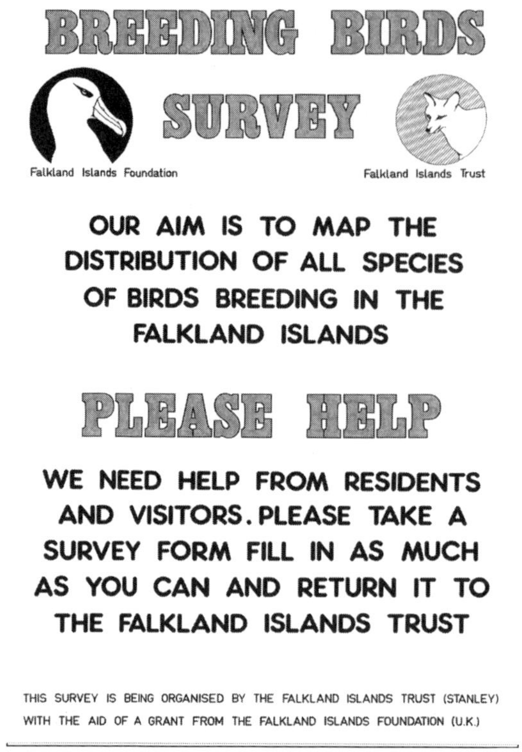

Figure 2

these species were described in the FI Foundation *Newsletter* (Woods 1985b). A list of squares grouped according to settlements was produced and a poster advertising the Breeding Birds Survey was printed by the FI Foundation (Figure 2). This new publicity material was used from early in the 1985/86 season. Posters were displayed in the Post Office and at Stanley Airport with adjacent display pockets holding supplies of survey forms. Results for the second full season were good, with 51 observers returning records for 91 squares.

Tom Davies retired early in 1986, and Shane Wolsey coordinated the survey and published a Newsletter for the FI Trust, in which he gave progress reports. In 1986/87, 34 observers returned survey forms for 164 squares. Analysis of the reports showed that settlement squares and the easily accessible places had been well covered, often several times, while the more remote areas were barely touched. Shane and I discussed the situation and decided that a full time ornithologist was required to improve coverage. Shane obtained a donation to FI Trust from a local company and the FI Foundation appointed Mike Riddy and paid his expenses for the 1987/88 season. Mike covered 60 squares on East Falkland and spent an average of 1·3 days in each square. He found up to 34 species in each square and confirmed breeding for 55% of species recorded. Altogether seven observers returned records for 87 squares in that season. By early 1988, 326 survey forms had been returned for 157 of the 255 10-km squares containing land. There was sufficient information for East Falkland in 1988 to detect differences in distribution between closely related species, such as the Upland Goose and Kelp Goose or the Grass Wren and Cobb's Wren and to relate some variations to differences between parts of this island in topography or vegetation.

Using a master record sheet system, collating material and adding to it annually was becoming complex. Therefore, a powerful relational database program (Ashton-Tate, 1st Software dBASEII) was obtained for the Amstrad PCW8512, and a database was designed for the survey. Records were entered in a way that allowed for more than one count or estimate for each species in any square. From mid-1988, lists of all records for each species in all squares could be obtained. Records from the survey forms had been added to species maps by hand, using small, medium and large red dots for the three classes of records and the species print-outs were checked against these maps. Total species present, probably breeding and confirmed breeding could also be listed for each square and league tables of species were published in the FI Foundation *Newsletter* (Woods 1989, 1990). A summary showing the total number of species present in each square was used to compile lists of squares still requiring field work. Publicity material prepared for the 1988/89 season included maps and lists showing the unsurveyed squares and those in which one to five and six to ten species had been recorded. This was sent to the FI Trust Secretary in Stanley and to some birdwatchers who made contact before they went to the Falklands.

One of these was Donna O'Daniel, a graduate ornithologist from Texas, who funded her own visit to West Falkland in 1987/88 to help with the survey. She recorded an average of over 15 species present from 15 squares around Fox Bay and on Weddell Island, but coverage of West Falkland was still sparse. The author discussed the progress of the survey with Roger Wilson, then Secretary of FI Foundation. It was agreed that Mike Riddy should be asked to survey parts of West Falkland and he flew out in mid-November, this time with a small salary.

By mid-February 1989 he had surveyed 58 squares, recording an average of 22 species present and an average of 12 confirmed breeding in each square. In two seasons, Mike covered much of East and West Falkland with the help of a motorcycle, tent and radio and hospitality from isolated farms, though lack of time meant that he was unable to visit offshore islands. His fieldwork has made a huge difference to the value of the survey. Since 1989, further records have been obtained for several seasons through correspondence with local people, civilian and military personnel based at Mount Pleasant Airport, visiting scientists, tour leaders and tourists.

Falkland birds in the future

Following the invasion by the armed forces of Argentina in April 1982 and the subsequent conflict ending in June 1982, environmental factors have changed the pressures on Falkland birds. There has been a substantial military presence. Access has improved with the opening of Mount Pleasant Airport in 1987 and the start of a camp road building programme. The large farms previously owned by absentee landlords have been sub-divided and almost all of the land is in local ownership, including some acquired from the FI Company by the Government in 1991 (Summers & McAdam 1993). Wool prices have fluctuated and some new owners of small farms have needed financial help from the government. Pedigree sheep have been imported to improve wool quality and an experimental herd of goats for cashmere wool was established in 1990. International commercial fisheries for squid and finfish developed rapidly after 1982. Concern that seabirds could be directly affected by over-fishing led the FI Foundation to initiate a preliminary study from October 1986 into the diet of six species on New Island, and later, Keppel Island and Steeple Jason (Thompson 1989). The sale of fishing licences by the FI Government since the 150 mile Interim Conservation and Management Zone was established in 1987 has brought greatly increased revenue though this varies from year to year according to the number of foreign vessels wishing to fish in Falkland waters. The effects of commercial fishing on the seabird populations are still not fully understood. Development of the potential oil resources of the continental shelf may pose greater problems for all the birds which depend on the marine environment.

The Seabird Monitoring Programme has continued but relatively little interest has been shown in the landbirds. A tree nursery started in 1986 may eventually contribute to far more wooded habitat for passerine birds if trees and shrubs become established at the new farm settlements. The importance of the tussac grass habitat to Falkland birds was demonstrated in an ecological study of Kidney Island between 1958 and 1963. On this island of only 32ha (80 acres), at least 5,000 pairs of 28 bird species were breeding (Woods 1970a). Research into factors affecting the establishment and early growth of tussac grass (McAdam & Walton 1990) may encourage replanting on a larger scale and a farmers' forum on tussac and sand grass was held in 1994 to share knowledge gained through experience. One speaker stated that 'the only successful Tussac grass grows in association with wildlife, especially birds' and that 'research should concentrate on the link between birds and Tussac grass' (Kerr 1994). There is also increased awareness of the damage caused by introduced predators to the native bird populations. Following a preliminary investigation into the feasibility of eradicating predators on a sample of islands, several landowners interested in conservation formed a Tussac Islands Restoration Group. However, large scale planting of trees and shrubs and replanting of tussac are too recent to have affected landbird populations and most Falkland habitats remain much as they were a century ago.

The Falkland Islands, with two endemic species, Cobb's Wren *Troglodytes cobbi* and the Falkland Steamer Duck *Tachyeres brachydactyla*, is recognised as an Endemic Bird Area (EBA). Tierra del Fuego and the Falklands share another EBA as the Tussacbird *Cinclodes antarcticus*, Ruddy-headed Goose *Chloephaga rubidiceps*, Striated Caracara *Phalcoboenus australis* and the Black-throated (Canary-winged) Finch *Melanodera melanodera* are present in both regions (BirdLife International 1992).

Stattersfield (1995) has reviewed the importance of the Falkland Islands for birds following the publication by BirdLife International of a world list of threatened birds (Collar *et al.* 1994). One Falkland species, Cobb's Wren is considered to be '**Vulnerable**' because it has a small population and has suffered from the introduction of predators and loss of habitat. Five species are described as '**Near-threatened**', which means that their populations deserve to be monitored for any signs of deterioration in status. Three of these are included in this Atlas: the Ruddy-headed Goose and Striated Caracara are listed because they have a restricted geographic range and the Falklands hold most of the world population. The Black-throated Finch is considered to be declining seriously in Tierra del

Fuego, and the endemic Falkland race comprises most of the world population. Of the other two, the Hudsonian Godwit *Limosa haemastica* is a scarce arctic-breeding shorebird with most of the population wintering in San Sebastian Bay, Tierra del Fuego though a few have occurred in the Falklands and the Fuegian (or Cordilleran) Snipe *Gallinago stricklandii* is a little known species (Marchant *et al.* 1986) that was reputed to breed in the Falklands (Bennett 1926). There are no documented breeding records or recent sightings and there is some doubt over the identity of the local specimen, lost when the first Stanley Museum was destroyed by fire in 1944.

The resident human population has not increased significantly since 1982 and wool production has remained an important industry. Wildlife tourism has grown and there has been concern about tourist pressure on some popular sites. Attitudes towards predatory and scavenging birds and to those species thought to be rare are slowly changing through greater knowledge of their status in the Falklands and abroad. It is certain that the next century will bring changes to the Falklands, some of which may benefit the native birds and some of which may be detrimental. This Atlas will provide a baseline from which the effects of these changes can be at least partly assessed.

Presentation of results

How much of the Falkland Islands was surveyed ?

It is difficult to estimate the degree of coverage achieved. Records have been received from 234 (92%) of the 255 10-km squares containing land and Figure 3 shows the total number of species recorded as present in each of these. Fifty or more species were recorded in TD40 (West Point Island), UB58 (Sea Lion Island) and VC47 (eastern Stanley and Port William). Absence of records does not necessarily mean that the species does not occur in a particular square, only that it has not been found during the ten years of the survey. The observers had varying amounts of experience in ornithology and skill in identification. The length of time spent in any square has varied from a few hours by a visiting birdwatcher to several years by some residents. Other atlas projects have found that more time spent in a square tends to produce more records, particularly of the less noticeable or smaller species.

Fortunately, with only 11 species of small songbirds, there are few inconspicuous landbirds in the Falklands. Most of the breeding birds are conspicuous or large, easily seen because of the general lack of tall vegetation, apart from dense tussac, and they are relatively tame compared with similar species in other countries. Confusion probably led to under-recording for the Flying Steamer Duck while the two species of wren have been difficult for some observers to separate. Many of the smaller islands, which comprise the only land in a particular square, have not been surveyed thoroughly. The nocturnal shearwaters, petrels, storm-petrels, diving petrels and prions, which generally nest underground on the outer islands, remain poorly known. This is due largely to the shortage of experienced ornithologists and readily available transport, the unpredictable weather and the difficulties of landing on uninhabited islands to stay overnight.

The Universal Transverse Mercator Grid code letters for 100-km squares and the numbers for 10-km squares are shown on Figure 4. These provide the labels used in this survey for each 10-km grid square. Lists of those regions of the Falklands where survey work was either lacking, or only partly covered the land area or the species likely to be present, can be found in Appendix B. Thirteen of the 21 unsurveyed squares (Table 10) include only a very small point of East or West Falkland or the tip of an island. Of the eight squares containing substantial amounts of land, one is west of Port Edgar, one comprises the western end of Dunnose Head and one covers the land-mined area around Eliza Cove near Stanley. Five squares from Mount Usborne to Mount Simon in the Wickham Heights, north of Mount Pleasant, are used as a military practice area and are not accessible to the public.

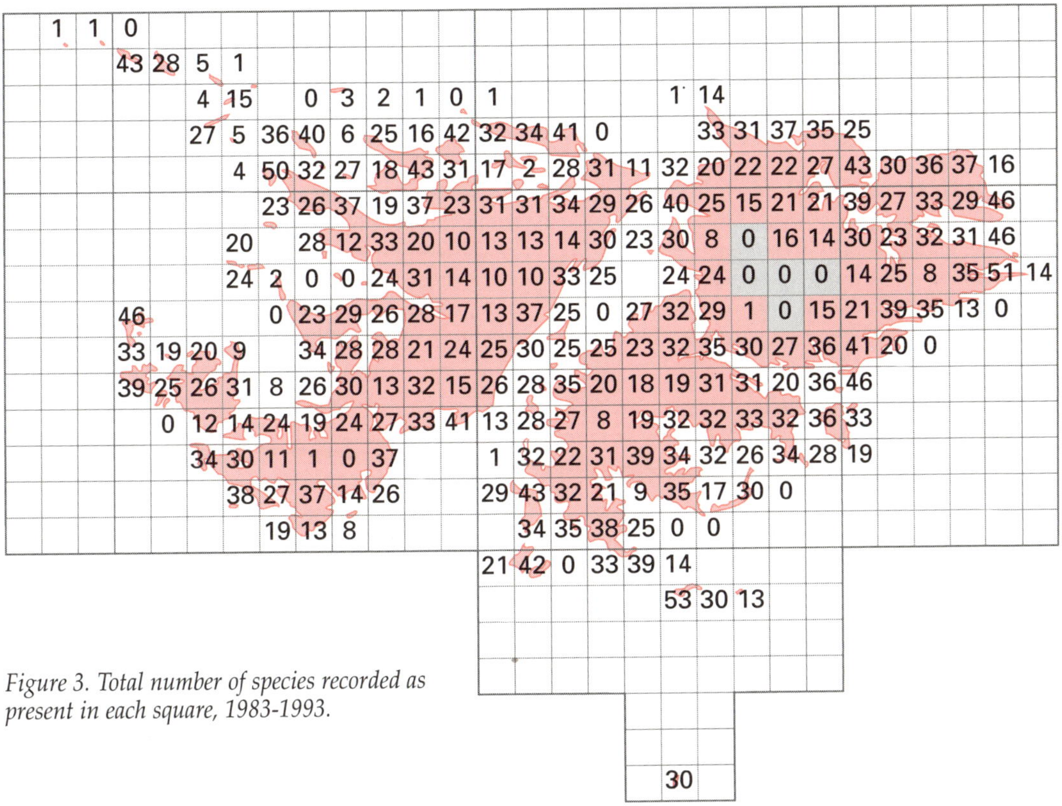

Figure 3. Total number of species recorded as present in each square, 1983-1993.

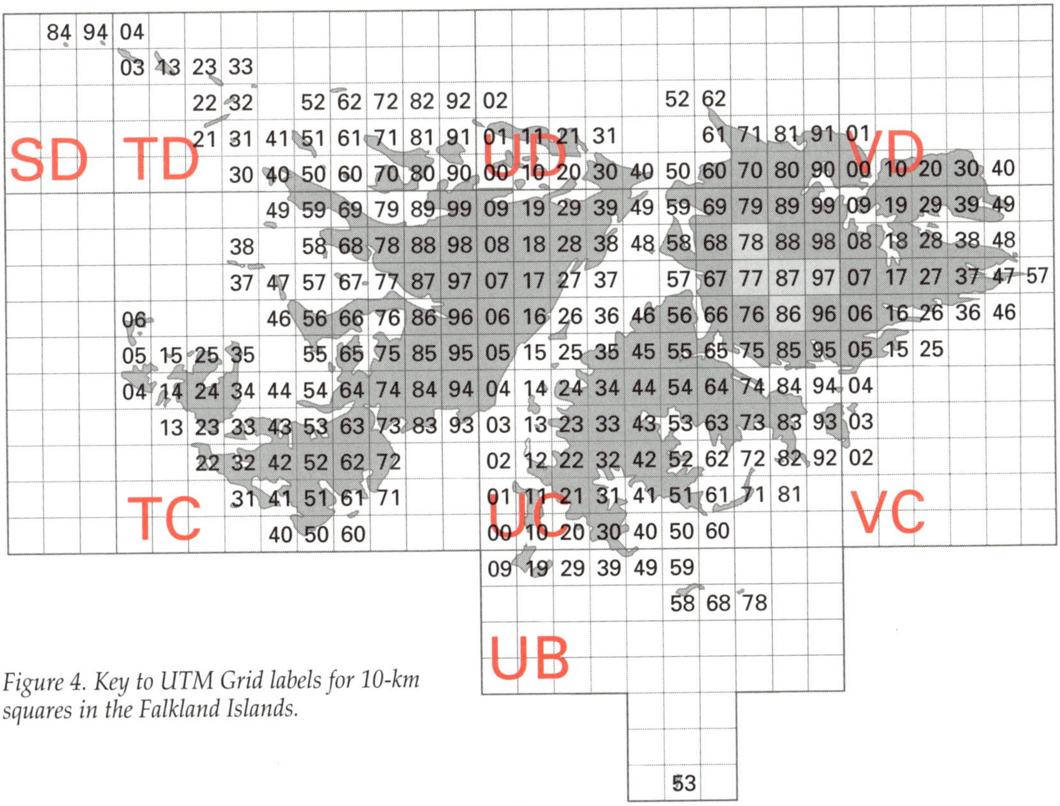

Figure 4. Key to UTM Grid labels for 10-km squares in the Falkland Islands.

Species texts, maps and data

Information for species 1-65 inclusive is presented in the sequence used in *Handbook of the Birds of the World* (del Hoyo *et al.* 1992) which differs slightly from the survey form. The text for each bird starts with the most recently accepted English name, with symbols indicating endemism where relevant, and the scientific name, which is usually in two parts but has a third name if geographic races have been described. Argentine and Chilean popular names are given, followed by any local Falkland Island names.

Under **History**, if the first scientifically-named specimen of the species was known to have come from the Falklands, the date, author and reference are quoted. A summary of early records follows, with comparisons between the reports of observers, particularly noticing comments on abundance made at different times, from 1764 to the start of the Breeding Birds Survey.

The section **Habitat and Breeding** summarises information already known on habitats used, with any seasonal changes, and on breeding behaviour.

Under **Breeding Birds Survey: analysis of data**, general assessments of the records are given and particularly interesting comments and counts from observers are quoted. Most of these remarks and records came from survey forms or letters while others have been published elsewhere. References to published and unpublished papers are in the Bibliography under observers' names. Where some habitat preferences were identified, attention is drawn to the evidence from the statistical comparison of records with categories of three topographical parameters. For some species, discussion of population estimates and comparisons with related species are included.

A brief summary from published sources in **World Distribution** includes notes on differences between Falkland races and others in South America, where these are known.

In the section **Implications for Conservation**, the possible effects of changes in the environment on the status of the species in the Falklands are considered. Where relevant, there are comments on classification under existing wildlife legislation, particularly where further protection is required, and suggestions for field work are given for some species.

References to the page numbers for the species texts in *The Birds of the Falkland Islands* (TBFI - Woods 1975) and *Guide to Birds of the Falkland Islands* (GBFI - Woods 1988) are given to enable readers to find other information, particularly on identification.

Species 66 to 75 are treated differently. There are no distribution maps and in place of 'Habitat and Breeding' and 'analysis of data', a new heading, **Breeding Birds Survey: records and comments** is used. These are the birds for which no evidence of breeding was found during the ten seasons of the survey. Some have bred in the Falklands in the past or were thought to breed, and some may possibly breed in the future.

There are **maps** for all of the breeding species and for two others, the Red Shoveler and Rufous-collared Sparrow, which were recorded in several years and may have bred. These maps show records for each 10-km square indicating that the species was 'present, but no evidence of breeding was found' (small dot), 'probably breeding' (medium dot), or that 'breeding was confirmed' (large dot), and totals are given for each category. If a particular square is blank, it means that the species was not recorded there during the survey. Figure 5a shows how records received for each species are presented, using data for Black-crowned Night Heron *Nycticorax nycticorax*. In Figure 5b the results of population calculations for this species are given, using the methods and data described for Tables 3 and 4 on pages 18 and 19.

Figure 5a

Number of squares in which recorded		
Present	76	
Probable	8	
Confirmed	52	38%
Total	136	58%

Figure 5b

Population (pairs)		
Minimum	572	Estimate
Maximum	4,892	1,900-3,600
Average	2,732	*Counts or estimates of population were given for 59 (43%) of the 136 squares in which recorded*
30% variation	820	

In the **data tables** 1-4 and 9, the scientific names of twelve species described as endemic Falkland races are prefixed by a single asterisk, '*' and scientific names of the two species recognised as endemic to the Falklands are prefixed by two asterisks, '**'. Where there is some doubt about endemism, a '?' precedes the asterisk. The species listed as having a restricted range are marked by '#' in tables and text.

Detectability

In the Falklands, apart from dense tussac grass, bird habitats are generally open to view from long distances. Because there are no native trees or woodland it is fairly easy to see many of the birds, but several factors affect the chances of seeing any species. These include how common the bird is, its size, how approachable it is, the sound produced as songs or calls, whether it lives and breeds in colonies or in solitary pairs, and the types of habitat used. Table 1 rates the detectability of each Falkland breeding species.

The first number before the bird name shows its position in this Atlas and the second is the Checklist number from *Guide to Birds of the Falkland Islands* (Woods 1988), to help with identification or when cross-checking information. The rating, on a five-point scale from very low to very high detectability, is a guide to the chances of finding the particular bird, assuming that it is present in the birdwatching area.

Table 1. Detectability

1 Very Low; 2 Low; 3 Moderate; 4 High; 5 Very High

Atlas	GBFI	Species			Detectability 1	2	3	4	5
01	1	King Penguin		*Aptenodytes patagonicus*					■
02	3	Gentoo Penguin		*Pygoscelis papua*					■
03	6	Rockhopper Penguin		*Eudyptes chrysocome*					■
04	7	Macaroni Penguin		*Eudyptes chrysolophus*			■		
05	10	Magellanic Penguin		*Spheniscus magellanicus*					■
06	11	White-tufted Grebe	?**	*Rollandia rolland*			■		
07	13	Silvery Grebe		*Podiceps occipitalis*			■		
08	16	Black-browed Albatross		*Diomedea melanophris*					■
09	24	Southern Giant Petrel		*Macronectes giganteus*					■
10	37	Thin-billed Prion		*Pachyptila belcheri*		■			
11	38	Fairy Prion		*Pachyptila turtur*		■			
12	40	White-chinned Petrel		*Procellaria aequinoctialis*		■			
13	42	Great Shearwater		*Puffinus gravis*	■				
14	43	Sooty Shearwater		*Puffinus griseus*		■			
15	46	Wilson's Storm-Petrel		*Oceanites oceanicus*	■				
16	49	Grey-backed Storm-Petrel		*Garrodia nereis*	■				
17	48	Black-bellied Storm-Petrel		*Fregetta tropica*	■				
18	50	Common Diving Petrel	*	*Pelecanoides urinatrix*	■				
19	52	Rock Shag/Cormorant		*Phalacrocorax magellanicus*					■
20	53	Imperial/King Shag		*Phalacrocorax atriceps*					■
21	60	Black-crowned Night Heron	*	*Nycticorax nycticorax*					■
22	66	Black-necked Swan		*Cygnus melancoryphus*			■		
23	71	Domestic Goose (feral)		*Anser anser*					■
24	69	Upland Goose	*	*Chloephaga picta*					■
25	70	Kelp Goose	*	*Chloephaga hybrida*					■
26	67	Ashy-headed Goose		*Chloephaga poliocephala*		■			
27	68	Ruddy-headed Goose	#	*Chloephaga rubidiceps*				■	
28	74	Flying Steamer Duck		*Tachyeres patachonicus*			■		
29	73	Falkland Steamer Duck	#**	*Tachyeres brachydactyla*					■
30	78	Chiloe Wigeon		*Anas sibilatrix*				■	
31	77	Speckled Teal		*Anas flavirostris*					■
32	72	Crested Duck		*Anas specularioides*					■
33	80	Yellow-billed Pintail		*Anas georgica*			■		
34	81	Silver Teal		*Anas versicolor*				■	
35	82	Cinnamon Teal		*Anas cyanoptera*			■		
36	83	Red Shoveler		*Anas platalea*		■			
37	87	Turkey Vulture		*Cathartes aura*					■
38	89	Red-backed Hawk		*Buteo polyosoma*				■	
39	92	Striated Caracara	#	*Phalcoboenus australis*					■
40	93	Crested Caracara		*Caracara plancus*					■
41	94	Peregrine Falcon		*Falco peregrinus*			■		
42	103	Magellanic Oystercatcher		*Haematopus leucopodus*					■

1 Very Low; 2 Low; 3 Moderate; 4 High; 5 Very High

Atlas	GBFI	Species			Detectability 1	2	3	4	5
43	102	Blackish Oystercatcher		*Haematopus ater*				■	
44	105	Two-banded Plover	?*	*Charadrius falklandicus*				■	
45	106	Rufous-chested Dotterel		*Charadrius modestus*					■
46	115	Magellanic Snipe	?*	*Gallinago magellanica*				■	
47	127	Falkland Skua		*Catharacta antarctica*					■
48	131	Dolphin Gull		*Larus scoresbii*					■
49	134	Kelp Gull		*Larus dominicanus*					■
50	136	Brown-hooded Gull		*Larus maculipennis*			■		
51	140	South American Tern		*Sterna hirundinacea*					■
52	146	Barn Owl		*Tyto alba*	■				
53	150	Short-eared Owl	*	*Asio flammeus*			■		
54	155	Tussacbird	#*	*Cinclodes antarcticus*					■
55	161	Dark-faced Ground-tyrant	*	*Muscisaxicola macloviana*					■
56	168	Chilean Swallow		*Tachycineta leucopyga*			■		
57	173	Falkland Pipit	*	*Anthus correndera*				■	
58	174	Falkland Grass Wren	*	*Cistothorus platensis*			■		
59	175	Cobb's Wren	#**	*Troglodytes cobbi*					■
60	178	Falkland Thrush	*	*Turdus falcklandii*					■
61	184	Rufous-collared Sparrow		*Zonotrichia capensis*			■		
62	182	Black-throated Finch	#*	*Melanodera melanodera*				■	
63	185	Long-tailed Meadowlark	*	*Sturnella loyca*					■
64	180	Black-chinned Siskin		*Carduelis barbata*				■	
65	179	House Sparrow		*Passer domesticus*			■		

Data: Records

Table 2 summarises all records received. The mean number of species recorded in each square was 25·1, with 13·6 species confirmed as breeding, 2·2 species probably breeding and 9·2 species recorded only as present. Considering all records, breeding was confirmed for 54·4% of the species recorded, and a further 8·8% were probably breeding.

Table 2. Summary of records received during the Breeding Birds Survey, 1983-1993

Column **1**. Total number of 10-km grid squares in which the species was recorded
2. Percentage of the 234 squares in which the species was recorded
3. Squares where the species was recorded only as present
4. Squares where the species was recorded as present and probably breeding
5. Number of squares where breeding was confirmed
6. Percentage of the total number of squares in which the species was recorded where breeding was confirmed

				1	2	3	4	5	6
01	King Penguin		*Aptenodytes patagonicus*	16	6·8	5	1	10	62·5
02	Gentoo Penguin		*Pygoscelis papua*	86	36·7	6	0	80	93·0
03	Rockhopper Penguin		*Eudyptes chrysocome*	31	13·2	3	0	28	90·3
04	Macaroni Penguin		*Eudyptes chrysolophus*	13	5·5	1	1	11	84·6
05	Magellanic Penguin		*Spheniscus magellanicus*	107	45·7	6	4	97	90·7
06	White-tufted Grebe	?**	*Rollandia rolland*	90	38·5	36	15	39	43·3
07	Silvery Grebe		*Podiceps occipitalis*	49	20·9	29	4	16	32·7
08	Black-browed Albatross		*Diomedea melanophris*	36	15·4	26	0	10	27·8
09	Southern Giant Petrel		*Macronectes giganteus*	134	57·3	109	3	22	16·4
10	Thin-billed Prion		*Pachyptila belcheri*	23	9·8	16	0	7	30·4
11	Fairy Prion		*Pachyptila turtur*	1	0·4	0	0	1	100
12	White-chinned Petrel		*Procellaria aequinoctialis*	18	7·7	16	0	2	11·1
13	Great Shearwater		*Puffinus gravis*	13	5·5	11	1	1	7·7
14	Sooty Shearwater		*Puffinus griseus*	26	11·1	20	0	6	23·1
15	Wilson's Storm-Petrel		*Oceanites oceanicus*	17	7·3	15	0	2	11·8

				1	2	3	4	5	6
16	Grey-backed Storm-Petrel		*Garrodia nereis*	7	3·0	5	2	0	0
17	Black-bellied Storm-Petrel		*Fregetta tropica*	1	0·4	1	0	0	0
18	Common Diving Petrel	*	*Pelecanoides urinatrix*	16	6·8	9	1	6	37·5
19	Rock Shag		*Phalacrocorax magellanicus*	168	71·8	34	4	130	77·4
20	King Shag		*Phalacrocorax atriceps*	121	51·7	54	3	64	52·9
21	Black-crowned Night Heron	*	*Nycticorax nycticorax*	136	58·1	76	8	52	38·2
22	Black-necked Swan		*Cygnus melancoryphus*	46	19·7	30	1	15	32·6
23	Domestic Goose		*Anser anser*	27	11·5	10	0	17	63·0
24	Upland Goose	*	*Chloephaga picta*	203	86·7	15	2	186	91·6
25	Kelp Goose	*	*Chloephaga hybrida*	181	77·3	41	8	132	72·9
26	Ashy-headed Goose		*Chloephaga poliocephala*	14	6·0	12	0	2	14·3
27	Ruddy-headed Goose	#	*Chloephaga rubidiceps*	175	74·8	39	3	133	76·0
28	Flying Steamer Duck		*Tachyeres patachonicus*	62	26·5	33	6	23	37·1
29	Falkland Steamer Duck	#**	*Tachyeres brachydactyla*	182	77·8	14	4	164	90·1
30	Chiloe Wigeon		*Anas sibilatrix*	89	38·0	50	6	33	37·1
31	Speckled Teal		*Anas flavirostris*	158	67·5	35	6	117	74·0
32	Crested Duck		*Anas specularioides*	181	77·3	31	3	147	81·2
33	Yellow-billed Pintail		*Anas georgica*	49	20·9	28	2	19	38·8
34	Silver Teal		*Anas versicolor*	63	26·9	28	10	25	39·7
35	Cinnamon Teal		*Anas cyanoptera*	8	3·4	5	1	2	25·0
36	Red Shoveler		*Anas platalea*	8	3·4	8	0	0	0
37	Turkey Vulture		*Cathartes aura*	163	69·7	106	9	48	29·4
38	Red-backed Hawk		*Buteo polyosoma*	150	64·1	52	9	89	59·3
39	Striated Caracara	#	*Phalcoboenus australis*	63	26·9	31	8	24	38·1
40	Crested Caracara		*Caracara plancus*	119	50·9	58	8	53	44·5
41	Peregrine Falcon		*Falco peregrinus*	90	38·5	52	8	30	33·3
42	Magellanic Oystercatcher		*Haematopus leucopodus*	172	73·5	33	13	126	73·3
43	Blackish Oystercatcher		*Haematopus ater*	162	69·2	67	37	58	35·8
44	Two-banded Plover	?*	*Charadrius falklandicus*	149	63·7	27	3	119	79·9
45	Rufous-chested Dotterel		*Charadrius modestus*	175	74·8	26	8	141	80·6
46	Magellanic Snipe	?*	*Gallinago magellanica*	158	67·5	72	27	59	37·3
47	Falkland Skua		*Catharacta antarctica*	152	64·9	90	3	59	38·8
48	Dolphin Gull		*Larus scoresbii*	123	52·6	83	2	38	30·9
49	Kelp Gull		*Larus dominicanus*	172	73·5	119	7	46	26·7
50	Brown-hooded Gull		*Larus maculipennis*	83	35·5	66	4	13	15·7
51	South American Tern		*Sterna hirundinacea*	130	55·5	82	10	38	29·2
52	Barn Owl		*Tyto alba*	8	3·4	2	2	4	50·0
53	Short-eared Owl	*	*Asio flammeus*	31	13·2	19	3	9	29·0
54	Tussacbird	#*	*Cinclodes antarcticus*	79	33·8	37	13	29	36·7
55	Dark-faced Ground-tyrant	*	*Muscisaxicola macloviana*	174	74·3	51	28	95	54·6
56	Chilean Swallow		*Tachycineta leucopyga*	9	3·8	8	0	1	11·1
57	Falkland Pipit	*	*Anthus correndera*	175	74·8	28	53	94	53·7
58	Falkland Grass Wren	*	*Cistothorus platensis*	100	42·7	37	40	23	23·0
59	Cobb's Wren	#**	*Troglodytes cobbi*	25	10·7	8	6	11	44·0
60	Falkland Thrush	*	*Turdus falcklandii*	162	69·2	31	9	122	75·3
61	Rufous-collared Sparrow		*Zonotrichia capensis*	6	2·6	2	4	0	0
62	Black-throated Finch	#*	*Melanodera melanodera*	194	82·9	32	58	104	53·6
63	Long-tailed Meadowlark	*	*Sturnella loyca*	180	76·9	36	39	105	58·3
64	Black-chinned Siskin		*Carduelis barbata*	82	35·0	36	10	36	43·9
65	House Sparrow		*Passer domesticus*	28	12·0	10	0	18	64·3
Total records				5869	–	2158	520	3191	–

Data: Population estimates

Answers to the question, 'How many birds?' varied between observers with most people using code letters indicating a range, rather than figures resulting from counts. The proportion of each grid square surveyed was not always recorded, though several observers, notably Mike Riddy, tried to visit the different habitats in each square. Therefore the data for each species contain a mixture of range estimates, counts of pairs or individuals or no information on numbers.

Table 3 shows the number of squares in which each species was recorded, followed by the number of squares for which no counts or estimates were given. The third column shows the number of squares with counts or estimates and the fourth gives the percentage of the species total where there were counts or estimates. Columns A to D show the number of squares with estimates of up to 10 (A), 100 (B), 1,000 (C) and more than 1,000 pairs (D).

Table 3. Population counts and estimates

				Total squares	No est.	Count /est.	%	A	B	C	D
01	King Penguin		*Aptenodytes patagonicus*	16	2	14	87	-	-	-	-
02	Gentoo Penguin		*Pygoscelis papua*	86	14	72	84	-	2	28	16
03	Rockhopper Penguin		*Eudyptes chrysocome*	31	7	24	77	-	2	2	12
04	Macaroni Penguin		*Eudyptes chrysolophus*	13	1	12	92	3	2	-	-
05	Magellanic Penguin		*Spheniscus magellanicus*	107	31	76	71	5	18	29	21
06	White-tufted Grebe	?**	*Rollandia rolland*	90	40	50	55	26	8	-	-
07	Silvery Grebe		*Podiceps occipitalis*	49	17	32	65	9	2	-	-
08	Black-browed Albatross		*Diomedea melanophris*	36	22	9	29	-	1	2	2
09	Southern Giant Petrel		*Macronectes giganteus*	134	98	33	25	8	11	3	-
10	Thin-billed Prion		*Pachyptila belcheri*	23	16	4	20	-	-	1	2
11	Fairy Prion		*Pachyptila turtur*	1	-	-	-	-	-	-	-
12	White-chinned Petrel		*Procellaria aequinoctialis*	18	8	2	20	-	1	1	-
13	Great Shearwater		*Puffinus gravis*	13	4	1	20	1	-	-	-
14	Sooty Shearwater		*Puffinus griseus*	26	13	4	23	1	2	-	1
15	Wilson's Storm-Petrel		*Oceanites oceanicus*	17	12	1	7	-	-	-	1
16	Grey-backed Storm-Petrel		*Garrodia nereis*	7	5	1	17	-	1	-	-
17	Black-bellied Storm-Petrel		*Fregetta tropica*	1	-	-	-	-	-	-	-
18	Common Diving Petrel	*	*Pelecanoides urinatrix*	16	14	2	12	-	-	1	1
19	Rock Shag		*Phalacrocorax magellanicus*	168	48	120	71	7	49	42	3
20	King Shag		*Phalacrocorax atriceps*	121	65	56	46	7	9	19	10
21	Black-crowned Night Heron	*	*Nycticorax nycticorax*	136	77	59	43	26	14	1	-
22	Black-necked Swan		*Cygnus melancoryphus*	46	22	24	52	7	2	-	-
23	Domestic Goose		*Anser anser*	27	7	20	74	11	6	-	-
24	Upland Goose	*	*Chloephaga picta*	203	35	168	83	9	107	35	10
25	Kelp Goose	*	*Chloephaga hybrida*	181	54	127	70	25	85	5	1
26	Ashy-headed Goose		*Chloephaga poliocephala*	14	8	6	43	1	1	-	-
27	Ruddy-headed Goose	#	*Chloephaga rubidiceps*	175	48	126	72	32	74	11	-
28	Flying Steamer Duck		*Tachyeres patachonicus*	62	30	32	52	12	1	-	-
29	Falkland Steamer Duck	#**	*Tachyeres brachydactyla*	182	38	145	79	32	100	7	-
30	Chiloe Wigeon		*Anas sibilatrix*	89	35	54	61	18	2	2	-
31	Speckled Teal		*Anas flavirostris*	158	51	107	68	28	63	5	-
32	Crested Duck		*Anas specularioides*	181	43	138	76	33	92	4	-
33	Yellow-billed Pintail		*Anas georgica*	49	18	31	63	9	1	-	-
34	Silver Teal		*Anas versicolor*	63	25	38	60	8	6	1	-
35	Cinnamon Teal		*Anas cyanoptera*	8	2	6	75	1	-	-	-
36	Red Shoveler		*Anas platalea*	8	2	6	75	-	-	-	-
37	Turkey Vulture		*Cathartes aura*	163	89	74	45	36	24	-	-
38	Red-backed Hawk		*Buteo polyosoma*	150	93	57	38	34	-	-	-
39	Striated Caracara	#	*Phalcoboenus australis*	63	37	25	40	11	2	-	-
40	Crested Caracara		*Caracara plancus*	119	80	39	33	24	-	-	-
41	Peregrine Falcon		*Falco peregrinus*	90	49	40	45	18	3	-	-
42	Magellanic Oystercatcher		*Haematopus leucopodus*	172	48	124	72	44	58	8	-
43	Blackish Oystercatcher		*Haematopus ater*	162	74	88	54	54	21	2	-
44	Two-banded Plover	?*	*Charadrius falklandicus*	149	35	114	76	52	45	7	-

				Total squares	No est.	Count /est.	%	A	B	C	D
45	Rufous-crested Dotterel		*Charadrius modestus*	175	52	123	70	27	79	10	-
46	Magellanic Snipe	?*	*Gallinago magellanica*	158	103	55	35	16	29	2	-
47	Falkland Skua		*Catharacta antarctica*	152	88	62	41	17	25	5	-
48	Dolphin Gull		*Larus scoresbii*	123	76	47	38	14	21	2	-
49	Kelp Gull		*Larus dominicanus*	172	107	64	37	5	28	14	3
50	Brown-hooded Gull		*Larus maculipennis*	83	67	16	19	3	7	1	-
51	South American Tern		*Sterna hirundinacea*	130	81	47	37	10	15	7	-
52	Barn Owl		*Tyto alba*	8	1	7	87	-	-	-	-
53	Short-eared Owl	*	*Asio flammeus*	31	18	13	42	9	-	-	-
54	Tussacbird	#*	*Cinclodes antarcticus*	79	38	41	52	10	9	11	4
55	Dark-faced Ground-tyrant	*	*Muscisaxicola macloviana*	174	78	96	55	31	51	6	-
56	Chilean Swallow		*Tachycineta leucopyga*	9	1	0	100	-	-	-	-
57	Falkland Pipit	*	*Anthus correndera*	175	71	104	59	28	67	5	-
58	Falkland Grass Wren	*	*Cistothorus platensis*	100	67	33	33	18	7	-	-
59	Cobb's Wren	#**	*Troglodytes cobbi*	25	11	14	56	6	6	2	-
60	Falkland Thrush	*	*Turdus falcklandii*	162	61	101	62	34	55	5	-
61	Rufous-collared Sparrow		*Zonotrichia capensis*	6	0	3	100	-	-	-	-
62	Black-throated Finch	#*	*Melanodera melanodera*	194	92	102	53	22	64	5	-
63	Long-tailed Meadowlark	*	*Sturnella loyca*	180	86	94	52	30	51	7	-
64	Black-chinned Siskin		*Carduelis barbata*	82	39	43	52	15	20	3	-
65	House Sparrow		*Passer domesticus*	28	10	18	64	1	13	-	-
Total records				5869							

Data: Population calculations

Table 4 lists most species in Atlas order, ranks them according to their estimated population from most to least numerous and gives the total number of squares in which each was recorded. Averages obtained from squares with counts or estimates (Table 3) were used to calculate figures for those squares without numerical data but where the species was present. The minimum, maximum and average figures were calculated from the counts, estimates and deduced figures. The arbitrary 30% variation from the average was thought to be sufficient to allow for most of this uncertainty. Population estimates were obtained by subtracting and adding 30% from the average, and rounding the figures to the nearest ten, hundred or thousand. Species 10-18 are listed at the end of the table as insufficient data were received for calculations.

Table 4. Population - calculated in pairs

	Rank	Total squares	Minimum	Maximum	Average	30% var.	Population estimate
01 King Penguin *Aptenodytes patagonicus*	49	16	147	155	151	45	100-200
02 Gentoo Penguin *Pygoscelis papua*	4	86	41246	164430	102838	30851	72000-134000
03 Rockhopper Penguin *Eudyptes chrysocome*	1	31	512827	587300	550063	165019	385000-715000
04 Macaroni Penguin *Eudyptes chrysolophus*	50	13	69	96	82	25	60-110
05 Magellanic Penguin *Spheniscus magellanicus*	3	107	23153	194613	108883	32665	76000-142000
06 W-tufted Grebe ?***Rollandia rolland*	37	90	195	1950	1073	322	750-1400
07 Silvery Grebe *Podiceps occipitalis*	40	49	258	1137	698	209	500-900
08 B-b Albatross *Diomedea melanophris*	2	36	319511	386800	353155	105947	247000-459000
09 S'n Giant Petrel *Macronectes giganteus*	22	134	3139	11592	7365	2210	5000-10000

	Rank	Total squares	Minimum	Maximum	Average	30% var.	Population estimate
19 Rock Shag *Phalacrocorax magellanicus*	7	168	10218	80105	45161	13548	32000-59000
20 King Shag *Phalacrocorax atriceps*	6	121	19313	110470	64891	19467	45000-84000
21 Blk-c. Night Heron **Nycticorax nycticorax*	30	136	572	4892	2732	820	1900-3600
22 Black-necked Swan *Cygnus melancoryphus*	46	46	112	751	431	129	300-600
23 Domestic Goose *Anser anser*	45	27	97	813	455	136	300-600
24 Upland Goose **Chloephaga picta*	5	203	11260	119711	65485	19645	46000-85000
25 Kelp Goose **Chloephaga hybrida*	12	181	2894	25292	14092	4228	10000-18000
26 Ashy-headed Goose *Chloephaga poliocephala*	51	14	24	110	67	20	50-90
27 Ruddy-h. Goose *#Chloephaga rubidiceps*	10	175	3761	37535	20648	6194	14000-27000
28 Flying St. Duck *Tachyeres patachonicus*	47	62	75	577	326	98	200-400
29 Falkland S.Duck *#**Tachyeres brachydactyla*	13	182	1748	22720	12234	3670	9000-16000
30 Chiloe Wigeon *Anas sibilatrix*	42	89	222	1140	681	204	500-900
31 Speckled Teal *Anas flavirostris*	20	158	1441	15598	8520	2556	6000-11000
32 Crested Duck *Anas specularioides*	18	181	1528	17231	9379	2814	7000-12000
33 Yellow-billed Pintail *Anas georgica*	38	49	184	1407	795	238	600-1000
34 Silver Teal *Anas versicolor*	36	63	292	2016	1154	346	800-1500
35 Cinnamon Teal *Anas cyanoptera*	52	8	8	26	17	5	12-22
36 Red Shoveler *Anas platalea*	53	8	9	15	12	4	8-16
37 Turkey Vulture *Cathartes aura*	29	163	495	5045	2770	831	1900-3600
38 Red-backed Hawk *Buteo polyosoma*	39	150	159	1302	731	219	500-1000
39 Striated Caracara *#Phalcoboenus australis*	43	63	268	1041	655	196	500-900
40 Crested Caracara *Caracara plancus*	44	119	124	1060	592	178	400-800
41 Peregrine Falcon *Falco peregrinus*	41	90	146	1232	689	207	500-900
42 Mag. Oystercat'r *Haematopus leucopodus*	16	172	1816	18691	10253	3076	7000-13000
43 B. Oystercatcher *Haematopus ater*	25	162	1084	11112	6098	1829	4000-8000
44 Two-banded Plover *?*Charadrius falklandicus*	17	149	1397	18413	9905	2971	7000-13000
45 R-c. Dotterel *Charadrius modestus*	11	175	2668	29523	16095	4829	11000-21000
46 Magellanic Snipe *?*Gallinago magellanica*	24	158	1326	12962	7144	2143	5000-9000
47 Falkland Skua *Catharacta antarctica*	23	152	1443	12978	7211	2163	5000-9000

	Rank	Total squares	Minimum	Maximum	Average	30% var.	Population estimate
48 Dolphin Gull *Larus scoresbii*	28	123	819	8115	4467	1340	3000-6000
49 Kelp Gull *Larus dominicanus*	8	172	6137	61965	34051	10215	24000-44000
50 Brown-hooded Gull *Larus maculipennis*	32	83	605	3471	2038	611	1400-2600
51 South American Tern *Sterna hirundinacea*	19	130	1729	16065	8897	2669	6000-12000
52 Barn Owl *Tyto alba*	54	8	7	8	7	-	7-10
53 Short-eared Owl **Asio flammeus*	48	31	32	277	155	47	100-200
54 Tussacbird *#*Cinclodes antarcticus*	9	79	1852	41315	21583	6475	15000-28000
55 D-f. Gr'd-Tyrant **Muscisaxicola macloviana*	27	174	1076	10527	5801	1740	4000-8000
56 Chilean Swallow *Tachycineta leucopyga*	56	9	1	-	-	-	?
57 Falkland Pipit **Anthus correndera*	14	175	1926	21738	11832	3550	8000-15000
58 Falkland Grass Wren **Cistothorus platensis*	34	100	299	3283	1791	537	1300-2300
59 Cobb's Wren *#**Troglodytes cobbi*	33	25	101	3652	1876	563	1300-2400
60 Falkland Thrush **Turdus falcklandii*	26	162	1007	11004	6005	1802	4000-8000
61 Rufous-c. Sparrow *Zonotrichia capensis*	55	6	3	6	4	-	3-6
62 Black-throated Finch *#*Melanodera melanodera*	15	194	1438	19734	10586	3176	7000-14000
63 L-t. Meadowlark **Sturnella loyca*	21	180	1378	14734	8056	2417	6000-10000
64 Black-chinned Siskin *Carduelis barbata*	31	82	347	3964	2155	647	1500-2800
65 House Sparrow *Passer domesticus*	35	28	192	2298	1245	373	900-1600

Species for which no counts and few estimates were available

	Total squares					
10 Thin-billed Prion *Pachyptila belcheri*	23	-	-	-	-	50000-100000?
11 Fairy Prion *Pachyptila turtur*	1	-	-	-	-	10000?
12 White-chinned Petrel *Procellaria aequinoctialis*	18	-	-	-	-	1000-5000?
13 Great Shearwater *Puffinus gravis*	13	-	-	-	-	50-100?
14 Sooty Shearwater *Puffinus griseus*	26	-	-	-	-	10000-20000?
15 Wilson's Storm-Petrel *Oceanites oceanicus*	17	-	-	-	-	5000?
16 Grey-backed Storm-Petrel *Garrodia nereis*	7	-	-	-	-	1000-5000?
17 Black-bellied Storm-Petrel *Fregetta tropica*	1	-	-	-	-	?
18 Common Diving Petrel **Pelecanoides urinatrix*	16	-	-	-	-	5000-10000?

How do some topographical features of the Falklands affect bird populations ?

Dry/Wet analysis

The amount of standing freshwater in each of the 255 10-km squares was assessed from the 1:50,000 DOS maps by counting tetrads. Tetrads (2x2 km squares or 4 sq km) are not printed on the maps, but their coordinates are shown in the margins and they can be identified by using a transparent overlay showing a grid of the 25 tetrads. Tetrads were counted as 'wet' if at least one pond or part of a lake was shown and as 'dry' if no ponds or lakes were marked. This gives only a rough idea of the wetness of squares as streams and rivers were not included but there are no other desk surveys of standing water and it is a useful guide.

The DOS maps were drawn from aerial photographs taken in October and November (1956), the spring months when most evaporation of surface water occurs. Since then some ponds have probably disappeared while others may have formed. If ponds were visible at the time of the aerial survey, they would certainly be present during the winter months, although some may evaporate by January or February in a dry summer. The counts of tetrads showing standing water are probably least accurate where only small pools were detected in 1956 and the wetness of squares without large permanent ponds or lakes may have been over-estimated. The numbers of 10-km squares containing from 1-25 wet tetrads are shown in Table 5. For example, 28 squares have only two wet tetrads, seven squares have 13 and only one square has 22 wet tetrads.

Squares were grouped according to the categories shown in Table 6 to allow records to be analysed. The table shows that 95% or more of the squares in four of the five categories were surveyed for the Atlas. The 12 squares with no standing water which could not be surveyed, consisted of small headlands or very small islands and they contained only about 68 sq km of land (the equivalent of two-thirds of an inland square). In the category of 1-5 wet tetrads, four of the unsurveyed squares together contained only about 76 sq km of land. The five unsurveyed squares in the Wickham Heights included two in the 1-5 category, two in the 6-10 category and one in the 11-15 category of wet tetrads.

Inland/Coastal analysis

The coastline of both main islands is deeply indented with many sheltered harbours. Analysis of the 1:250,000 maps shows that: only 22 squares (9%) are completely inland; 15 squares (6%) have 90% or more land with only a small creek or estuary; while 218 squares (85%), have a considerable length of coastline. Using an overlay carrying symbols to mark squares in the first two categories, it was possible to examine the map for each species and count the records in 'inland', 'almost completely inland' and 'coastal' squares. The result is shown in the Inland/Coastal squares analysis for each species. Table 7 shows the number of squares in each category and the proportions surveyed. Coverage was good for two of the three categories, but almost a quarter of the inland squares were in the Wickham Heights on East Falkland and could not be surveyed.

Upland/Lowland analysis

West Falkland is generally more hilly than East Falkland though no mountains on any island are higher than 705m (2,300 feet). On West Falkland, the Hornby Mountains behind Port Howard run parallel to Falkland Sound and the Mount Robinson/Mount Adam range runs east to west across the northern part of the island. There is some mountainous ground east of Port Stephens and another similar area south of Chartres. Weddell Island, the largest offshore island, rises to about 380m (1,200 feet). On East Falkland, north of Choiseul Sound, which almost bisects the island, the land rises gently to the Wickham Heights (up to 600m or 2,000 feet) extending westwards from Stanley. Lafonia, the southern half of East Falkland, is low and undulating ground, rarely rising above 60m (200 feet). The 1:250,000 maps, which are contoured in feet, rather than metres, show that only 12 10-km squares (5%) have at least half of the land higher than 500 feet, 26 (10%) have between one quarter and half of the land above 500 feet, 73 (29%) have less than one quarter of their area above 500 feet and 144 (56%) are lowland squares with no land above 500 feet (152m). Another overlay containing symbols for these four categories made it possible to assess species distribution according to the altitude category of each square. Four of the squares in the highest category of upland could not be surveyed because public access was prohibited.

Table 5. Tetrads with standing water in 255 10-km grid squares

Wet tetrads	0	1	2	3	4	5	6	7	8	9	10	11	12
Grid squares	50	33	28	18	13	18	17	8	10	10	9	6	4

Wet tetrads	13	14	15	16	17	18	19	20	21	22	23	24	25
Grid squares	7	3	5	5	4	3	1	0	2	1	0	0	0

Table 6. Proportions of 10-km squares in categories of wetness

Categories		Surveyed		Not surveyed		Total
1.	No standing water	38	76%	12	24%	50
2.	1-5 wet tetrads	104	95%	6	5%	110
3.	6-10 wet tetrads	52	96%	2	4%	54
4.	11-15 wet tetrads	24	96%	1	4%	25
5.	16-22 wet tetrads	16	100%	0	—	16
	Total	234	92%	21	8%	255

Table 7. Proportions of Inland and Coastal 10-km squares

Categories		Surveyed		Not surveyed		Total
I:	Completely inland	17	77%	5	23%	22
IC:	≥ 90% inland	14	93%	1	7%	15
C:	Coastal squares	203	93%	15	7%	218
	Total	234	92%	21	8%	255

Table 8. Proportions of 10-km squares by Altitude

Categories		Surveyed		Not surveyed		Total
High	≥1/2 of land >500 ft asl	7	58%	5	42%	12
MH	≥1/4 and <1/2 of land >500 ft asl	25	96%	1	4%	26
ML	<1/4 of land >500 ft asl	73	100%	0	0	73
Low	No land >500 ft asl	129	90%	15	10%	144
	Total	234	92%	21	8%	255

Table 9. Comparisons between species records and three topographical parameters

Key for table and species data

10-km: 10km x 10km square (100km^2) of the Universal Transverse Mercator Grid
>: more than
≥: equal to or more than
<: less than
asl: above sea level
500': 500 feet or 152 metres
♦: A strong tendency towards a high count of records in any category, but not statistically significant
*: A high count in a category, which shows a difference between distribution of the species records and the parameter considered that is significant at the 5% level; such a difference is unlikely to occur by chance more than once in 20 samples. In statistical terms this difference is described as significant at the 5% level ($p<0{\cdot}05$)
**: A highly significant difference that is unlikely to occur by chance more than once in a hundred samples. This is described as significant at the 1% level ($p<0{\cdot}01$)
***: A very highly significant difference that is unlikely to occur by chance more than once in a thousand samples; it is described as significant at the 0·1% level ($p<0{\cdot}001$)

Dry/Wet analysis: standing water by 10-km square on the 1:50,000 maps

1: None of the 25 2km x 2km (4km^2) tetrads in a square shows standing water
2: 1-5 wet tetrads,
3: 6-10
4: 11-15
5: 16-22

Inland/Coastal analysis: coastline proportion in each 10-km square on the 1:250,000 maps

I: Inland squares
IC: Inland/Coastal squares with very little (less than 10%) coastline
C: Coastal squares with several kilometres (10% or more) of coastline

Upland/Lowland analysis: proportion of land over 500 feet in each 10-km square on the 1:250,000 maps

H: High squares with ≥½ of the land over 500 feet
MH: Moderately High squares with ≥¼ and <½ of the land over 500 feet
ML: Moderately Low squares with <¼ of the land over 500 feet
L: squares with no land over 500 feet

Species	Squares	Dry/Wet 1	2	3	4	5	Inland/Coastal I	IC	C	Upland/Lowland H	MH	ML	L
01 King Penguin *Aptenodytes patagonicus*	16												
02 Gentoo Penguin *Pygoscelis papua*	86								**			*	
03 Rockhopper Penguin *Eudyptes chrysocome*	31											**	
04 Macaroni Penguin *Eudyptes chrysolophus*	13												
05 Magellanic Penguin *Spheniscus magellanicus*	107								**				♦
06 White-tufted Grebe ?** *Rollandia rolland*	90			***	***	***							♦
07 Silvery Grebe *Podiceps occipitalis*	49			***	***	***		♦					♦
08 Black-browed Albatross *Diomedea melanophris*	36											♦	
09 Southern Giant Petrel *Macronectes giganteus*	134								***				
19 Rock Shag *Phalacrocorax magellanicus*	168								***			♦	
20 King Shag *Phalacrocorax atriceps*	121								***			♦	
21 Blk-crwnd Night Heron **Nycticorax nycticorax*	136								**				
22 Black-necked Swan *Cygnus melancoryphus*	46				***	***							

Species		Squares	Dry/Wet 1	2	3	4	5	Inland/Coastal I	IC	C	Upland/Lowland H	MH	ML	L
23	Domestic Goose *Anser anser*	27												
24	Upland Goose * *Chloephaga picta*	203												
25	Kelp Goose * *Chloephaga hybrida*	181								***				
27	Ruddy-headed Goose # *Chloephaga rubidiceps*	175												
28	Flying Steamer Duck *Tachyeres patachonicus*	62				**	**		**					
29	Falkland Steamer Duck #** *Tachyeres brachydactyla*	182								***				
30	Chiloe Wigeon *Anas sibilatrix*	89			***	***	***							♦
31	Speckled Teal *Anas flavirostris*	158			**	**	**							
32	Crested Duck *Anas specularioides*	181								♦				
33	Yellow-billed Pintail *Anas georgica*	49				**	**							♦
34	Silver Teal *Anas versicolor*	63			***	***	***							
37	Turkey Vulture *Cathartes aura*	163												
38	Red-backed Hawk *Buteo polyosoma*	150												
39	Striated Caracara # *Phalcoboenus australis*	63	♦	♦						♦				
40	Crested Caracara *Caracara plancus*	119												
41	Peregrine Falcon *Falco peregrinus*	90								♦				
42	Mag. Oystercatcher *Haematopus leucopodus*	172								***				
43	Blackish Oystercatcher *Haematopus ater*	162								***				
44	Two-banded Plover ?* *Charadrius falklandicus*	149								♦				
45	Ruf.-chested Dotterel *Charadrius modestus*	175												
46	Magellanic Snipe ?* *Gallinago magellanica*	158												
47	Falkland Skua *Catharacta antarctica*	152								**				
48	Dolphin Gull *Larus scoresbii*	123								**				
49	Kelp Gull *Larus dominicanus*	172								*				
50	Brown-hooded Gull *Larus maculipennis*	83								*				
51	South American Tern *Sterna hirundinacea*	130								*				
53	Short-eared Owl * *Asio flammeus*	31			**	**								
54	Tussacbird #* *Cinclodes antarcticus*	79								**				♦
55	Dark-faced Gr'd-Tyrant * *Muscisaxicola macloviana*	174												

Species	Squares	Dry/Wet 1	2	3	4	5	Inland/Coastal I	IC	C	Upland/Lowland H	MH	ML	L
57 Falkland Pipit *Anthus correndera	175			*	*	*							
58 Falkland Grass Wren *Cistothorus platensis	100												
59 Cobb's Wren #**Troglodytes cobbi	25	♦											
60 Falkland Thrush *Turdus falcklandii	162											♦	
62 Black-throated Finch #*Melanodera melanodera	194												
63 Long-tailed Meadowlark *Sturnella loyca	180												
64 Black-chinned Siskin *Carduelis barbata*	82								♦			♦	
65 House Sparrow *Passer domesticus*	28												

Species for which no comparisons were made due to lack of information

Species	Squares
10 Thin-billed Prion *Pachyptila belcheri*	23
11 Fairy Prion *Pachyptila turtur*	1
12 White-chinned Petrel *Procellaria aequinoctialis*	18
13 Great Shearwater *Puffinus gravis*	13
14 Sooty Shearwater *Puffinus griseus*	26
15 Wilson's Storm-Petrel *Oceanites oceanicus*	17
16 Grey-backed Storm-Petrel *Garrodia nereis*	7
17 Black-bellied Storm-Petrel *Fregetta tropica*	1
18 Common Diving Petrel *Pelecanoides urinatrix	16
26 Ashy-headed Goose *Chloephaga poliocephala*	14
35 Cinnamon Teal *Anas cyanoptera*	8
36 Red Shoveler *Anas platalea*	8
52 Barn Owl *Tyto alba*	8
56 Chilean Swallow *Tachycineta leucopyga*	9
61 Rufous-collared Sparrow *Zonotrichia capensis*	6

Species Accounts

01 King Penguin

Aptenodytes patagonicus

A. & C. Pingüino rey

History
This species was first named as 'The Patagonian Pinguin', by Thomas Pennant in 1769, though the specimen apparently came from the Falkland Islands. Early records suggest that it was never plentiful here. A keeper from the Zoological Society of London, Adolphe Lecomte, visited Volunteer Lagoon, East Falkland in December 1867 and captured about a dozen King Penguins from a group of about 20, among a large colony of Gentoo Penguins behind Volunteer Beach (Sclater 1868). By 1870, it was almost exterminated in the Falklands for its oil and feathers. Vallentin (in Boyson 1924) believed that none had been recorded since 1898 but Cobb (unpublished) noted five individuals in captivity early this century; one was at Darwin and four in Stanley, the two at Sulivan House having been caught in Fitzroy camp in the 1906/07 season. Bennett (1926) noted that it was occasionally seen in the archipelago and might even breed. Cawkell & Hamilton (1961) reported that a small colony existed in the 1920s at Cow Bay and that nesting was proved there in 1945 and 1946 though only one bird was seen in 1953. In December 1971, Sewall Pettingill visited a well-established colony behind Volunteer Beach, 'on the periphery of a thriving colony of some 200 pairs of Gentoos', with about 30 courting or incubating adults, nine huge 'teddy-bear' chicks and ten moulting adults some distance away. Two months later, they saw 22 adults, four with small chicks and seven with eggs; the nine large chicks had left the colony (Pettingill 1975).

Habitat and Breeding
It is the largest penguin known to breed in the Falkland Islands and is found on flat ground on the coast, usually in association with Gentoo Penguins. The complete breeding cycle takes longer than one year, which means that a pair can only rear two chicks in three years. It makes no nest; a single large pale greenish-white egg is laid in November or December and is incubated on the feet. The egg is covered so completely, by a loose fold of skin and the tail, that a casual observer would not know that the bird was incubating. King Penguins are seen near their colonies throughout the year and occasional moulting individuals may appear on sand beaches.

Breeding Birds Survey: analysis of data
The main colony is at Volunteer Point where on 21 January 1989, Phil Gregory recorded c250 adults, about 120 of them having eggs or chicks. By 1992/93, breeding had been proved at nine other sites, all close to large colonies of Gentoo Penguins, giving a breeding population of up to 200 pairs. Individuals had been seen at five other Gentoo colonies and a pair was probably breeding at Wood Cove, Port Stephens in 1987.

World Distribution
The King Penguin has a circumpolar distribution and also breeds at South Georgia, the Prince Edward Islands, Iles Crozet and Kerguelen, Heard and Macquarie Islands. The estimated world population is at least 1·23 million pairs and numbers continue to increase at all sites. The presence of many more birds moulting on the coasts of Tierra del Fuego suggests that breeding may soon start there (Croxall *et al.* 1994).

Implications for Conservation
The small Falkland population of King Penguins is vulnerable because the site of the largest colony is a well-known tourist attraction. Other isolated pairs nest at Gentoo colonies which are also easily accessible, but with good joint management of visits by tour operators and landowners, it is expected that the population will continue to increase.

References
TBFI (1975) p67-8, GBFI (1988) p93-4

01 King Penguin *Aptenodytes patagonicus*

Detectability: **5 Very High**

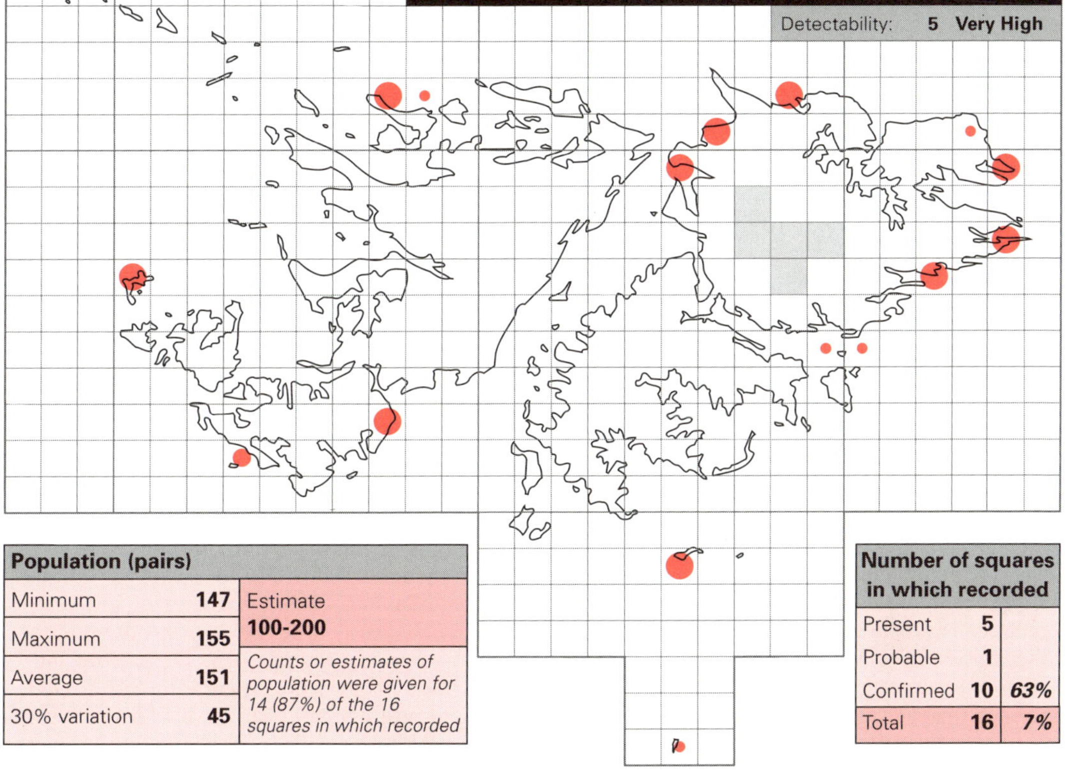

Population (pairs)		
Minimum	**147**	Estimate **100-200**
Maximum	**155**	
Average	**151**	*Counts or estimates of population were given for 14 (87%) of the 16 squares in which recorded*
30% variation	**45**	

Number of squares in which recorded		
Present	**5**	
Probable	**1**	
Confirmed	**10**	***63%***
Total	**16**	***7%***

Comparisons between species records and topography	Dry/Wet Number of wet tetrads					Inland/Coastal Amount of coast			Upland/Lowland Proportion of square > 500′				
	Nil	**1-5**	**6-10**	**11-15**	**16-22**	**None**	**<10%**	**≥10%**	**≥½**	**≥¼**	**<¼**	**None**	**Total**
Surveyed squares	38	104	52	24	16	17	14	203	7	25	73	129	234
% of surveyed squares	16	45	22	10	7	7	6	87	3	11	31	55	100
Grid squares in which recorded	1	11	1	2	1	0	0	16	0	1	7	8	16
% records in each category	6	69	6	13	6	0	0	100	0	6	44	50	100
Significant variation													

02 Gentoo Penguin

Pygoscelis papua papua

A. Pingüino de pico rojo
C. Pingüino papua

History
The Gentoo Penguin was first described in 1781 by IR Forster from a specimen brought to London from the Falklands by JS Hausmann (Hellmayr & Conover 1948). Clayton (1774) reported Gentoos on Saunders Island but gave no indications of abundance. Brooks (1917) learned from the captain of a schooner that his crew had annually killed about 70,000 Gentoos by corralling them in thousands, clubbing them and rendering them down in try-pots and he noted that, 'A penguin furnished about a pint of oil, a gallon of which brought 2/6d'. An earlier report from Adolphe Lecomte (Sclater 1868) mentioned four vessels collecting 50,700 gallons of penguin oil in 1867. Vallentin (in Boyson 1924) and Murphy (1936) assumed this referred to the Rockhopper Penguin, but Sclater did not name the species and Gentoos would have been easier to corral than Rockhoppers.

In 1932/33, AG Bennett (1933) as Acting Government Naturalist, estimated that there were over 127,600 pairs (70,070 on West Falkland and 45,950 on East with a further 10% for several islands not counted). Croxall *et al.* (1984) noted Strange's surveys between 1965 and 1970 and estimated a breeding population of about 100,000 pairs. Eggs of this species have long been collected for human consumption and, according to Hamilton (in Cott 1953), were favoured over those of Rockhopper and Magellanic Penguins. Cobb (unpublished) noted about 1922 that eggs were sold 'at eight shillings per 100 in Stanley'. Cawkell & Hamilton (1961) said that large numbers were still taken, in theory under Government licence, and cited 10,000 from Stop Cove on Weddell Island in one season 'in living memory' noting that the colony had now disappeared. Cawkell (in Cott 1954) said that excessive egg collecting had led to a great decrease in numbers and he gave the price of Gentoo eggs in Stanley in 1952 as two shillings for 25 (the price quoted by Cobb 30 years earlier). He noted that the colony at Sparrow Cove near Stanley was protected and the eggs there were dyed so that anyone caught with such an egg was liable to prosecution. In 1962, 15,000 eggs were said to be collected from Sea Lion, Speedwell and Bleaker Islands and Albemarle and eggs were still being collected for food under licence in 1983.

Habitat and Breeding
The Gentoo breeds in groups of smaller colonies, or large colonies of up to several thousands, on fairly flat ground, up to about 200m above sea level usually near the coast and on offshore islands (Pettingill 1964). Colonies are sometimes several kms inland; birds travel between the sea and their nests using traditional routes, sometimes walking circuitously farther than is apparently necessary. At mainland colonies, nests are made of available materials, mainly diddle-dee which is torn up and piled in a heap. Each year, colonies change their sites slightly and excreta adds nitrogenous material to the soil, encouraging the growth of grass and improving the pasture. On Beauchêne Island, where there is no diddle-dee, nests are rudimentary heaps of pebbles with mud lumps or albatross bones. Two large round white eggs are laid from the first week of October, occasionally late September. Smaller replacement eggs may be laid up to December, after earlier clutches have been robbed (Murphy 1936). Incubation takes about five weeks and fledging to independence a further three months.

Breeding Birds Survey: analysis of data
In 27 grid squares estimates of 100-1,000 pairs were reported and in a further 16 squares, 1,000+ pairs. In six others there were estimates of between 2,500 and 6,000 pairs. Thompson (1990) at Beaver Island found nine subcolonies containing a total of about 1,200 pairs while at Bull Point, six colonies contained a total of at least 6,000 pairs. A few counts give conflicting impressions about the stability of the population. Josie Larsen reported a large increase on Speedwell Island from 1985 to 1986. David Osborn (1991) reported estimates of Gentoos on Sea Lion Island by Dave Gray and David Hosking in 1988 as about 2,500 pairs. By the beginning of the 1990 season the same observers felt that the population had declined to about 800 pairs, a decrease of nearly 70%. David Osborn speculated that excessive pressure from visitors may have affected breeding success on Sea Lion Island. Records show a highly significant bias towards the coastal squares, which held all but one of the 86 records ($p = 0{\cdot}006$) and a significant tendency ($p = 0{\cdot}03$) for more records from squares with less than one quarter of the land above 500 feet.

02 Gentoo Penguin *Pygoscelis papua*

Detectability: **5 Very High**

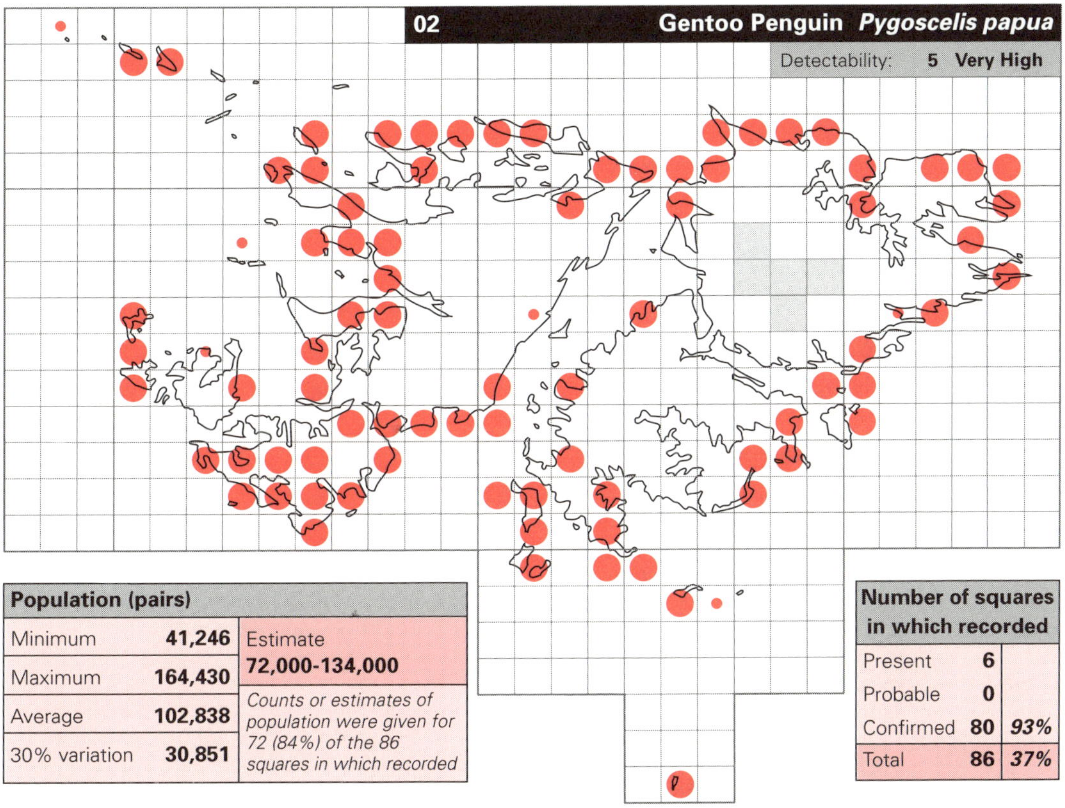

Population (pairs)		
Minimum	**41,246**	Estimate **72,000-134,000**
Maximum	**164,430**	
Average	**102,838**	*Counts or estimates of population were given for 72 (84%) of the 86 squares in which recorded*
30% variation	**30,851**	

Number of squares in which recorded		
Present	**6**	
Probable	**0**	
Confirmed	**80**	***93%***
Total	**86**	***37%***

Comparisons between species records and topography	**Dry/Wet** Number of wet tetrads					**Inland/Coastal** Amount of coast			**Upland/Lowland** Proportion of square > 500′				
	Nil	**1-5**	**6-10**	**11-15**	**16-22**	**None**	**<10%**	**≥10%**	**≥1/2**	**≥1/4**	**<1/4**	**None**	**Total**
Surveyed squares	38	104	52	24	16	17	14	203	7	25	73	129	234
% of surveyed squares	16	45	22	10	7	7	6	87	3	11	31	55	100
Grid squares in which recorded	10	50	17	6	3	0	1	85	2	3	40	41	86
% records in each category	12	58	20	7	3	0	1	99	2	3	47	48	100
Significant variation								**			*		

World Distribution

The Gentoo Penguin has a circumpolar distribution. The larger race *Pygoscelis p. papua* breeds on many subantarctic islands including the Falklands. The smaller race *P. p. ellsworthi* breeds on the Antarctic Peninsula south to about 65°S and adjacent islands. Gentoo Penguins generally remain near their colonies throughout the year, though some have been reported on the coast of Argentina to about 43°S. The minimum world population was recently estimated at 314,000 pairs (Woehler 1993); South Georgia and the Falklands both have about 100,000 pairs. Populations appear to fluctuate substantially from year to year though data from most areas indicate essentially stable population sizes (Woehler 1993).

References

TBFI (1975) p68-9, GBFI (1988) p94-5.

Implications for Conservation

The Falklands hold one of the world's most important populations of this species. Dietary studies under the Seabird Monitoring programme of Falklands Conservation have shown that the Gentoo Penguin can exploit a wide range of prey sizes and types and that it takes *Loligo* squid of a size that are fished commercially and blue whiting smaller than those that could be harvested. Counts by local residents for Falklands Conservation at some East Falkland colonies indicate a decline from 1987/88 to 1993/94 averaging about 3% per year (Bingham 1994). Continuing studies are required to monitor the effects of increased fishing activity and the development of tourism. Egg collecting at vulnerable sites may also have to be considered as a potential threat, though comparisons with the few estimates preceding this survey suggest that the Falkland population is about the same size as it was 60 years ago.

03 Rockhopper Penguin

Eudyptes chrysocome chrysocome

A. & C. Pingüino de penacho amarillo

Local name: Rocky

History

The Rockhopper Penguin was first named scientifically from Falkland Islands specimens by JF Miller in 1784 (Hellmayr & Conover 1948). Clayton (1774) had already noted the 'very small' size of the 'Jumping Jacks' on Saunders Island and their nesting association with Black-browed Albatrosses. In 1797, Fanning (1833) visited New Island and found both species breeding. He also noticed sea lions capturing and eating Rockhoppers as they approached their landing places. Rockhopper eggs were a favourite food for early travellers and settlers in the islands. Fanning described in detail how he preserved the eggs by soaking them in seal oil and packing them in barrels of sand. Under the heading 'Rockhopper Penguin', both Vallentin (in Boyson 1924) and Murphy (1936) mention Sclater's (1868) report that in 1867 there were four vessels engaged in the 'Penguin and Seal fishery' around this archipelago, that they had made 50,700 gallons of penguin oil and that eight penguins were estimated to make one gallon of oil. Murphy comments, 'this implied the destruction of nearly half a million birds during a single season'. However, the species of penguin involved is not stated in Sclater's notes; Gentoos would appear to be a much more likely target for penguin oil collectors, as noted by Brooks (1917). In 1915, Beck visited Kidney Island and reported that more than 25,000 eggs were collected there in December that year. Many thousands were preserved for winter use and, because some colonies were seldom visited he thought it was likely that 'this excellent and inexpensive food will long continue to form a welcome change from the steady diet of mutton' (Beck 1917).

In the 1932/33 season, AG Bennett organised a census of penguins by asking observers to report the number of nests from which he felt the approximate number of adults could be obtained by doubling, though he considered the figures given were conservative. He quoted the number of nests for each camp as follows: Port Howard 15,000; West Point Island 50,000; Passage Islands 10,000; Port Stephens 2,000; Pebble Island 6,000; Jasons 2,625,000; Roy Cove 7,200; New Island 130,000; Saunders Island 3,400; Kidney Island etc 12,000; Port Louis 3,000; Sea Lions 150,000; Johnsons Harbour 150,000; San Carlos 500 and Bleaker Island 5,000. With an additional 10% for those colonies 'uncounted' Bennett suggested that, excluding Beauchêne Island, there was a breeding population of almost 3·5 million pairs (Bennett 1933). The Rockhoppers on Beauchêne were first counted in 1980, when Smith & Prince (1985), using sample counts and calculations of the area covered, estimated about 300,000 pairs.

Habitat and Breeding

Rockhoppers return in early October from winter pelagic wanderings to breed in close-packed colonies on cliffs, often in association with Black-browed Albatrosses and King Shags. Two pear-shaped, whitish, limy eggs are laid between the last days of October and mid-November, in a shallow depression on the ground. The first egg is much smaller and this chick is rarely reared to fledging. Immature and nonbreeding birds moult from mid-January, breeding adults a little later; most have completed moult by late March and the colonies are deserted by late April.

Breeding Birds Survey: analysis of data

Croxall *et al.* (1985) commented that the Rockhopper was the most abundant species of penguin in the Falklands and therefore the most important avian consumer of marine resources in Falkland waters. The survey shows that it is also the most numerous bird species, with at least half a million pairs breeding at about 30 colonies, the majority around the coasts of West Falkland. All records of this species came from grid squares with a considerable length of coastline. A significant ($p < 0{\cdot}01$) proportion of these squares have less than one quarter of the land above 500 feet and the others have no land above 500 feet.

Historical data on Populations

Published and unpublished records from three localities show a decline over at least the past 60 years. Some information lacks precision such as that given by Bennett for the 'Jasons', where he quoted 2,625,000 nests but did not list the islands included. In 1988, observers for the Seabird Monitoring Programme calculated 190,000 pairs on Steeple Jason alone, but it is impossible to detect any trend from comparing these figures.

The colony at Kidney Island has declined, as the following table shows. The figures for 'eggs' are the reported numbers collected, but it is not known whether they are from one collecting trip

03 Rockhopper Penguin *Eudyptes chrysocome*

Detectability: **5 Very High**

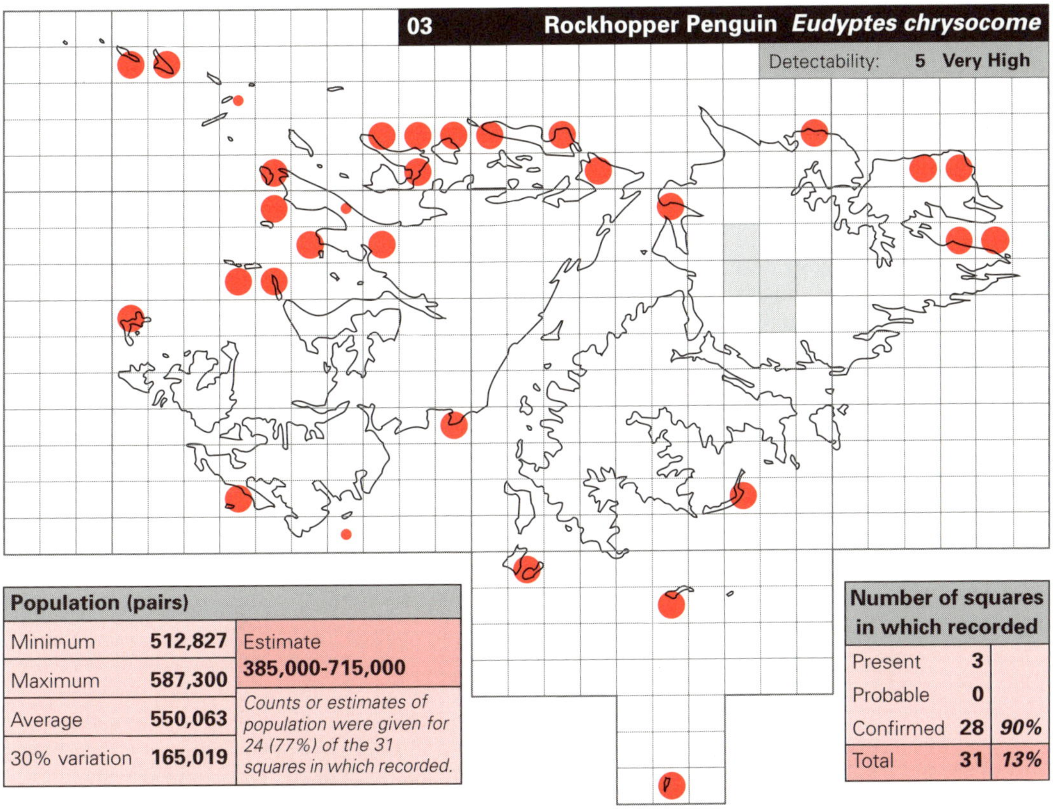

Population (pairs)		
Minimum	**512,827**	Estimate **385,000-715,000**
Maximum	**587,300**	
Average	**550,063**	*Counts or estimates of population were given for 24 (77%) of the 31 squares in which recorded.*
30% variation	**165,019**	

Number of squares in which recorded		
Present	**3**	
Probable	**0**	
Confirmed	**28**	***90%***
Total	**31**	***13%***

Comparisons between species records and topography	Dry/Wet: Number of wet tetrads					Inland/Coastal: Amount of coast			Upland/Lowland: Proportion of square > 500′				
	Nil	**1-5**	**6-10**	**11-15**	**16-22**	**None**	**<10%**	**≥10%**	**≥1/2**	**≥1/4**	**<1/4**	**None**	**Total**
Surveyed squares	38	104	52	24	16	17	14	203	7	25	73	129	234
% of surveyed squares	16	45	22	10	7	7	6	87	3	11	31	55	100
Grid squares in which recorded	5	17	6	1	2	0	0	31	0	0	20	11	31
% records in each category	16	55	19	3	7	0	0	100	0	0	65	35	100
Significant variation											**		

or several, nor whether all eggs were taken. Therefore, they are not strictly comparable.

Year	Numbers	Source
1911	85,000 eggs	EM Cawkell (24 August 1953) in Cott (1954)
1915	25,000 eggs	RH Beck (1917)
1917	15,000 eggs	B Roberts (5 November 1951) in Cott (1953)
1930	9,500 eggs	B Roberts (5 November 1951) in Cott (1953)
1932/33	12,000 nests	AG Bennett (1933, unpublished)
1934	7,000 eggs	B Roberts in Cott (1953)
1936	5,000 eggs	B Roberts in Cott (1953)
1952	1,000 eggs	EM Cawkell in Cott (1954)
1960	3,000 pairs	RW Woods (1970a)
1983	600 nests	RW Woods
1990	24 nests	P Gregory (*Penguin News* Vol.3 No.1, 25 Jan.1991)
1995	100 pairs	RW Woods

Comparing the numbers of nests over the 63-year period from 1932/33 to 1995/96, the colony has apparently been reduced by about 11,900 pairs (99·2%) or about 1·6% per year. However, comparison between the lowest count in 1990 and the 1995 figure suggests a slight recovery.

There are also some figures for West Point Island:

1932/33	50,000 nests	AG Bennett (1933, unpublished)
1960	8,000 pairs	RB Napier
1993	2,475 pairs	Falklands Conservation, monitoring group
1995	4,000 pairs	M Riddy (Falklands Conservation)

The comparison of the nests/pairs over the 63-year period from 1932/33 to 1995/96 shows a very large reduction of 92%, or about 1·5% per year. However, there are again signs that the colonies are recovering.

The similarity of the long-term decline in population at Kidney and West Point Islands suggests that powerful influences are operating. The effect of collecting eggs for food is thought to have been serious in the 19th century and up to the middle of the 20th century. Hamilton (in Cott 1953) believed that colonies near Stanley had been depleted by excessive collecting and that less than 10,000 eggs were taken annually in the late 1940s. Egg collecting has declined considerably in the last thirty or so years. There has been evidence from Campbell Island south of New Zealand, that changes in ocean currents have led to warming of surface waters, which has changed the availability of the marine animals upon which Rockhoppers feed (Moors 1986). The Seabird Monitoring programme of Falklands Conservation has shown that Rockhoppers depend heavily on euphausiid Crustacea with squid apparently being more important during chick rearing and fish being of relatively minor importance. In general, it seems that commercial fishing activities have little effect on Rockhoppers during the breeding season. However, diet samples from adults returning to moult at West Point Island in March 1990 contained about 11% squid remains, of which the majority were of the commercially fished *Loligo* species (Thompson, 1989). Large numbers of Rockhoppers were found dead or dying in the latter part of the 1985/86 season. The evidence collected from corpses and from samples obtained in February 1987, suggested that starvation due to a shortage of their main source of food caused the unusually high mortality. The shortage happened to coincide with the post-breeding moult, when they are most susceptible to additional stress (Keymer 1988).

World Distribution
The Rockhopper Penguin has a circumpolar distribution in the subantarctic zone, with three races recognised; *E. c. chrysocome* breeds in the Falklands and islands off Tierra del Fuego; *E. c. moseleyi* breeds at Tristan da Cunha, Gough, St Paul and Amsterdam Islands; *E. c. filholi* breeds at other southern Indian Ocean and southern Pacific Ocean islands south of New Zealand. At least two pairs of *E. c. chrysocome* were found breeding amongst Macaroni Penguin colonies on South Georgia (Prince & Payne 1979) and moulting immatures have occurred with a large Macaroni Penguin colony (Prince & Croxall 1996). An adult *E. c. moseleyi*, with distinctive long head plumes, was present at Macbride Head, East Falkland in December and January 1995/96 and was photographed by Mike Morrison. The extent of winter pelagic movements is unknown but Rockhoppers have reached 35°S off Argentina. The world population for all three races was recently estimated to be at least 3,687,600 pairs (Woehler 1993) suggesting that it is the third most numerous penguin. The Falklands may still hold the world's largest breeding population.

Implications for Conservation
The magnitude of the decline in population of Rockhoppers in the Falklands is similar to those suggested for the Auckland, Antipodes and Campbell Islands (Croxall 1994). The causes of these long-term and very significant reductions in numbers are likely to be complex. Whatever the operating factors, it is important that ecological studies of Rockhopper Penguins are continued and extended, in order to gain a better understanding of their dietary requirements and other factors affecting their survival.

References
TBFI (1975) p69-71, GBFI (1988) p96-7.

04 Macaroni Penguin

Eudyptes chrysolophus

A. Pingüino anaranjado
C. Pingüino macaroni

History
The Macaroni Penguin was first recorded in 1860 by Captain Abbott who saw 15 among a colony of about 20,000 Rockhoppers in the North Camp of East Falkland and he commented later (1861) that it was by no means common. Sclater (1860) noted that Captain Pack had sent skins to the British Museum and later (1868) quoted Lecomte's remark that the Macaroni did not make its own colonies but was found sparingly mixed among the Rockhoppers. Wace (1917 unpublished) recorded that it bred, but Dabbene in his spanish language version of Wace's paper (1921) confusingly added that it was common while Bennett (1926) wrote that it 'only appears as a straggler in the Falklands'. In his 1932/33 report on penguins for the Falkland Islands Government, Bennett (1933) commented on the small number of Macaronis, 'probably immature birds' and excluded them from the survey. Cawkell & Hamilton (1961) stated that they bred amongst the Rockhoppers 'in extremely small numbers', that the small rookery at Eagle Point,

04 Macaroni Penguin *Eudyptes chrysolophus*

Detectability: **3 Moderate**

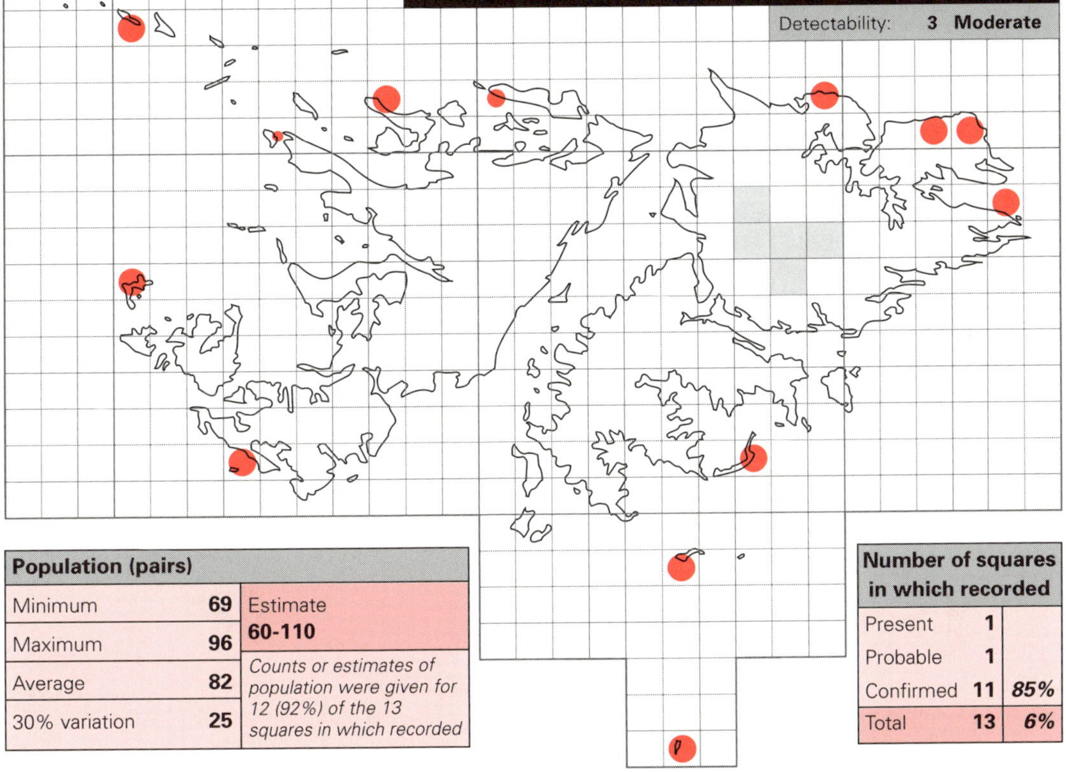

Population (pairs)		
Minimum	69	Estimate **60-110**
Maximum	96	
Average	82	*Counts or estimates of population were given for 12 (92%) of the 13 squares in which recorded*
30% variation	25	

Number of squares in which recorded		
Present	1	
Probable	1	
Confirmed	11	*85%*
Total	13	*6%*

Comparisons between species records and topography	Dry/Wet: Number of wet tetrads					Inland/Coastal: Amount of coast			Upland/Lowland: Proportion of square > 500′				
	Nil	1-5	6-10	11-15	16-22	None	<10%	≥10%	≥1/2	≥1/4	<1/4	None	Total
Surveyed squares	38	104	52	24	16	17	14	203	7	25	73	129	234
% of surveyed squares	16	45	22	10	7	7	6	87	3	11	31	55	100
Grid squares in which recorded	2	6	3	0	2	0	0	13	0	0	7	6	13
% records in each category	>15	46	23	0	>15	0	0	100	0	0	54	46	100
Significant variation													

Johnson's Harbour in the 1920s had now gone and that it had also been recorded at Port Stephens, West Point, Kidney and Bleaker Islands in recent years. At Seal Bay in Port Louis camp, Bruce Withers reported a few pairs among Rockhoppers in November 1958 and on 16 January 1960, Tony Nelson counted at least 32 adults nesting in the same area. A single Macaroni was found in a Gentoo rookery on Bleaker Island by Archie Short in March 1959 and in the 1961/62 season he saw five birds, the highest number recorded since 1949. A single pair was seen each season from 1960 to 1962 on Kidney Island.

Habitat and Breeding

It is migratory but there is very little information on arrival or departure dates or other aspects of breeding in the Falklands. It is similar to the Rockhopper in laying two white eggs in November and usually only the larger second egg is incubated successfully.

Breeding Birds Survey: analysis of data

It was found in small numbers at Rockhopper colonies, well-spread around the coasts of both East and West Falkland and all records come from grid squares with substantial amounts of coastline. Mixed pairs of Macaroni x Rockhopper Penguins were recorded at Sea Lion Island by Shane Wolsey and at Kidney Island by Tui de Roy and Mark Jones in 1986/87 but no hybrid offspring have been reported. However, in December 1995 Ian Strange pointed out an incubating adult at New Island with some bill and head plumage characteristics of both species.

There were unsubstantiated reports of a small colony of Royal Penguins *Eudyptes (chrysolophus) schlegeli* in the 1991/92 season, somewhere in the West Falkland area but no confirmatory information has been provided nor descriptions published. This penguin, only known to breed at Macquarie Island south of New Zealand, has a range of facial colouration which makes specific separation difficult; it is often regarded as a subspecies of the Macaroni Penguin (Marchant & Higgins 1990).

World Distribution
The Macaroni is the most numerous penguin species with a world population of about 9,000,000 pairs (Croxall 1994). The Falklands are at the northern edge of its breeding range. Colonies with up to one million pairs are found at several islands or groups between 45°S and 65°S in the southern Atlantic and Indian Oceans. South Georgia has many large colonies with a total population estimated at about 2·7 million pairs.

Implications for Conservation

There appears to be no noticeable decline in Macaroni Penguin numbers in the Falklands. It is likely to be as susceptible to environmental change as the Rockhopper Penguin and further study of its breeding habits here would be interesting.

References
TBFI (1975) p71-3, GBFI (1988) p97.

---

05 Magellanic Penguin

Spheniscus magellanicus

A. & C. Pingüino de Magallanes
Local name: Jackass

History
It was first named by Forster in 1781 but it is not known whether the specimen came from the Falklands or the Magellanic region (Hellmayr and Conover 1948). Clayton (1774) called it the 'Holey' Penguin as it burrowed under the sand hills like rabbits. Interestingly, Abbott (1860) notes that near Stanley the burrows 'run in to a considerable distance, on account, I conclude, of their being so often robbed of their eggs, which are taken out by means of a piece of iron-hoop fastened to the end of a pole. ... At a distance from the settlement, these penguins breed in holes close to the surface of the ground'. Vallentin (in Boyson 1924) thought they were increasing in numbers, at least up to the end of 1911 when he left the islands and there is no evidence that these birds were killed for their oil. Bennett (1926) stated that it was a common breeding species and though in his survey of penguins in 1932/33 it could not be counted because it nested in burrows, he speculated that it was not less numerous than the Gentoo (127,600 pairs). Cobb (1933) commented that 'acres of good ground are spoiled by these burrowing penguins' and added that, 'on ground badly riddled with holes, horses generally refuse to proceed, which adds another difficulty to sheep-gathering in these parts'. Cobb said that the eggs were more like hens eggs than those of the other penguins but they were 'difficult to collect in large numbers, and the collector runs the risk of being smothered in Jackass fleas'. Magellanic Penguins defend their eggs vigorously with the body blocking the burrow and the hooked heavy bill, which can inflict a nasty bite, facing outwards. Hamilton considered that the annual crop of eggs collected was 'rather less than 5,000' (Cott 1953).

Habitat and Breeding
It generally returns to coastal regions in early September and leaves about mid-April, though a few may be seen during the winter. It breeds in large or small colonies, burrowing into soft soil or peat on slopes facing the sea. Most colonies are on ground that formerly carried dense tussac grass and burrows are still found in the very soft waterlogged peat that accumulates beneath mature tussac on offshore islands. The burrows slope downwards, are up to 2m (6 feet) deep and have a nest-chamber slightly higher than the adjacent tunnel floor allowing rain water to collect away from the eggs. Two white eggs are laid between mid-October and mid-November on a nest of diddle-dee or tussac and they rapidly become stained by the damp peaty soil.

Breeding Birds Survey: analysis of data
Analyses of records confirm its very high occurrence in grid squares with a considerable length of coastline. There was also a strong but non-significant tendency for more records to come from squares with no land over 500 feet. In the concurrent Seabird Monitoring programme of Falklands Conservation, it has been found that this species largely takes small crustaceans (lobster-krill), small fish and mostly smaller species of squid than are fished commercially. Evidence has also been obtained of variability in diet between the colonies on Keppel Island and New Island (Thompson 1989). Samples from a study area at

05 Magellanic Penguin *Spheniscus magellanicus*

Detectability: **5 Very High**

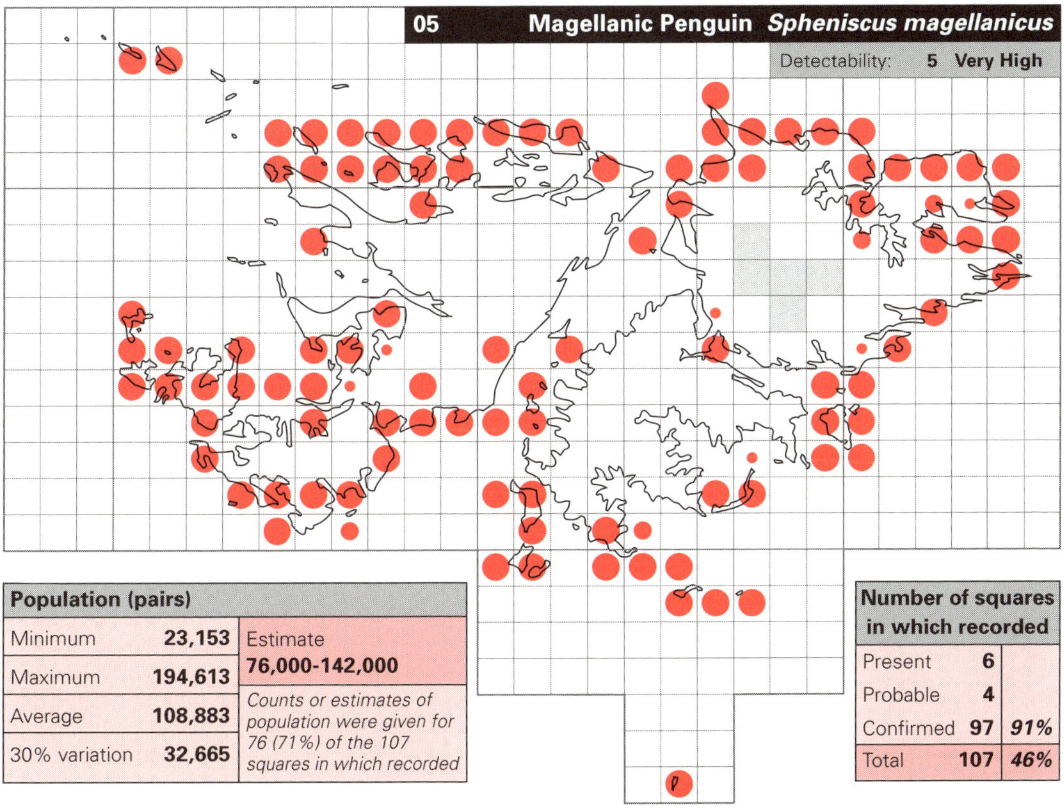

Population (pairs)		
Minimum	**23,153**	Estimate **76,000-142,000**
Maximum	**194,613**	
Average	**108,883**	*Counts or estimates of population were given for 76 (71%) of the 107 squares in which recorded*
30% variation	**32,665**	

Number of squares in which recorded		
Present	**6**	
Probable	**4**	
Confirmed	**97**	***91%***
Total	**107**	***46%***

Comparisons between species records and topography	Dry/Wet: Number of wet tetrads					Inland/Coastal: Amount of coast			Upland/Lowland: Proportion of square > 500′				
	Nil	**1-5**	**6-10**	**11-15**	**16-22**	**None**	**<10%**	**≥10%**	**≥1/2**	**≥1/4**	**<1/4**	**None**	**Total**
Surveyed squares	38	104	52	24	16	17	14	203	7	25	73	129	234
% of surveyed squares	16	45	22	10	7	7	6	87	3	11	31	55	100
Grid squares in which recorded	15	57	19	10	6	0	1	106	0	5	38	64	107
% records in each category	14	53	18	9	6	0	<1	>99	0	5	35	60	100
Significant variation								**				♦	

Volunteer Point in 1989/90 contained much less fish than the New Island samples of 1986/87 and they included more than 10% of the commercially fished patagonian squid (Thompson 1990). Roddy Napier reported in 1993 that the population at West Point Island has been declining in recent years and sample counts made by Falklands Conservation at Volunteer Point and Bull Point since 1990 indicate a possible decline of up to 10% for each year. These studies suggest that shortage of food when adults are feeding chicks has led to a low survival rate for the chicks (Bingham 1994).

World Distribution

The Magellanic Penguin breeds on the Pacific coast of South America from about 37°S in Chile and on the Atlantic coast from about 41°S in Argentina to the islands south of Tierra del Fuego. It ranges north in winter to about 30°S off Chile and to southern Brazil, occasionally reaching Rio de Janiero (23°S). There are seven records of vagrants at South Georgia, all in the austral autumn (Prince & Croxall 1996) and vagrants have reached New Zealand. The world population has been roughly estimated as between 4·5 and 10 million birds. It is not globally threatened; some colonies have declined and some new ones have been noted on the coast of Argentina, including Punta Tombo where there are about 225,000 pairs and Punta Clara with about 70,000 pairs (del Hoyo *et al.* 1992).

Implications for Conservation

The Falkland population of this penguin, possibly less than 10% of the world population, is not thought to be under any particular threat. However, these birds are susceptible to disturbance when nesting and potentially vulnerable to accidental oil spills or discharge of waste by vessels offshore.

References

TBFI (1975) p73-4, GBFI (1988) p98-9.

06 White-tufted Grebe

Rollandia rolland rolland

A. Macá común

C. Zambullidor pimpollo

Local names: Brown, Black, Golden or Rolland's Grebe, Hell Diver

History
The White-tufted (or Rolland's) Grebe was first described by MM. Quoy and Gaimard (in Freycinet 1824), from specimens collected in the Falkland Islands by M. Rolland, a master gunner of the *Uranie*. Garnot (1826) gave a brief description and said that it could be found on the sea and the edges of rivers while Darwin collected specimens in March 1833 or 1834 and stated that it was 'common at the head of the tortuous bays which intersect these islands' (Gould & Darwin 1838-41). Abbott (1861) felt that it was rather common on fresh and salt water, 'though more frequent on the freshwater streams'. Vallentin (1904) described it as generally distributed at least over northern East Falkland and later, after his visit from 1909-11, he noted (in Boyson 1924) that it was found in all the larger ponds and pools on West Falkland. Brooks (1917) remarked that it was 'fairly common by hearsay but quite local actually' and recorded it as breeding on varying sized ponds not far from the sea but in winter it was 'found on salt water generally near the shore'. Bennett's (1926) comment that it was 'quite common, though not abundant' was perhaps clarified by Cawkell & Hamilton (1961) who stated that it was widely distributed but nowhere plentiful, commonest in Lafonia, 'but the population is quite small'. Pettingill (1960) described Rolland's Grebe as one of the two species, with the Logger duck, considered endemic in the Falklands.

Habitat and Breeding
When breeding, it particularly favours freshwater ponds with emergent vegetation and is found on streams and small rivers. It also occurs in creeks and coastal waters, more frequently in winter. It breeds between mid-October and January laying 1-3 eggs which are ivory white at first but rapidly become stained. The floating nest of waterweeds and grass is placed under the overhanging bank of a pond or stream, sometimes in rushes, and is difficult to detect.

Breeding Birds Survey: analysis of data
Rolland's Grebe is mainly found in the lowlying parts of the Falklands, particularly in Lafonia and south of the Wickham Heights along the northern side of Choiseul Sound. Some offshore islands (Pebble, Saunders, Speedwell, Bleaker and Lively) support small colonies and it also breeds at ponds in the northern part of East Falkland and central West Falkland north of Fox Bay. Although this grebe is not particularly conspicuous and some birds may have been overlooked, it can probably only be described as locally common with a small total population. The highest single count in one square (UC30, Fanny Cove, south of North Arm) was of nine pairs. Only one of the comparisons with the topographical analyses shows a highly significant result ($p = 0{\cdot}0002$): a very high proportion of the squares in which it was recorded had between 6 and 22 wet tetrads. There were no significant results relating to coastal or inland squares but there was a tendency for more records to come from squares with no land above 500 feet.

World Distribution
Rolland's Grebe is so much larger than the two mainland races of *R. rolland* that Wetmore (1926) felt it was sufficiently distinct to be recognised as a separate species. Fjeldså & Krabbe (1990) describe the Falkland form as 'gigantic', currently classified as a megasubspecies or distinct taxon which does not intergrade with the other smaller races described. The smallest form *R. r. chilensis* occurs in South American lowlands from 10°S in Peru through Argentina to Tierra del Fuego. *R. r. morresoni*, intermediate in size, breeds in the Andes of central Peru at Lake Junin.

Implications for Conservation
Further investigation is needed to establish whether it is an endemic species, distinct from the mainland forms. Rolland's Grebe is rare in the Falklands and possibly less numerous than formerly. Studies of the wetland habitats important for its survival, may be useful in safeguarding its future.

References
TBFI (1975) p74-5, GBFI (1988) p99-100.

06 ?** White-tufted Grebe *Rollandia rolland*

Detectability: 3 Moderate

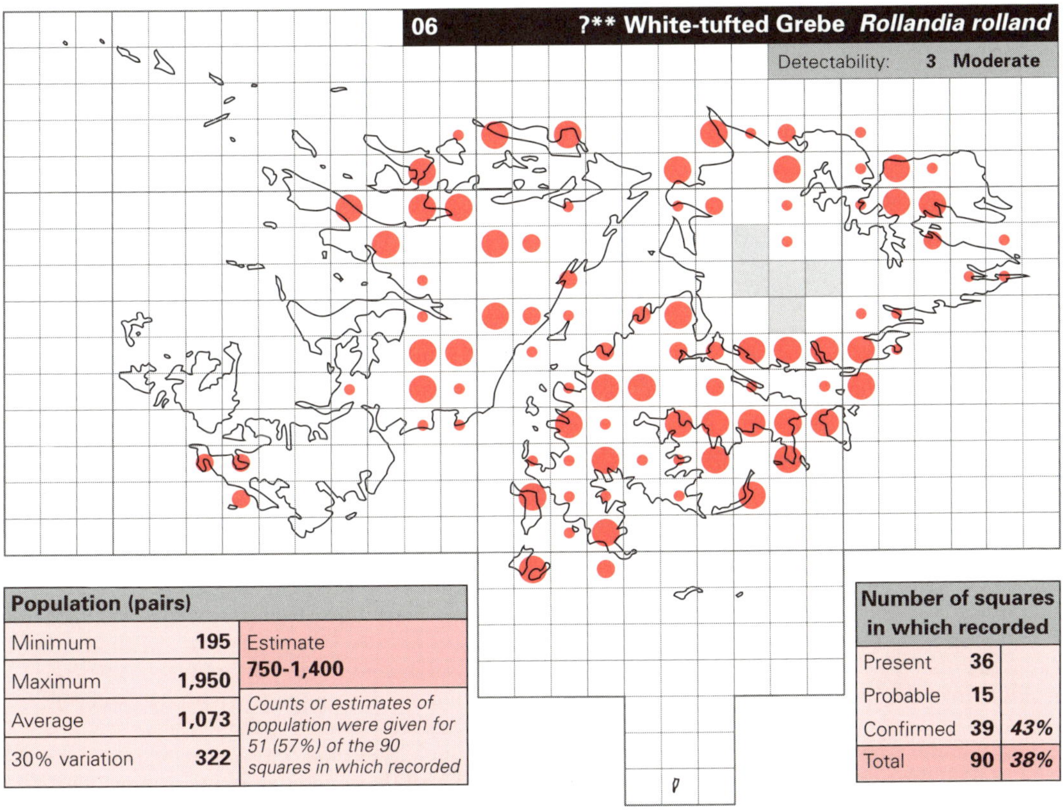

Population (pairs)		
Minimum	195	Estimate **750-1,400**
Maximum	1,950	
Average	1,073	*Counts or estimates of population were given for 51 (57%) of the 90 squares in which recorded*
30% variation	322	

Number of squares in which recorded		
Present	36	
Probable	15	
Confirmed	39	*43%*
Total	90	*38%*

Comparisons between species records and topography	Dry/Wet Number of wet tetrads					Inland/Coastal Amount of coast			Upland/Lowland Proportion of square > 500′				
	Nil	1-5	6-10	11-15	16-22	None	<10%	≥10%	≥1/2	≥1/4	<1/4	None	Total
Surveyed squares	38	104	52	24	16	17	14	203	7	25	73	129	234
% of surveyed squares	16	45	22	10	7	7	6	87	3	11	31	55	100
Grid squares in which recorded	1	32	29	18	10	9	6	75	1	9	26	54	90
% records in each category	1	36	32	20	11	10	7	83	1	10	29	60	100
Significant variation			***	***	***							♦	

07 Silvery Grebe

Podiceps occipitalis occipitalis

A. Macá plateado

C. Zambullidor blanquillo

Local name: White Grebe

History
The Silvery Grebe was first named scientifically as *P. occipitalis* by Garnot in 1826, from a specimen obtained in November/December 1822 probably in the Port Louis area. Lesson and Garnot later named it *P. kalipareus* in 1827. Darwin noted that he saw 'very few individuals and these only on one small fresh water lake' (Gould & Darwin 1838-41). Abbott (1861) found it only in the interior of East Falkland, 'on the small inland ponds'. Vallentin (in Boyson 1924) noted that it was 'very rare' and he apparently only saw it on Byron Sound lagoon, West Falkland. Bennett (1926) felt that it was a less common breeding resident than the White-tufted Grebe, an opinion repeated by Cawkell (1960) who stated that the White-tufted Grebe was 'by far the more common' of the two species. In Cawkell & Hamilton (1961) however, it was stated that the Silvery Grebe was 'more numerous than *P. rolland*', which must have been a typographical error because no other observers have claimed that there were more Silvery Grebes.

Habitat and Breeding
This species is often found on the same ponds as the White-tufted Grebe. Though individually small and inconspicuous, pairs are more easily noticed when breeding because they are often gregarious. At Garden Point pond, North Arm in 1995, following the survey, adults were sitting on 16 nests in a band about 50m long, parallel to and about 20m in from the bank, within thick waterweeds. Pairs are in territory from September and 2 blue-white eggs are laid in the floating weed nest between November and January. Silvery Grebes may be found in coastal kelp patches in autumn and winter.

Breeding Birds Survey: analysis of data
Breeding was confirmed in fewer than one third of the grid squares in which it was recorded. Most records were from the region between the southern side of the Wickham Heights and central Lafonia, with others in the generally lowlying parts of East Falkland. Flocks of up to 35 have been seen in March and April at ponds south of Mount Pleasant Airport but there is no evidence that the Falkland population migrates to the continent for the winter. The records from West Falkland include the well-known wetlands on the eastern half of Pebble Island and the Hawks Nest ponds area north of Fox Bay. It was also recorded from Saunders Island, South Jason and Carcass Island but seems to be absent from southern West Falkland. The population estimate suggests that there are about three pairs of White-tufted Grebes to two pairs of Silvery Grebes. The distribution of Silvery Grebe records is very strongly biased ($p = 0{\cdot}0002$) towards squares with 6-22 wet tetrads, suggesting that standing freshwater is a very important factor. There are also non-significant tendencies for it to occur more often in squares with very little or no coastline rather than coastal squares and to be found slightly more often in the squares with no land above 500 feet.

World Distribution
Two races of the Silvery Grebe have been described. *P. o. occipitalis* also occurs in South America from Tierra del Fuego to about 25°S in Chile and Argentina, usually at altitudes below 1300m (Fjeldså & Krabbe 1990). It is migratory in the south, arriving in Tierra del Fuego in late September and leaving by May. Hellmayr & Conover (1948) remark that the two poor Falkland skins they examined were slightly larger than the average of continental birds, otherwise they did not seem to be different. *P. o. juninensis* (with brown plumes and white foreneck) occurs in the Andes of northern Chile, northwestern Argentina, Bolivia, Ecuador and Peru. This andean race should perhaps be classed as a distinct species, on the basis of differences in body structure and behaviour (del Hoyo *et al.* 1992).

07 Silvery Grebe *Podiceps occipitalis*

Detectability: 3 Moderate

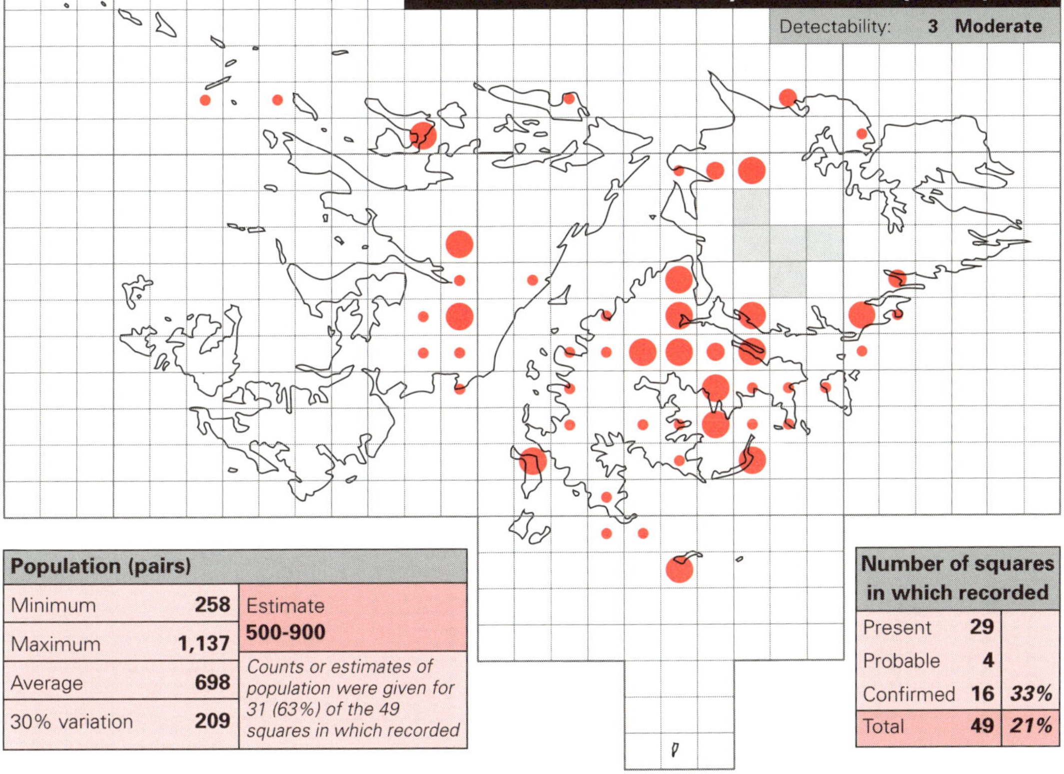

Population (pairs)		
Minimum	258	Estimate **500-900**
Maximum	1,137	
Average	698	*Counts or estimates of population were given for 31 (63%) of the 49 squares in which recorded*
30% variation	209	

Number of squares in which recorded		
Present	29	
Probable	4	
Confirmed	16	*33%*
Total	49	*21%*

Comparisons between species records and topography	Dry/Wet Number of wet tetrads					Inland/Coastal Amount of coast			Upland/Lowland Proportion of square > 500′				
	Nil	1-5	6-10	11-15	16-22	None	<10%	≥10%	≥1/2	≥1/4	<1/4	None	Total
Surveyed squares	38	104	52	24	16	17	14	203	7	25	73	129	234
% of surveyed squares	16	45	22	10	7	7	6	87	3	11	31	55	100
Grid squares in which recorded	0	13	18	10	8	4	6	39	1	3	10	35	49
% records in each category	0	27	37	20	16	8	12	80	2	6	>20	>71	100
Significant variation			***	***	***		♦					♦	

Implications for Conservation

Only a very small proportion of the world population is found in the Falklands. A taxonomic investigation is necessary to determine whether it forms a distinct insular race. There do not appear to be any special threats in the Falklands, though oil-spills in coastal waters could affect it during the winter. The Silvery Grebe is largely dependent on the most extensive wetland areas and it may be necessary to give special protection to certain regions in the future.

References

TBFI (1975) p75, GBFI (1988) p101.

08 Black-browed Albatross

Diomedea melanophris

A. Albatros chico

C. Albatros de ceja negra

Local name: Mollymawk

Clayton (1774) was the first to describe this albatross in the Falklands, when he found hundreds on Saunders Island. He commented on their appearance, nesting behaviour and their eggs which were 'excellent nourishing food'. In 1797, Fanning saw them nesting among Rockhopper Penguins at New Island and stated that the first eggs were laid about 10th October (Fanning 1833). Snow (1857) mentioned a colony on Keppel Island in 1855 and Abbott (1861) said that this albatross 'is very seldom seen in East Falkland, but breeds in large numbers in the adjacent islands,' and he noted that eggs 'are collected in large numbers and brought to Stanley for sale'. Vallentin (in Boyson 1924) visited a large rookery at the north end of Saunders Island in 1910 which continued to thrive, 'in spite of the raids which are made every spring on its eggs, for which there is always a steady demand'. In late January 1916, Beck (in Murphy 1936) visited several colonies which were on the steep seaward slopes of Pebble, Saunders and West Point Islands.

Bennett (1926) merely commented that it was common on some of the outer islands while Cawkell & Hamilton (1961) described it as abundant and nesting on Steeple, Grand, Elephant and East Jason Islands, West Point, Saunders, Keppel, New, North, Saddle, Diddle-dee (Smylie's Channel, south of Weddell Island), Bird (very large) and Beauchêne Islands. There were also small colonies on Death's Head and Calm Head on the West Falkland mainland coast but not on Pebble Island, as reported by Beck in 1916. On West Point Island in 1962, about 3,400 chicks were counted before fledging, suggesting a breeding population of 5,000 to 7,000 pairs (Woods 1962). In an article in *Animals* magazine, Strange (1965) estimated that over 2½ million pairs of Black-browed Albatrosses were breeding on Beauchêne Island. An independent calculation of the area required for each nest, based on census work at West Point Island in 1963, led to the conclusion that Beauchêne held about 130,000 pairs (Woods 1966). Prince (1982) surveyed the Beauchêne Island colony in December 1980 and calculated that it held 140,000 to 170,000 pairs. Before the Breeding Birds Survey, it was also known that Steeple Jason in the northwest of the archipelago and Bird Island off the southwest coast both held very large colonies. More Black-browed Albatrosses have been ringed in the Falklands than any other species, due to initiatives by the Bird Banding Program of the United States Antarctic Research Program and the enthusiasm of Roddy and Lily Napier of West Point Island, supported by a few friends. In February 1952, 18 nestlings and 21 adults were ringed and between 1960 and 1968 about 11,000 nestlings were ringed just prior to fledging. Recoveries were made off the eastern coast of South America north to Cape Frio, Brazil (23°S) with a few reaching the west coast of South Africa between two and a half and six months after leaving the nest. From all recoveries, it seems that birds ringed as nestlings in the Falklands normally remain within the South Atlantic as they develop to maturity (Tickell 1967). Longevity of 30 years has been reached by several birds on West Point Island and a strong attachment to the colony of birth has also been shown. In 1989/90, of 84 ringed birds aged 20 years or more at one colony, only seven were known to have been ringed elsewhere on the island.

Habitat and Breeding

This species may be seen throughout the year in Falkland waters though it is more numerous during the breeding season. Between September and April, Black-browed Albatrosses are at their colonies on steep slopes or rocky plateaux of several islands, mainly off West Falkland. The nest is a solid pillar up to 50cm (20") high of mud and guano with some tussac grass and seaweed incorporated and is re-used annually. A single very large whitish egg variably marked with red-brown is laid in early October and fledged young birds leave the nest between mid-March and early April.

Breeding Birds Survey: analysis of data

Population estimates were obtained for all but three grid squares where colonies were reported. These included a survey of Steeple Jason Island in December 1987, where about 200,000 pairs (± 10% error) were found to be breeding (Thompson &

08 Black-browed Albatross *Diomedea melanophris*

Detectability: **5 Very High**

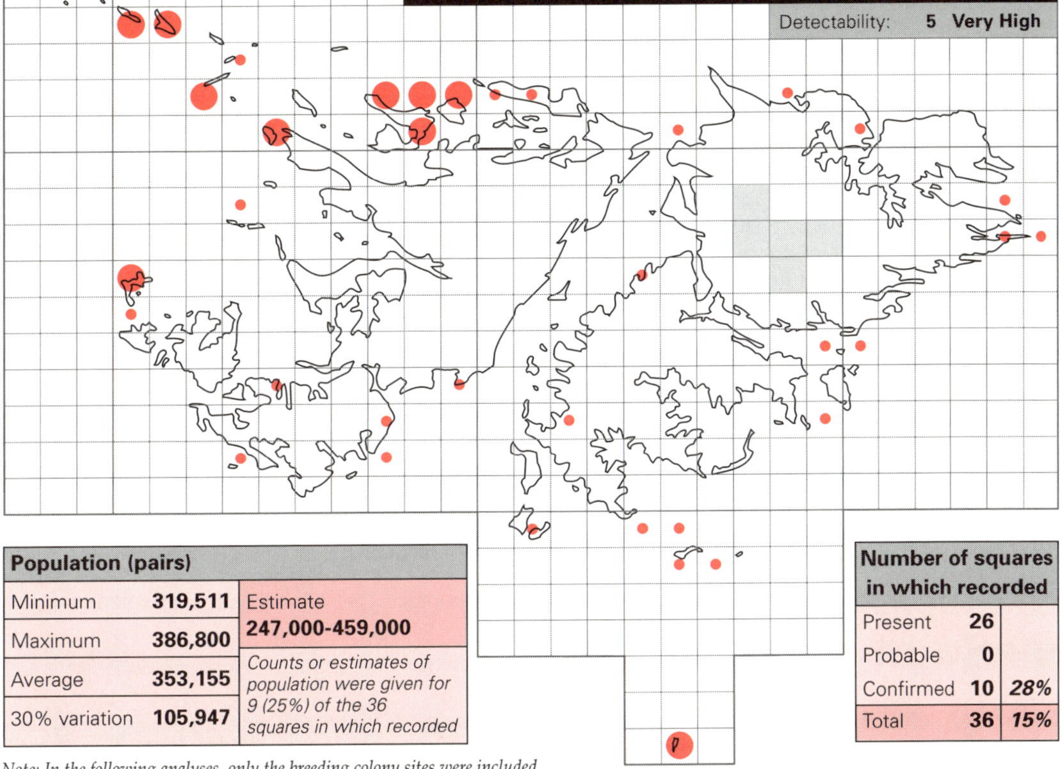

Population (pairs)		
Minimum	**319,511**	Estimate **247,000-459,000**
Maximum	**386,800**	
Average	**353,155**	*Counts or estimates of population were given for 9 (25%) of the 36 squares in which recorded*
30% variation	**105,947**	

Number of squares in which recorded		
Present	**26**	
Probable	**0**	
Confirmed	**10**	***28%***
Total	**36**	***15%***

Note: In the following analyses, only the breeding colony sites were included.

Comparisons between species records and topography	Dry/Wet: Number of wet tetrads					Inland/Coastal: Amount of coast			Upland/Lowland: Proportion of square > 500′				
	Nil	**1-5**	**6-10**	**11-15**	**16-22**	**None**	**<10%**	**≥10%**	**≥1/2**	**≥1/4**	**<1/4**	**None**	**Total**
Surveyed squares	38	104	52	24	16	17	14	203	7	25	73	129	234
% of surveyed squares	16	45	22	10	7	7	6	87	3	11	31	55	100
Grid squares in which recorded	2	8	1	0	0	0	0	11	0	0	10	1	11
% records in each category	18	73	9	0	0	0	1	99	0	0	91	9	100
Significant variation											♦		

Rothery 1991), making it the largest known colony in the Falklands. The population at West Point Island has increased substantially and may have doubled since the early 1960s. In November 1989, a total of 12,050 breeding pairs was counted (Thompson 1990) and Roddy Napier reported that Falklands Conservation had resurveyed the colonies in December 1993 and estimated that they held about 15,200 pairs. The population estimate indicates that this is almost certainly the second most abundant Falkland breeding bird. Away from the breeding colonies, it was under-recorded as it is known to occur widely in coastal waters. Records from northern West Falkland probably reflect the presence of steep west to northwest-facing slopes, which generate updrafts used for taking flight. As expected, all colonies were in squares with considerable coastline.

There is evidence that this species has benefited at least temporarily, from the development of large-scale commercial fishing for patagonian squid particularly around Beauchêne Island. Thompson (1992) found that waste amounted to about 5% of the total reported catch and that albatrosses scavenged about 50% of this waste. The quantity taken was equivalent to 10 to 15% of the total food requirements of the breeding albatrosses on Beauchêne when the fishery was operating. The long-term effect of this may be detrimental because it has been calculated that about 46% of the estimated peak stock of patagonian squid was caught in the same year.

World Distribution

The Black-browed Albatross has a circumpolar distribution in subantarctic waters. The Falklands are the most important breeding site, probably supporting 75% of the world population (Croxall *et al.* 1984). The race *D. m. melanophris* also breeds at 51°S off the coast of Chile (Clark 1985), at Diego Ramirez (56°37'S), Ildefonso to the west of Cape Horn, Staten Island off Tierra del Fuego and several other subantarctic islands including South Georgia, Heard, Kerguelen and Antipodes Islands. The race *D. m. impavida*, which may be specifically distinct, breeds only at Campbell Island, south of New Zealand. Black-browed Albatrosses generally range between 65°S and 20°S in the southern oceans, but reach 10°S off the coast of Peru in the Humboldt Current and at least 16°S off the coast of Angola.

References

TBFI (1975) p77-8, GBFI (1988) p103-4.

Implications for Conservation

Fluctuations in the amount of squid fishing could reduce or increase the availability of this source of food for the albatrosses. It is not known whether longline fishing methods are causing increased mortality, though there are signs that the largest threat to southern hemisphere albatrosses comes from incidental mortality associated with such fisheries (Prince *et al* 1994). Egg collecting for human consumption is limited to a few hundred annually and is controlled by licensing; it does not seem to have reduced populations in the recent past and if it continues at the same level, is unlikely to have an adverse effect. It is important that those breeding colonies which have not been investigated, such as Bird Island, should be censused and that all colonies should be monitored regularly, using methods similar to those already employed at West Point, Beauchêne and Steeple Jason Islands.

---

09 Southern Giant Petrel

Macronectes giganteus

A. Petrel gigante

C. Petrel gigante antártico

Local name: Stinker

Northern Giant Petrel

Macronectes halli

A. Petrel gigante picocorto

C. Petrel gigante subantártico

History

Garnot (1826) appears to have been the first person to remark on the presence of Giant Petrels in the Falklands, though he made no comment on their abundance. Abbott (1861) reported that the 'Stinkard' was common along the shores of East Falkland and that it bred on many of the adjacent islets but he did not name any sites. Vallentin (1904) stated that it was only a visitor, blown into harbours and inlets by strong winds; he had seen occasional birds eagerly devouring slaughterhouse refuse from the sea in Stanley. During his later visit (1909-1911) he classed it as a frequent visitor to the harbours and coves of the archipelago, though numbers had decreased since Abbott's time. He was told that it only nested at New Island, where it was not abundant (in Boyson 1924). Beck (1917) found about 100 pairs nesting on Sea Lion Island in 1915 and Bennett (1926) stated that it was very common and bred in many places. Cawkell (in Cott 1954) however, believed that it was declining in the Falklands. Many eggs had been taken in the past, notably from Elephant Cays, and eggs were still collected for food, including from Sea Lion Island. Eggs had been deliberately smashed on New Island because the birds attacked fallen sheep and the once-flourishing colony had been reduced to three pairs in 1953. Cawkell and Hamilton (1961) recorded eight known colonies, including Elephant Cays at the southern end of Falkland Sound with 'hundreds of birds' and Sea Lion Island, where the colony was 'not much smaller'. They noted that the 'large eggs, weighing about nine ounces (255g) had the best flavour of the "wild eggs" eaten in the Falklands. Normally they are seldom collected but in the period 1950-54, large numbers were taken'.

The two species of Giant Petrel were described as distinct in 1966 by Bourne & Warham, who felt that the Giant Petrels in the Falklands could be intermediate between the Northern and Southern

09 Southern Giant Petrel *Macronectes giganteus*

Detectability: **5 Very High**

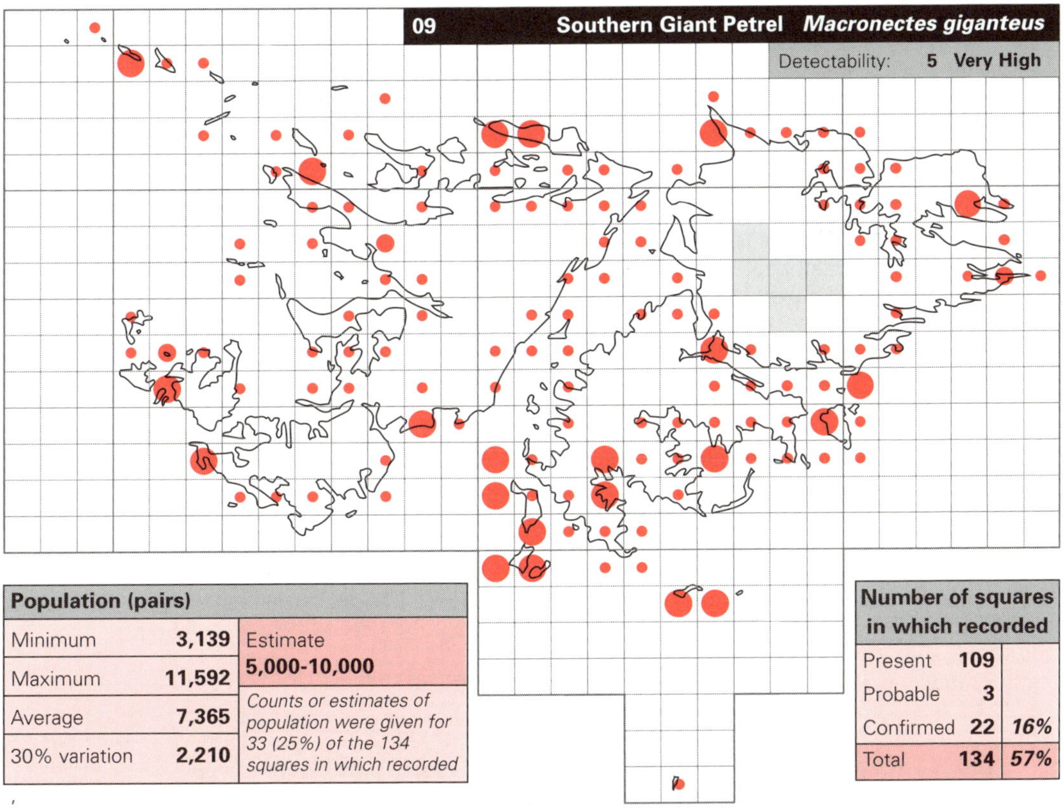

Population (pairs)		
Minimum	3,139	Estimate **5,000-10,000**
Maximum	11,592	
Average	7,365	*Counts or estimates of population were given for 33 (25%) of the 134 squares in which recorded*
30% variation	2,210	

Number of squares in which recorded		
Present	109	
Probable	3	
Confirmed	22	*16%*
Total	134	*57%*

Comparisons between species records and topography	Dry/Wet: Number of wet tetrads					Inland/Coastal: Amount of coast			Upland/Lowland: Proportion of square > 500'				
	Nil	1-5	6-10	11-15	16-22	None	<10%	≥10%	≥1/2	≥1/4	<1/4	None	Total
Surveyed squares	38	104	52	24	16	17	14	203	7	25	73	129	234
% of surveyed squares	16	45	22	10	7	7	6	87	3	11	31	55	100
Grid squares in which recorded	17	66	32	15	4	0	2	132	1	9	41	83	134
% records in each category	13	49	24	11	3	0	1	99	<1	7	31	62	100
Significant variation								***					

species. Devillers & Terschuren (1980) thought that both species could be breeding, but few observers have examined the breeding birds closely. Twelve colonies listed in 1982 by Steve Hunter from data supplied by Steve Whitley, were all said to be *M. giganteus* and recent evidence indicates that only Southern Giant Petrels breed in the Falklands (John Croxall pers. comm.). The Giant Petrels are difficult to distinguish in the field and there were only two sight records of the Northern species before 1984. Therefore only 'Giant Petrel (Stinker)' was included on the record form for the survey. There have been rumours that *M. halli* is breeding on Beauchêne Island where several were seen close inshore in February 1987 by Andy Douse. A few Giant Petrels seen off Carcass Island and Cape Pembroke, were probably of this species (Gregory 1994). Bourne & Curtis (1985) state that while *M. halli* was frequent among *M. giganteus* north of 54°S, they 'did not normally occur inshore around the Falklands'. At subantarctic islands where both breed, *M. halli* lays about six weeks earlier and tends to nest in loose colonies or alone. There were no reports of colonies in the Falklands having these characteristics.

Habitat and Breeding

Giant Petrels are familiar in Falkland coastal waters throughout the year, habitually entering harbours during daylight. The Southern species occurs in two colour phases, one mostly dark grey-brown becoming lighter with age and the other almost wholly white from the juvenile stage, forming about 10% of populations on the Antarctic Peninsula and 1% of those breeding at South Georgia. White phase birds are occasionally seen offshore, but only birds of the dark phase have been found breeding in the Falklands. It nests socially on low flat ground near shores. On George Island with about 30 pairs in 1961, the nests were mere scrapes in 'black ground' (bare decomposed peat, previously covered by tussac grass but eaten out by sheep and eroded by wind and rain), while on Sea Lion Island the eggs were laid in shingle or in hollows in the sand. A single large white egg is laid in late October, incubation takes about two months and fledging a further four months.

Breeding Birds Survey: analysis of data
Breeding colonies were recorded in 22 grid squares, with a tendency for a concentration in southern East Falkland. One island in the Elephant Cays held the largest colony of about 1,000 pairs, recorded by Shane Wolsey in 1986/87. Kate Thompson found 300 pairs on Steeple Jason in 1987/88 and Sally Poncet reported 250 pairs in the Beaver Island group in 1989. The George Island colonies had increased markedly; Ian McLeod estimated 100-1,000 pairs in 1985/86 and Ken Maule and Paul Fitzpatrick reported a similar number on 28-29 February 1992, with 20 young birds still on nests. John Reid counted 19 pairs in the colony on Pebble Island in 1986/87 but according to David Osborn, this had increased to about 50 pairs in December 1990. On Sea Lion Island Tim Duggleby and Shane Wolsey reported only five or six pairs in the 1986/87 season but in December 1990, David Osborn noted about 20 pairs, of which there were 'about 14 pairs on sand & shingle below the tussac'. In December 1987, Mike Riddy found a mainland colony of 26 pairs in Low Bay Rincon (TC62). The population estimate appears low, given the number of records of presence in Falkland waters. It is likely that many of these birds are on passage or dispersal from colonies elsewhere. In comparison with the topographical parameters, the Inland/Coastal analysis shows a very significant bias towards occurrence in squares with much coastline (p = 0·0008).

World Distribution
The Giant Petrels have a circumpolar distribution in southern oceans. Both species breed at South Georgia, Marion, Crozet, Kerguelen and Macquarie Islands. *M. giganteus* also breeds on the Antarctic Peninsula, on islands off Chubut province, Argentina, at Diego Ramirez and Isla Noir in Chile and at Gough Island. *M. halli* also breeds at subantarctic islands in the southern Indian Ocean and around New Zealand. The Falklands hold a significant proportion of the world population of Southern Giant Petrels, a species that is declining in many parts of its range (John Croxall pers. comm.).

Implications for Conservation
Giant Petrels are very sensitive to human disturbance during incubation and the tourist organisation has found it necessary to erect hides for viewing the colonies on Pebble and Sea Lion Islands from a safe distance. There is no evidence that egg collecting is now popular or widespread. The Falkland population does not appear to be under any current threat.

References
TBFI (1975) p80-1, GBFI (1988) p109-10.

10 Thin-billed Prion

Pachyptila belcheri

A. Petrel-ballena pico delgado

C. Petrel-paloma de pico delgado

Local name: Firebird

History
The first record of a 'blue petrel' which could have been this species, was from Garnot (1826) who was at Port Louis in 1822. There is no published evidence that the Blue Petrel *Halobaena caerulea* breeds or has ever bred in the Falklands, but Wace (1917) listed the Blue Petrel (*Prion coeruleus*) as a breeding species and added the name 'Firebird'. In the edited version of Wace's paper (1921) Dabbene noted the Blue Petrel *H. caerulea* as breeding and Bennett (1926) gave information about the Thin-billed Prion whilst believing that he was discussing the Blue Petrel. Confusion over the status of these two species in the Falklands continued though Hamilton (1951) stated that he had handled more specimens of Thin-billed Prions than he could remember and 'never one of *Halobaena*' and that the latter was certainly not known to breed. Cawkell & Hamilton (1961) said that the Thin-billed was the only prion to be

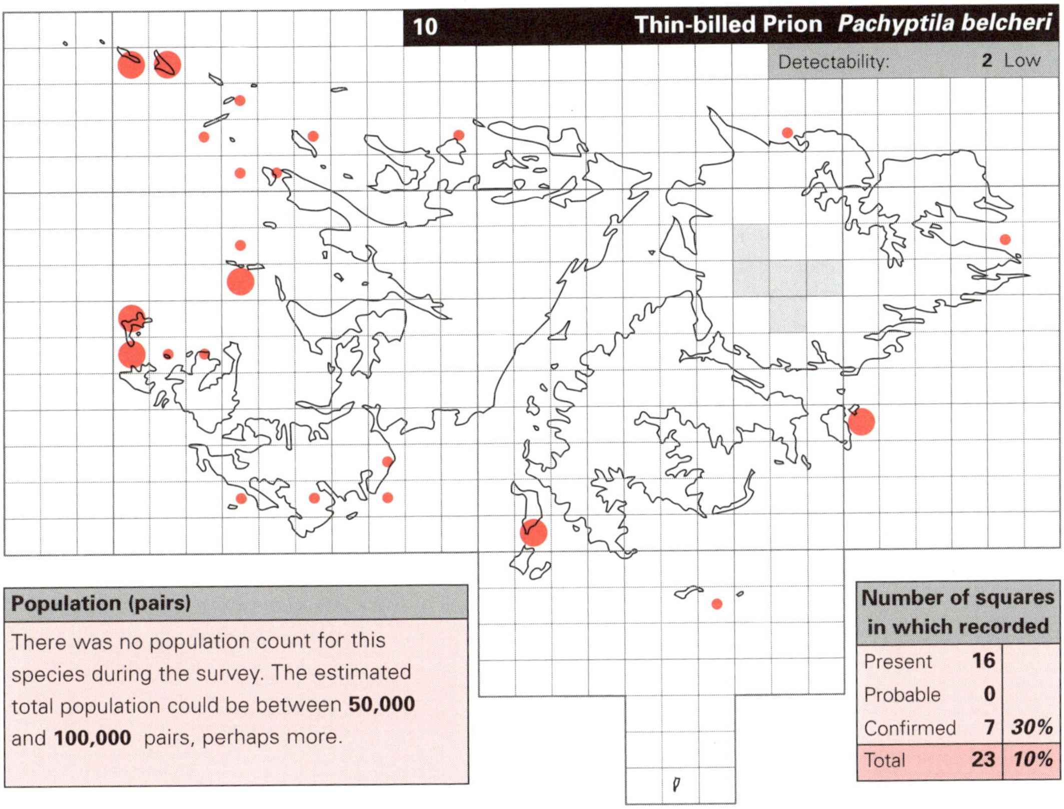

10 Thin-billed Prion *Pachyptila belcheri*

Detectability: 2 Low

Population (pairs)

There was no population count for this species during the survey. The estimated total population could be between **50,000** and **100,000** pairs, perhaps more.

Number of squares in which recorded		
Present	16	
Probable	0	
Confirmed	7	*30%*
Total	23	*10%*

found in any numbers in the Falklands and it was breeding in thousands at New Island and also on Grand Jason, Gibraltar Rock, the Passage and Tyssen Islands and at Macbride Head where the colony appeared to date from about 1944. They noted that the population had increased by many thousands in the last thirty years. In 1962, Roddy Napier and I investigated a known small colony of this species on West Point Island. All that could be found were some chewed remains of prions though there was some evidence that birds were still visiting the colony. Beck in 1916 noted 'the disastrous effect that both rats and mice have had upon many of the small petrels at the Falklands. Some islets once well populated by certain species, no longer have a trace of petrels and Beck found that most such are now overrun with either rats or mice' (Murphy 1936). On New Island this species was treated as a pest because its burrows were thought to ruin sheep pastures and it was listed, in the 1964 Wild Animals and Birds Protection Ordinance, as a species that could be destroyed at any time. A large flock of domestic cats was kept with the aim of reducing prion numbers and dogs were also known to dig out and kill some birds (Cawkell & Hamilton 1961). Croxall *et al.* (1984) however, described it as 'very abundant at New Island but apparently less so at other sites'. They commented that, 'no reliable estimates are available but the population almost certainly reaches one million pairs and the Falkland Islands are probably the most important of the three known breeding sites of this species'.

Habitat and Breeding

Adults usually return to colonies in late September though sometimes as early as mid-August and a single egg is laid between late October and mid-November. It nests in burrows of up to 2m (6 feet) long, excavated in soft peat or sandy soil. At New Island it also nests under rocks, diddle-dee or tussac clumps when the tunnels may be up to 4m (12 feet) long and contain several nesting chambers. Nests have been found from 15m to 190m (50 to 625 feet) above sea level (Cawkell & Hamilton 1961). Breeding birds feed over the continental shelf east and south of the Falklands during daylight. Most chicks fledge by late February and the population begins to leave from the beginning of March. Most are thought to migrate northwards during winter but some remain in Falkland waters with other prion species (Bourne & Curtis 1987).

Breeding Birds Survey: analysis of data

Breeding was confirmed at only seven colonies, mostly on islands off the western coasts of West Falkland and at Lively and Speedwell Islands. Only four estimates of population were received in the survey, two in range 'D' (1000+ pairs) at

New Island (TC06) from Keith Reddick and Mike Reeves in 1985/86 and Tui de Roy and Mark Jones in 1986/87. In 1992, Sally Poncet estimated 1000+ at Hecate Island and Seal Rocks off Beaver Island (TC05) and there was an estimate in range 'C' (100-1000 pairs) at South Jason Island (TD21) from Nicholas Beswick. There has been no evidence of a population approaching one million pairs. An interesting record from mainland East Falkland came from Robin Pitaluga of Salvador. He reported that numerous carcasses could usually be seen on the cliffs of Lion Point (UD81) at the east end of Concordia Beach. Preliminary work on this species in the Seabird Monitoring Project of Falklands Conservation in 1986/87 at New Island showed that the diet included small crustaceans, amphipods and small squid (Thompson 1989). This species was under-recorded during the survey and analysis of grid square records in comparison with topography would not have produced useful results.

World Distribution

The Thin-billed Prion also breeds at Kerguelen and the Crozet Islands in the subantarctic Indian Ocean and 'a very large breeding colony' was discovered at Isla Noir off southern Chile by Gerry Clark in February 1984 (Winser & Winser 1985). The colony at Kerguelen is very large but none of them is thought to hold as many pairs as the Falklands.

Implications for Conservation

Several offshore islands which could hold substantial colonies of prions have not been examined by ornithologists. Far more investigation of these islands is necessary before any definitive statements can be made about the size of the Falkland breeding population. There is no known conflict between the commercial fishing industry around the Falklands and the Thin-billed Prion but any serious pollution of its feeding grounds could have an immediate detrimental effect. It should be given protected status in future conservation legislation.

References

TBFI (1975) p85, GBFI (1988) p116-7.

---

11 Fairy Prion

Pachyptila turtur

A. Petrel-ballena picocorto

C. Petrel-paloma chico

History

Before 1967, the Fairy Prion was thought to be a vagrant in the Falklands, recorded once when a bird was found dead near Stanley on 11 March 1917 (Bennett 1926). It was discovered breeding on Beauchêne Island, 80km (50 miles) south of East Falkland in early January 1967 (Strange 1968). In December 1980 Smith & Prince (1985) found them to be quite widespread over the island and estimated the population to be several thousand pairs.

Habitat and Breeding

At Beauchêne, it nests in burrows or in cavities beneath rock slabs. A single egg is laid and adults were seen incubating in mid-December and with small chicks in early January. Beauchêne, the southernmost island in the archipelago has three topographical areas. The northern two-thirds is a broad plateau with sheer cliffs, 30-60m high on the eastern side, tilting gently to the west and mostly covered with tussac grass or tussac over very deep peat. The narrow centre is lower, with a tussac ridge along its length between an eastern coast of shelving rock platforms with stacks and a western coast of large rounded boulders. The southern end is higher and surrounded by vertical cliffs with caves and narrow inlets. It lacks vegetation on rock platforms which have rock slabs and loose boulders sometimes embedded in mud (Smith & Prince 1985). It is possible that the Fairy Prion is breeding at other islands in the archipelago with similar habitat and that the local population is relatively sedentary.

11 Fairy Prion *Pachyptila turtur*

Detectability: 2 Low

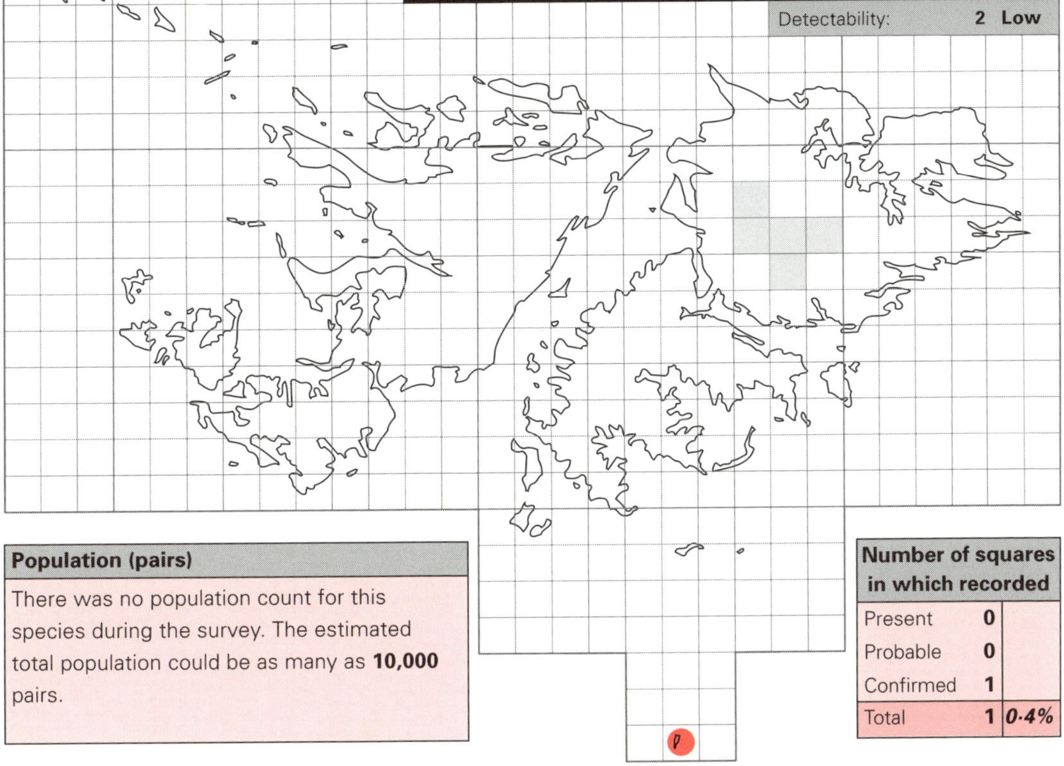

Population (pairs)

There was no population count for this species during the survey. The estimated total population could be as many as **10,000** pairs.

Number of squares in which recorded		
Present	0	
Probable	0	
Confirmed	1	
Total	1	*0·4%*

Breeding Birds Survey; analysis of data

Beauchêne Island provided the only breeding record. Small numbers were seen during the winter along the edge of the continental shelf, south of the Falklands in the years 1983-1985, (Bourne & Curtis 1985) but there were no breeding season records at sea during the survey. The topographical analysis was not possible with only one record.

World Distribution

Apart from the Falklands, the Fairy Prion breeds in small numbers at South Georgia, at Prince Edward and Marion Islands in the southern Indian Ocean and at islands off New Zealand and Australia. The total population may be several million birds and it is not globally threatened although some colonies have been particularly harmed by feral cats (del Hoyo *et al.* 1992).

Implications for Conservation

The one known colony on Beauchêne Island may be adequately protected because this island is Crown Property and recognised as an animal and bird sanctuary under the 1964 Wild Animals and Birds Protection Ordinance. The introduction of mammalian predators should be guarded against and unnecessary disturbance by visitors avoided. However, although Beauchêne Island is unlikely to be used for agriculture, the sanctuary status does not automatically protect resident birds from disturbance nor the tussac habitat from destruction. Much more scientific work should be carried out on Beauchêne Island, which has major biological interest because it has a 'fascinating simple ecosystem in which ecological hypotheses, particularly related to predator-prey interrelationships, may be tested' (Smith & Prince 1985). In particular, the breeding biology of this Fairy Prion population is little known and deserves study. Based on the diet of the species at South Georgia (Prince & Copestake 1990), it is unlikely that it depends on marine prey that are or could be harvested commercially.

References

TBFI (1975) p86, GBFI (1988) p117.

12 White-chinned Petrel

Procellaria aequinoctialis

A. Petrel negro

C. Fardela negra grande

Local names: Shoemaker, Cobbler, Nighthawk

History

The first suggestion that the White-chinned Petrel occurred in the Falklands came from Sclater (1860) who noted that Captain Abbott had sent eggs to the Zoological Society of London and John Gould thought that they could be of this species. Oustalet (1891) listed it as occurring in the Falklands area and in December 1908, Colonel AM Reid collected an adult specimen on Kidney Island, East Falkland, which is in the Natural History Museum collection at Tring. Wace (1921) noted that it had been observed in neighbouring seas and breeding was confirmed in November 1922 (Bennett 1926). Bennett collected a female on Kidney Island on 24 November 1929 and the British Grahamland Expedition collected a male and female there on 8 December 1937. Cawkell & Hamilton (1961) noted that it bred 'sparingly on New Island, in numbers on Kidney Island and perhaps elsewhere'.

Habitat and Breeding

Adults return to the breeding colony on Kidney Island in late September for courtship and burrow renovation. This colony is concentrated mainly along about 400m of the south-facing slope, which is covered with dense tussac grass, in a band about 15m wide from the beach. The single egg is laid in late October or November in a burrow about 2m (6 feet) long and usually curved, in soft tussac-peat. The nest itself is a slightly raised platform of peat and dry grass. In October and November, courting pairs and groups make their presence known by their ear-splitting trilling, which can usually be heard throughout the night. The last juvenile birds leave their burrows by the end of April.

Breeding Birds Survey: analysis of data

The only breeding records are from Kidney Island, where up to 1,000 pairs were estimated by Nicholas Beswick, David Lee and Carol O'Nians in December 1992, and New Island where Keith Reddick and Mike Reeves photographed some adults at a hilltop colony of up to 100 pairs in January 1986. Records in the vicinity of the Passage Islands, east of George Island, off Beauchêne and other islands could indicate other breeding colonies. 'Individuals are widespread out at sea throughout the year, though they are much commoner in the summer than the winter, with a maximum of about 80 birds seen per hour around the vessels fishing along the edge of the continental shelf' (Bourne & Curtis 1987). Comparison with topographical analyses would not be useful with so few records.

World Distribution

The Falkland population of the White-chinned Petrel is apparently a small isolated outlier from the only other breeding grounds in the South Atlantic at South Georgia, where 2,000,000 pairs were estimated by Croxall *et al.* (1984). It also breeds in large numbers at the Crozets, Kerguelen and Prince Edward Islands in the southern Indian Ocean and Antipodes, Campbell and Auckland Islands south of New Zealand (Marchant & Higgins 1990).

Implications for Conservation

This large petrel is probably less vulnerable to predation from rats than smaller burrowing petrels but, in the absence of other prey species, feral cats would cause considerable damage to colonies. At sea, this species is known to be especially vulnerable to mortality associated with longline fishing vessels (Ashford *et al.* 1995). Dietary studies at South Georgia (Croxall *et al.* 1995) suggest that it could be affected by commercial fishing for some species of fish, squid and large crustaceans.

References

TBFI (1975) p86-8, GBFI (1988) p118.

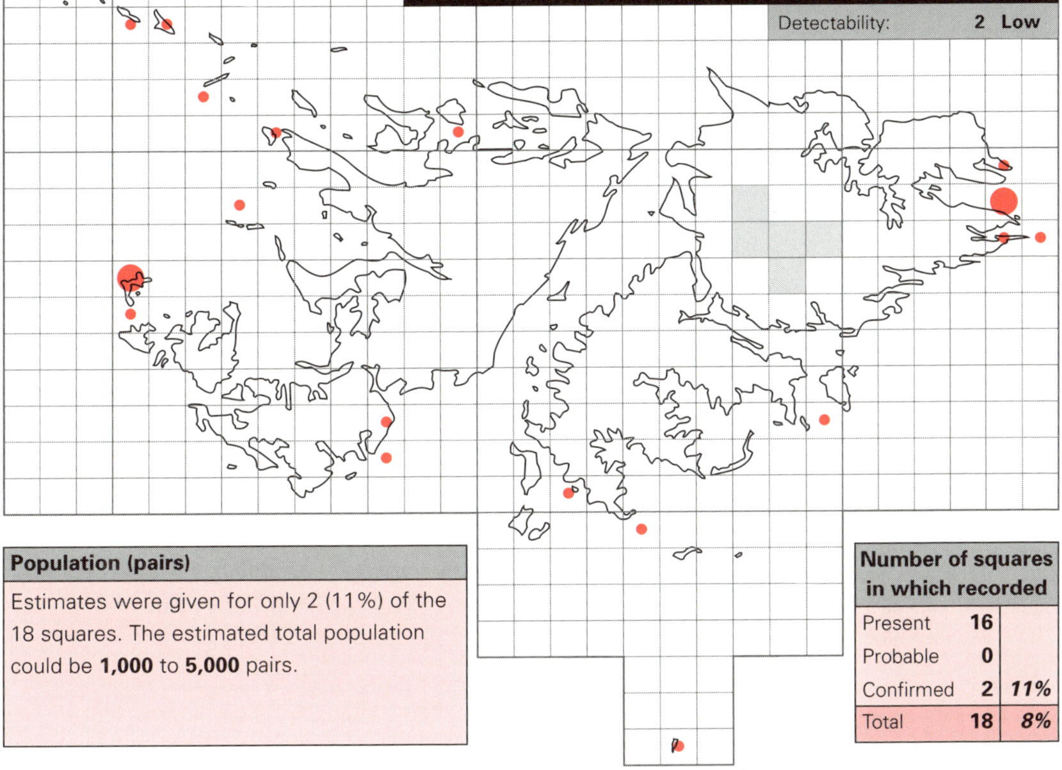

12 White-chinned Petrel *Procellaria aequinoctialis*

Detectability: 2 Low

Population (pairs)
Estimates were given for only 2 (11%) of the 18 squares. The estimated total population could be **1,000** to **5,000** pairs.

Number of squares in which recorded		
Present	16	
Probable	0	
Confirmed	2	*11%*
Total	18	*8%*

13 Great Shearwater

Puffinus gravis

A. Petrel pardo/Pardela capitorada
C. Fardela capitorada

History
The Great Shearwater was first reported as occurring in Falkland waters by Salvin (1896). Specimen(s) were said to have been obtained by the English Antarctic Expedition of 1842, according to Wace (1921) who stated that it occurred in seas near the Falklands. Bennett (1926) simply noted that there was a Falkland skin in the British Museum. Cawkell & Hamilton (1961) stated that it had been seen at sea off the Falklands in summer and winter and also that Dr and Mrs OS Pettingill had seen several large white-bellied shearwaters coming into Kidney Island at night in February 1954, but had been unable to collect a specimen. Breeding in the Falklands was confirmed in 1961 when a pair with an egg were found in a burrow at the western end of Kidney Island (Woods 1970b). In 1980, an increase in numbers off Cape Pembroke was noticed during April and in May, passage counts reached a peak of 30-40 an hour (Peatfield 1981).

Habitat and Breeding
Adults were present on Kidney Island on 18 September 1983, when at least two were seen in a raft of about 1,500 Sooty Shearwaters offshore and one was seen entering a burrow in tussac peat on 19 September. Burrows are about 1m long, usually curved and sloping downwards. The single egg is probably laid between late October and early November, with incubation taking about eight weeks. The egg found on 28 December 1961 was on the point of hatching. On Gough Island, juveniles apparently fledged during April and had left the island by mid-May (Swales 1965). Contrary to the report in Woods (1982), the Great Shearwater has not been found breeding on Sea Lion Island; this error was due to confusion in reporting or mis-identification.

Breeding Birds Survey: analysis of data
The Great Shearwater is in Falkland waters from September to April/May and is the commonest shearwater between the Falklands and South Georgia by April. The map shows a small concentration of inshore records, mainly near the two grid squares where breeding is probable or has been confirmed. Bill Curtis visited Kidney Island on 28 February 1987 and estimated that about 15 pairs were nesting there. Mike Morrison caught a Great Shearwater on 16 December 1988 in a burrow at the Sooty Shearwater colony near Wineglass Hill, west of Seal Bay, 35km (23 miles) northwest of Kidney Island and it is possible that one or more pairs are in this colony. It possibly breeds at other offshore tussac islands, judging by its abundance during the summer and later, in the autumn, when the Tristan da Cunha population is thought to be heading northwards on migration. More appear inshore in March and it becomes the commonest shearwater there after the Sooty Shearwaters leave. Bill Bourne saw 146 and 177 in 30 minutes off Cape Dolphin on 27 April and 6 May 1985 respectively. He later counted 588 in one hour about 115km (70 miles) northeast of East Falkland on 13 February 1986 and more than 100 were seen passing Cape Pembroke on 1 March 1987 by Shane Wolsey. Harro Müller reported 250 just east of Cape Pembroke on 28 March 1987 and Bill Bourne saw 40 between 110 and 150 nautical miles eastsoutheast of Cape Pembroke on 24 February 1990.

World Distribution
The Great Shearwater breeds in very large numbers (5,000,000 or more pairs; del Hoyo *et al.* 1992) at Inaccessible and Nightingale Islands of the Tristan da Cunha group and also at Gough Island. Post-breeding dispersal is to the northwest then north, past the eastern coast of South America into the North Atlantic to about 66°N, returning to the breeding grounds from September. Apart from the offshore records above it was observed inshore near Desolation Island off the western end of the Magellan Straits in early February and near Cape Horn Island in late December 1984 (Clark 1988), suggesting perhaps that there could be breeding colonies in the Magellanic region.

13 Great Shearwater *Puffinus gravis*

Detectability: 1 Very Low

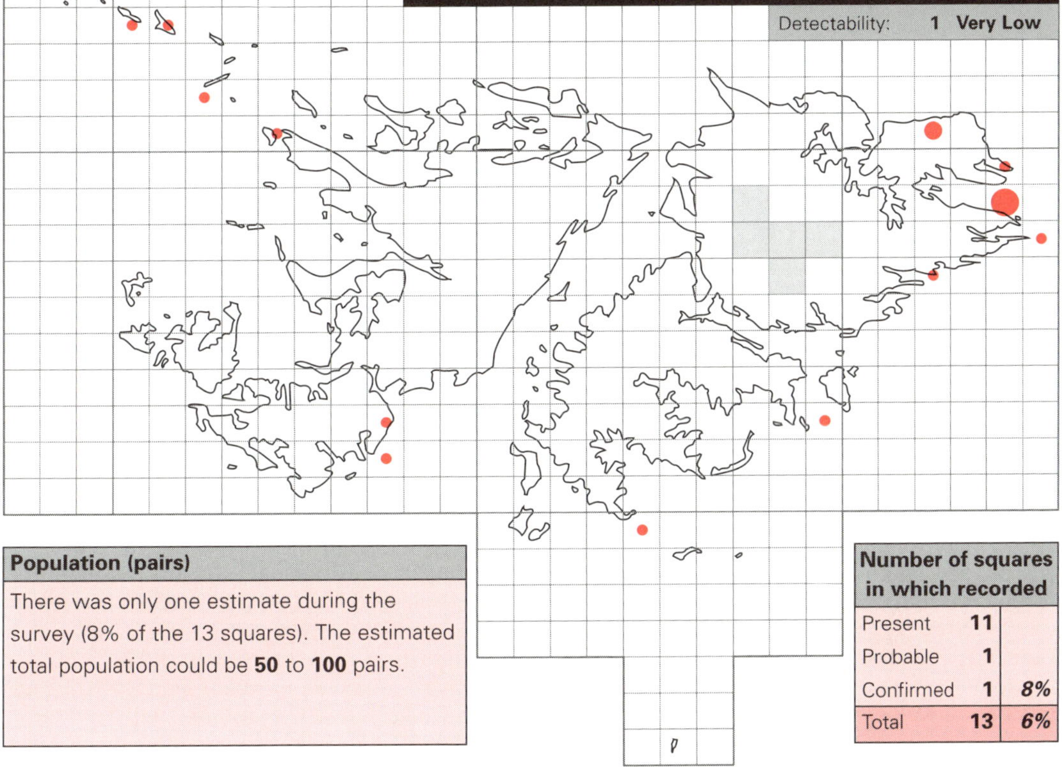

Population (pairs)

There was only one estimate during the survey (8% of the 13 squares). The estimated total population could be **50** to **100** pairs.

Number of squares in which recorded		
Present	11	
Probable	1	
Confirmed	1	*8%*
Total	13	*6%*

Implications for Conservation

Very little is known about the Great Shearwater in the Falklands. It would be useful to establish the size of the known colonies on Kidney Island and near Wineglass Hill and to discover the timing of the breeding cycle. There are many offshore islands, especially those with dense tussac and without introduced predators where it may be breeding. The dietary requirements of this species in Falkland waters are not known. In the North Atlantic winter quarters, it takes *Illex* squid, fish and crustaceans (Brown *et al.* 1981), indicating that its local diet could overlap with the species taken by fisheries.

References

TBFI (1975) p89, GBFI (1988) p119-20.

14 Sooty Shearwater

Puffinus griseus

A. Pardela oscura

C. Fardela negra

Local name: Shyer Bird

History

The first record of Sooty Shearwaters in the Falklands was published by Dabbene (in Wace 1921); a male and female captured near Stanley by AG Bennett on 11 May 1919. Later, Bennett (1926) stated that 'since then one or two have been taken from nesting-burrows on islands covered with tussock-grass, but so far without eggs. They undoubtedly do breed in the Falklands'. Hamilton had collected a specimen at Cape Pembroke on 11 March 1923, but breeding was not proved until 29 November 1936 when the British Grahamland Expedition in the *Penola* collected three eggs and five specimens on Kidney Island, East Falkland. Cawkell & Hamilton (1961) noted that it nested in large numbers at Kidney Island, 'and probably elsewhere'.

Habitat and Breeding

Kidney Island birds return in mid-September to excavate or renovate burrows in soft tussac peat, usually 1·3m (4 feet) or more deep, with a raised mat of tussac as a nest. Some eggs are laid in late November, but the extent of the egg laying period is not known. Adults probably leave in late March and juveniles from early April to early May. Wrecked birds (probably mostly juveniles) are sometimes found in Stanley after severe weather in the autumn. One grounded on 3 May 1962 was kept overnight and ringed the following morning as it seemed to be in very good condition. It was released in Stanley harbour, was attacked twice by a territorial male Falkland Steamer Duck and disappeared under water. Twenty-eight days later it was found dead entangled in a fishing net at St. Philip, Barbados, West Indies, having flown at least 9,200km (5,700 miles) around the horn of Brazil at an average speed of about 320km (200 miles) per day (Woods 1975).

Breeding Birds Survey: analysis of data

Sooty Shearwaters are numerous over the continental shelf from August and are present in Falkland waters throughout the year except in midwinter (Bourne & Curtis 1987). The best known colony is on Kidney Island, where several thousand pairs have honeycombed the western part of the island with their burrows. Other colonies were found at Sea Lion Island, where nests with young were seen by Alison Robson in January 1986, at the northern end of George Island by Paul Fitzpatrick and Ken Maule and on New Island and Beauchêne Island. The only mainland colony, thought to have 10 to 20 burrows in use, was found near Wineglass Hill, west of Seal Bay on northeastern East Falkland, by Mike Morrison in October 1984. These breeding records are spread around the Falklands and it is likely that other colonies will be discovered. Phil Gregory noted more than 4,000 Sooty Shearwaters near Kidney Island on 16 January 1989, but the largest number on record was estimated from visual and radar observations to be about 80,000 in a compact flock covering about one square mile. This enormous gathering in the southern end of Falkland Sound on 28-29 March 1987 may have originated from Falkland colonies or perhaps from breeding grounds in the Magellanic region (Curtis 1988). All records were from coastal squares; apart from this, analysis of grid square records in comparison with topography would not have been meaningful.

World Distribution

The Sooty Shearwater also breeds at islands off Tierra del Fuego and in large numbers on Isla Guafo, off the coast of Chile (43°36'S) and on Cape Horn Island (Clark 1988). Very large populations breed at islands south and east of New Zealand, including the Snares Islands, where 2,750,000 pairs were estimated in 1970/71 (Marchant & Higgins 1990). They migrate into the North Atlantic and North Pacific Oceans, reaching the coasts of Canada and Alaska. The world population is numbered in millions and it is not considered to be threatened globally.

14 Sooty Shearwater *Puffinus griseus*

Detectability: 2 Low

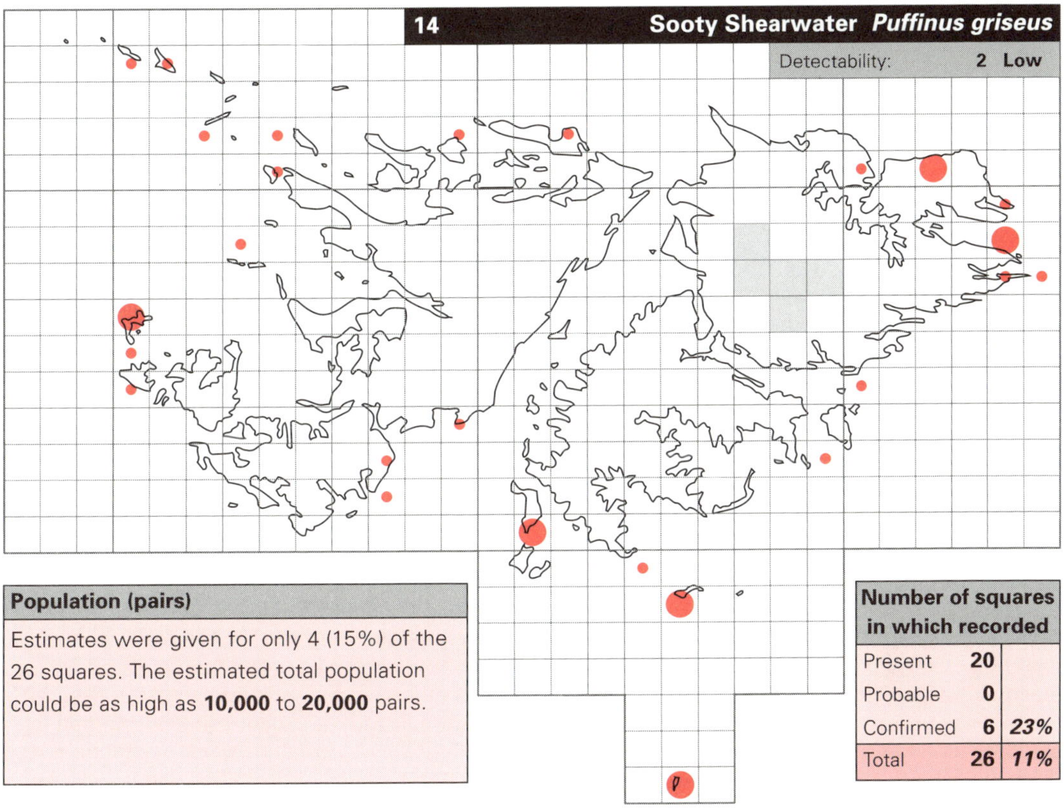

Population (pairs)

Estimates were given for only 4 (15%) of the 26 squares. The estimated total population could be as high as **10,000** to **20,000** pairs.

Number of squares in which recorded		
Present	20	
Probable	0	
Confirmed	6	*23%*
Total	26	*11%*

Implications for Conservation

Introduced mammalian predators, particularly cats and rats, can cause severe damage to breeding colonies. At the only mainland colony it is probable that feral cats keep numbers down. While it is impractical to eradicate them from the mainland, efforts should be made to eradicate cats and rats from tussac islands and to ensure that they are not introduced accidentally. A study of diet in North Atlantic winter quarters showed that euphausiid crustaceans were important prey, with squid and soft-bodied fish (Brown *et al.* 1981). It is not known whether commercial fishing in Falkland waters is affecting this species.

References

TBFI (1975) p90-3, GBFI (1988) p120.

15 Wilson's Storm-Petrel

Oceanites oceanicus

A. Paíño común

C. Golondrina de mar

History
Abbott (1861) was the first to record the presence of a nearly black species of Storm-Petrel in the Falklands. He believed that it bred and had heard that it was 'always to be found on Long Island, in Berkeley Sound'. He also reported specimens picked up dead on the shores of East Falkland. Wace (1921) stated that it was common in the seas near the islands and that it bred. Bennett (1926) discussed its occurrence and abundance in the antarctic and merely said that he had seen some eggs reputed to be of this species from a colony in the Falklands. Cawkell & Hamilton (1961) noted that Grand Jason Island was 'the known breeding station', 'but no estimate has ever been made of the numbers' and they suspected breeding on New Island, Kidney Island and 'probably elsewhere'. At Beauchêne Island in December 1980, Smith & Prince (1985) found nests with eggs in peat beneath tussac, saw many flying at night and caught large numbers in a mist net. They remarked that Croxall *et al.* (1984) had described it as a very local breeding species in the Falklands. Smith and Prince felt that it was probably more abundant on Beauchêne Island than elsewhere in this archipelago.

Habitat and Breeding
At antarctic colonies this species nests in deep crevices among rocks or makes burrows in soft soil and lays one egg, white with reddish-brown spots around the broad end. The breeding season in the Falklands probably extends from December to April, with Falkland waters usually being deserted by May after adults and young have migrated towards the northern oceans for the nonbreeding season. A few birds have been seen during the winter.

Breeding Birds Survey: analysis of data
Large numbers have been seen between November and April along the edge of the continental shelf south of the Falklands and off the entrances to some of the sounds, especially Falkland Sound (Bourne & Curtis 1987). Breeding was confirmed at Grand Jason (Osborn 1992) and in February 1987, Andy Douse estimated 1,000+ pairs around nest sites on Beauchêne Island. All records were from coastal squares; apart from this, analysis of grid square records in comparison with topography would not have been meaningful. Wilson's Petrels have been seen offshore or near several small islands, but the status of this species in the Falklands is poorly known and information on its breeding biology in the archipelago is minimal.

World Distribution
The Wilson's Storm-Petrel also breeds at islands off Tierra del Fuego, South Georgia, Kerguelen Island, at the South Shetlands, and at many other antarctic islands and sites on the coast of Antarctica. Migration is transequatorial; it is abundant between April and August in the North Atlantic and Indian Oceans but less common in the North Pacific. Small numbers are thought to breed in the Cape Horn area (Schlatter 1984) but the largest populations have been estimated at colder subantarctic and antarctic islands, including South Shetlands (1,000,000 pairs), South Georgia (600,000 pairs), South Orkneys (10,000 pairs), South Sandwich Islands (10,000-100,000 pairs, Croxall *et al.* 1984) and Iles Crozet (>10,000 pairs, Jouventin *et al.* 1984).

Implications for Conservation

The populations on Grand Jason and Beauchêne should be investigated and more work is needed to locate other Falkland colonies. There is no evidence of predation in the Falklands, but cats or rats would cause immense damage if introduced to islands with colonies.

References
TBFI (1975) p93-4, GBFI (1988) p122.

15 Wilson's Storm-Petrel *Oceanites oceanicus*

Detectability: **1 Very Low**

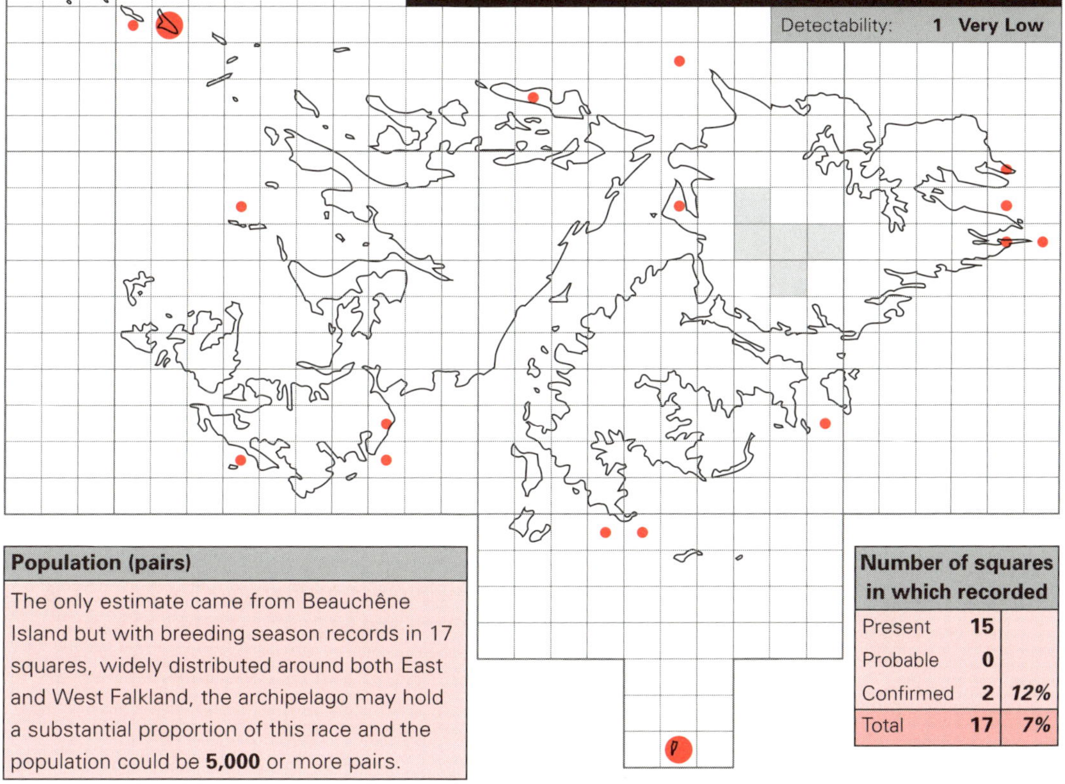

Population (pairs)
The only estimate came from Beauchêne Island but with breeding season records in 17 squares, widely distributed around both East and West Falkland, the archipelago may hold a substantial proportion of this race and the population could be **5,000** or more pairs.

Number of squares in which recorded		
Present	**15**	
Probable	**0**	
Confirmed	**2**	***12%***
Total	**17**	***7%***

16 Grey-backed Storm-Petrel

Garrodia nereis

A. Paíño gris
C. Golondrina de mar subantártica

History
The Grey-backed Storm-Petrel was first recorded in the Falklands by Abbott, who picked up a dead specimen in March 1858 (Sclater 1860, Abbott 1861). Another specimen was received by the Natural History Museum, London in 1888 from the Falkland Islands Company. Colonel A M Reid collected six specimens at Cape Pembroke Lighthouse in December 1908. In 1915, Beck saw many at sea around the Falklands and in November found birds incubating on Kidney Island and Cochon Island nearby. In mid-January 1916 he found four young birds showing contour plumage through their down in nests on Kidney (Murphy 1936). Members of the British Grahamland Expedition on the yacht *Penola* collected nine specimens from Kidney Island in the December of 1934 and 1936. Bennett (1926) had earlier stated that it bred on outer islets covered with tussac grass, a belief supported by Cawkell & Hamilton (1961). They felt that it was a common breeding species on some tussac islands, especially in the east of the Falklands. In 1961 and 1962 there was a substantial breeding population on Kidney Island in very dense, mature tussac (Woods 1970a).

Habitat and Breeding
This species has been found only on islands with tussac and is vulnerable to predation from introduced cats and rats. The breeding season appears to be extended. On Kidney Island some birds return by mid-September. Nest sites are not protected by being underground but are enlarged cavities between dead tussac stems, in the skirt of leaves above the basal pedestal or at ground level. One white egg with minute reddish dots is laid between mid-October and late December. Fledglings probably leave between early February and mid-April but some birds have been found there in late May. Many adults are caught by Short-eared Owls on Kidney Island and their remains are often found in owl pellets or on plucking sites. Müller (1990) had evidence of similar predation on Sea Lion Island in 1987.

Breeding Birds Survey: analysis of data
There were very few records, including only two of probable breeding, which both involved birds calling at nest crevices in tussac on Carcass and Kidney Islands. Comparison of grid square records with topography would not have been useful. This species is strictly nocturnal at the nest and as most birds do not return until two to three hours after sunset, it is difficult to record their presence. The shortage of records during the survey was probably due more to a lack of overnight observations than a low population level as it is widespread over the continental shelf in summer, often feeding over drifting kelp (Bourne & Curtis 1987). Its confirmed presence at Carcass, Beauchêne, Sea Lion, East Sea Lion and Kidney Islands, all known to be free of introduced mammalian predators, suggests that it could breed at other predator-free islands with mature tussac grass. More field work on the outer uninhabited islands could help to establish the distribution and population size of this attractive storm-petrel.

World Distribution
The Grey-backed Storm-Petrel has a circumpolar distribution and also breeds at South Georgia, Gough, Crozet, Kerguelen, Chatham, Auckland and Antipodes Islands. The pelagic range is not well-known, but it appears to stay fairly close to breeding colonies.

16 Grey-backed Storm-Petrel *Garrodia nereis*

Detectability: 1 **Very Low**

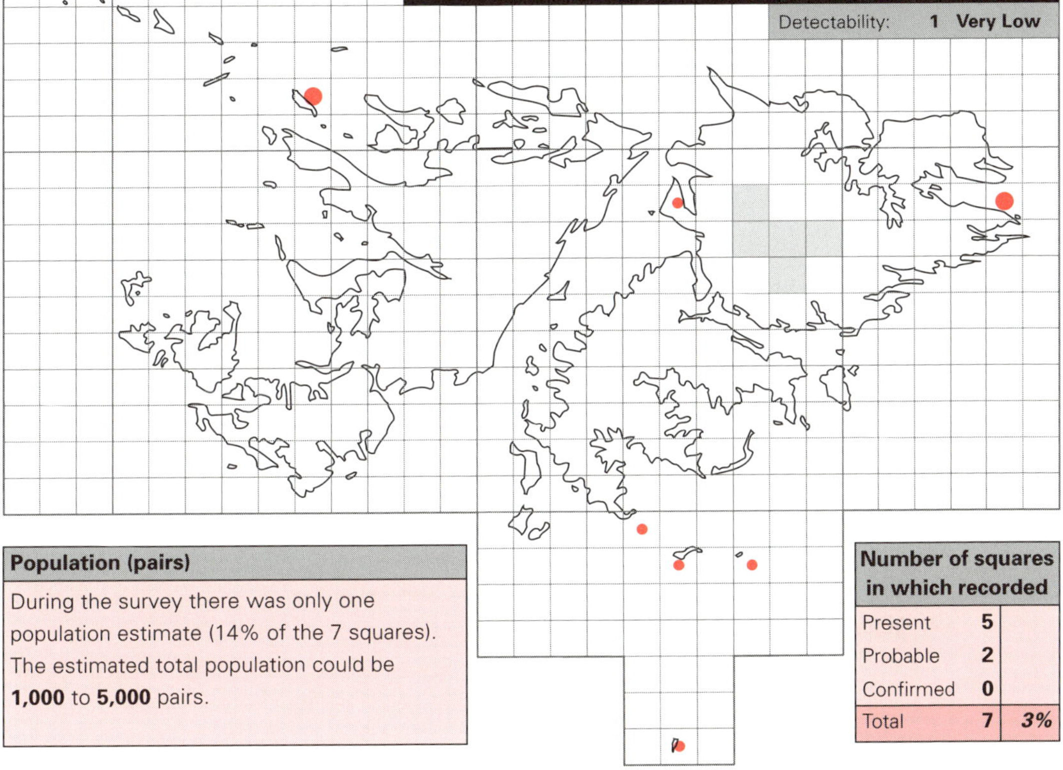

Population (pairs)

During the survey there was only one population estimate (14% of the 7 squares). The estimated total population could be **1,000** to **5,000** pairs.

Number of squares in which recorded		
Present	5	
Probable	2	
Confirmed	0	
Total	7	*3%*

Implications for Conservation

Though there is no direct evidence in the Falklands, the introduction of domestic cats, rats or house mice or grazing of tussac grass by sheep and cattle could be expected to have serious effects on breeding populations of this species. It is likely that the availability of islands with dense tussac and with no introduced mammals is necessary for its survival. In the Chatham Islands (44°S 176°W) 800km east of New Zealand, it was found that the removal of grazing animals from two small islands allowed the re-establishment of this storm-petrel, and other species, within about 20 years (Bell & Robertson 1994).

References

TBFI (1975) p94-6, GBFI (1988) p124-5.

17 Black-bellied Storm-Petrel

Fregetta tropica

A. Paíño ventrinegro

C. Golondrina de mar de vientre negro

History

It has long been on record that a single egg of this species was obtained in the Falklands by the *Challenger* Expedition of 1872-1876 (eg Wace 1921). In Appendix I of the *Challenger* report (Sclater 1881), it is stated that one egg of this species came from the Falklands and one from Kerguelen Island. A footnote records that, 'The eggs of the Falkland Islands were nearly all presented to the Expedition by Mr Deans, of Stanley, Falkland Islands'. Included in this list is an entry for the Elegant Crested-Tinamou *Eudromia elegans*, stating that two eggs from the Falklands were donated. Since 1881, no authors have claimed that this tinamou occurred in the Falklands, yet an exactly similar entry for one egg of the Black-bellied Storm-Petrel has led to repeated claims that it breeds in the archipelago, though none was recorded until a century later. It is possible that this egg also came from Kerguelen Island. Bennett (1926) remarked that it was said to breed in the Falklands and South Georgia and noted that he had found nests twice, on the South Orkneys and the South Shetlands. Cawkell & Hamilton (1961) stated that they had no evidence of its occurrence in the Falkland Islands. The first sightings date from December 1980 when Smith & Prince (1985) saw a few near Beauchêne Island and Bill Curtis saw two off Mengeary Point on 31 December 1982.

Habitat and Breeding

There are no records of breeding in the Falklands. At Signy Island in the South Orkneys, nests were well-hidden in small crevices among consolidated scree of small boulders about 20cm diameter (Beck & Brown 1971). At South Georgia it is a common, but rather local breeder (Prince & Croxall 1996). Peter Prince has also reported that its favoured habitat there is crevices of rocky cliffs with scattered tussac grass.

Breeding Birds Survey: analysis of data

During the survey, there were no records from the islands and very few were seen offshore. Bill Bourne saw one 48km (30 miles) north of Cape Dolphin on 19 November 1983 and Bill Curtis observed this species 22km (14 miles) and 60km (37 miles) north of the Falklands on 2 July 1984. One was seen 'just off the East Falkland coast' in February 1985 by Andy Douse and at least one was seen off Beauchêne Island by Callan Duck between 28 February and 6 March 1991 and reported by Mike Riddy. The records close to Beauchêne Island in 1980 and 1991 may indicate a breeding colony; suitable habitat is possibly present at the southern end of this isolated and ungrazed island.

World Distribution

The Black-bellied Storm-Petrel has a circumpolar distribution, on antarctic and the colder subantarctic islands. It breeds at South Georgia, the South Orkneys and South Shetlands, Kerguelen, Crozet, Auckland, Antipodes and Prince Edward Islands. In the nonbreeding season it is pelagic, occurring almost to the Equator in the Pacific Ocean and north of the Equator in the Atlantic and Indian Oceans.

Implications for Conservation

Further investigations of offshore islands may help to determine the status of this species in the Falklands and indicate any conservation measures that may be necessary.

References

TBFI (1975) p215, GBFI (1988) p123-4.

17 Black-bellied Storm Petrel *Fregetta tropica*

Detectability: **1 Very Low**

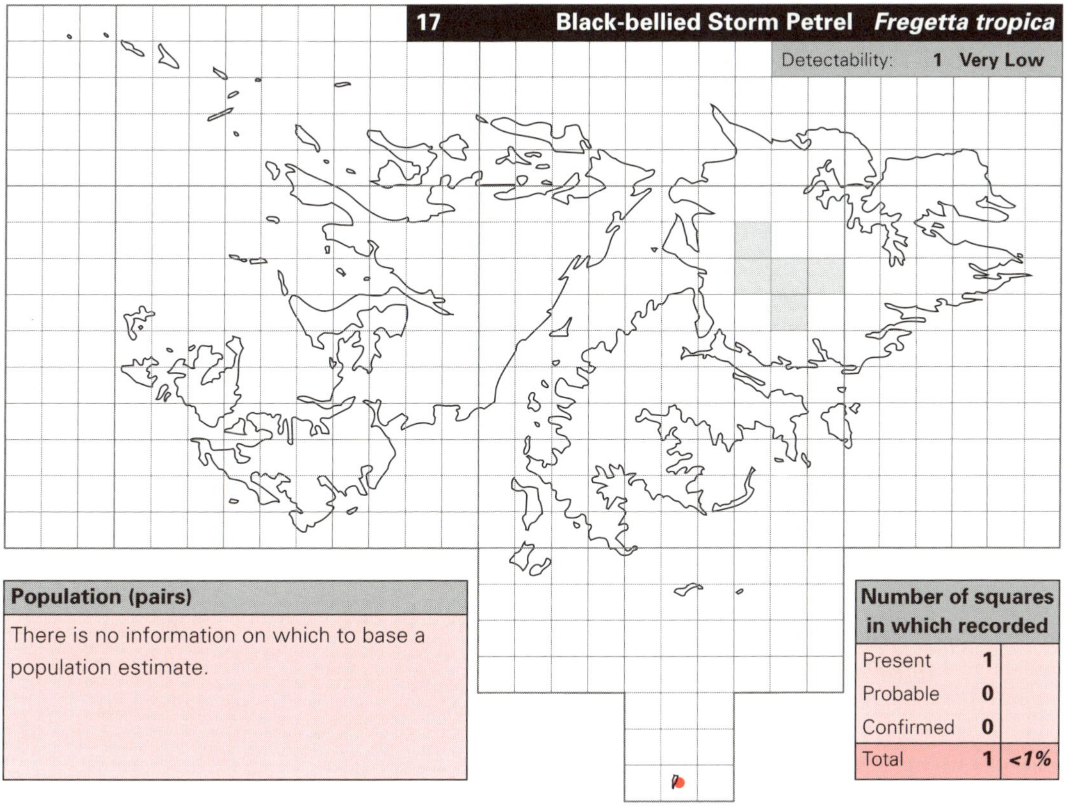

Population (pairs)
There is no information on which to base a population estimate.

Number of squares in which recorded		
Present	**1**	
Probable	**0**	
Confirmed	**0**	
Total	**1**	***<1%***

18 Common Diving Petrel

Pelecanoides (urinatrix) berard

A. Potoyunco malvinero

C. Yunco de los canales

Local names: Diver, Firebird

History
The Falkland race of the Common Diving Petrel was first named by Gaimard in 1823 as *Procellaria Bérard* from a specimen collected by MM. Quoy and Gaimard in 1820. It was called Bérard after a fellow officer from their expedition vessel, *Uranie*, which had sunk in Berkeley Sound. Later ornithologists decided that this was the same species as *Procellaria urinatrix* from South Island, New Zealand named by Gmelin in 1789. Gmelin's name pre-dated that of Gaimard, so the Falkland race was named *berard*. The generic name was also changed to the earlier one, *Pelecanoides*, published in 1799 by Lacépède (Hellmayr & Conover 1948).

Abbott (1860) said he had seen it only in Berkeley Sound, where it was supposed to breed in holes along the shores of the adjacent islands, and that it was not common. Murphy (1936) reported that Beck had found birds in burrows on Cochon and Kidney Islands in November 1915. Bennett (1926) noted that it bred commonly on the 'outer islands' but did not name any. Cawkell & Hamilton (1961) stated that it nested on 'the Jasons', Gibraltar Rock, Kidney and Cochon Islands, possibly New Island and almost certainly on some of the outer tussac islands, and Smith & Prince (1985) found that it was abundant on Beauchêne Island in December 1980.

Habitat and Breeding
Very little is known about the breeding season in the Falklands but Payne & Prince (1979) have described its breeding biology and that of the closely related *P. georgicus* at South Georgia. In the Falklands, some burrows have been found in peat beneath tussac grass on sloping ground. Adults probably return to the breeding colonies in late August or September and the single egg may be laid between late September and early October. With incubation and fledging both taking nearly two months on average, the juveniles would fly between late January and late February (based on data in Marchant & Higgins, 1990). Most populations at other subantarctic islands are thought to stay in the vicinity of their colonies throughout the year and it is certainly present in Falkland waters in all months, particularly over the edge of the continental shelf. Single birds in whirring flight are sometimes seen briefly offshore.

Breeding Birds Survey: analysis of data
Breeding was confirmed on Steeple Jason, New, Kidney and Bird Islands and it probably nested on other tussac islands where the soil or tussac peat is suitable for excavating burrows. Diving Petrels are strictly nocturnal when breeding and there has been very little careful investigation of outer, uninhabited tussac islands. The map probably shows only a fraction of its true distribution. Bourne & Curtis (1987) recorded up to 150 in an hour in the southern approaches to Falkland Sound on 27 May 1984 and a group of more than 300 was seen off New Island on 5 February 1989 by Phil Gregory. Phil also reported that a single bird was found dead near the Monument in Stanley in early May 1990, when several Sooty Shearwaters were wrecked. It is considered that this species was under-recorded during the survey and analysis of grid square records in comparison with topography would not have produced useful results. In December 1986, Tim Duggleby found some remains of diving petrels at a Peregrine Falcon eyrie on Second Passage Island. Since the end of the survey, evidence of presence on Motley Island in January 1995 was obtained, when bones of this species were found in a Short-eared Owl's pellet and wings from carcasses were identified (Woods 1995).

World Distribution
Many island races of the Common Diving Petrel have been described from southern oceans between about 35°S and 55°S and it is possible that some, even perhaps including *berard*, should be recognised as full species. However, the taxonomic relationships of these forms have not yet been investigated fully.

18 *Common Diving Petrel *Pelecanoides (urinatrix) berard*

Detectability: **1 Very Low**

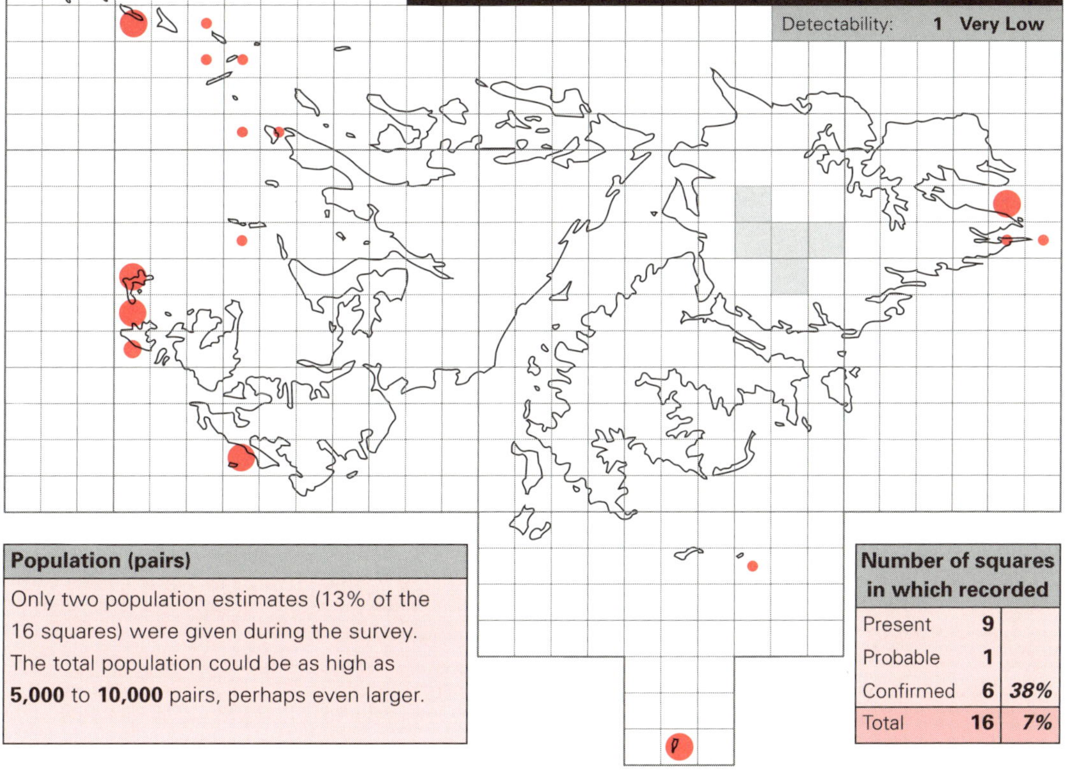

Population (pairs)

Only two population estimates (13% of the 16 squares) were given during the survey. The total population could be as high as **5,000** to **10,000** pairs, perhaps even larger.

Number of squares in which recorded		
Present	**9**	
Probable	**1**	
Confirmed	**6**	***38%***
Total	**16**	***7%***

Implications for Conservation

Some colonies are known, but there is uncertainty about their distribution and size. This position will continue unless serious efforts are made to investigate islands which have good cover of tussac grass. This species has been shown to be susceptible to predation by cats on Marion Island (Moors & Atkinson 1984) and would probably be seriously affected if cats or rats were introduced to islands with breeding colonies.

References

TBFI (1975) p96-7, GBFI (1988) p125-6.

19 Rock Shag

Phalacrocorax magellanicus

A. Viguá cuello negro

C. Cormorán de las rocas

Local name: Black Shag

History

This widespread species was noted by some of the earliest visitors but the variations in its plumage with age and season have caused problems in identification. Clayton (1774) at Saunders Island wrote of the 'Shelldrakes' which sat in rows on the rocks and cliffs and, 'every low water they leave their habitations to seek food' but he did not distinguish between this species and the King Shag. Garnot in 1822, equally confused, stated that three species of Cormorants occurred and then described four. One of these was interpreted as the Brown Booby *Sula fusca=leucogaster* by Oustalet (1891). Bennett (1926) on the same evidence, listed the 'White-bellied Booby' as an accidental visitor. Hellmayr & Conover (1948) felt that the bird described by Garnot was something quite different, and perhaps not even a Gannet. Bennett's (1926) comment that Garnot had misidentified immature Rock Shags in 'entirely black plumage' as examples of *P. vigua* must have been based on his reading of Wace (1921). This was an edited spanish version with additional species and comments by Dabbene of the Buenos Aires Natural History Museum. However, the original typescript list by Wace (1917) which was recently discovered at the Natural History Museum, Tring, names only the Rock Shag and King Shag. Dabbene therefore perpetuated Oustalet's misinterpretation of Garnot's descriptions and Bennett compounded the error.

Abbott (1861) noted that the Rock Shag was very common along the coasts all the year round and bred on the cliffs in communities. Vallentin (1904) only comments on, 'a large rookery of this species within easy walking distance of Stanley'. Bennett (1926) described it as 'an abundant resident'. Cawkell & Hamilton (1961) said that it was common, 'favouring the sheltered bays and estuaries' and that 'certainly since 1919', eggs have not been regularly gathered for food. In Cott (1954), however, Cawkell is quoted as stating in a letter dated 24 August 1953 that 'some eggs are still taken for eating, mainly by shepherds'.

Habitat and Breeding

The Rock Shag is a marine species which habitually feeds amongst kelp patches in inshore waters and may use jetties and hulks of beached sailing ships for resting and nesting. It usually nests colonially on ledges of steep cliffs in sheltered inlets at least 6m (20 feet) above the sea. Colonies vary from six to several hundred though it sometimes nests singly and occasionally on cliff tops, where it may be associated with nesting King Shags. Nests are constructed from tussac grass, fine seaweeds and sometimes diddle-dee twigs and are renovated or built from late September. Two to 5, usually 3, greenish-white eggs with chalky deposits are laid between early November and mid-December and the fledged juveniles leave the nests between mid-January and late February.

Breeding Birds Survey: analysis of data

The widespread distribution of this species around the coasts of the Falklands can be seen from the map and all but three of the records came from squares with a considerable length of coastline. A high proportion of records (77%) were confirmed as breeding, probably because it was often possible to look into nests on cliff sides. The largest numbers recorded were autumn counts of 500 on Beaver Island, 260 at Sea Lion Island and 200 at Keppel Island in March 1987 by Harro Müller. No records came from completely inland squares which supports the belief that the Rock Shag is a marine species, not normally inhabiting freshwater lakes or the upper reaches of larger rivers.

World Distribution

The Rock Shag also breeds on the coasts of southern South America from 37°S in Chile and 50°S in Argentina to Cape Horn. Continental birds are said to be mainly sedentary but some have occurred as far north as 33°S in Chile and to 35°S on the coast of Uruguay (Harrison 1983). It is particularly common around the coasts of Tierra del Fuego and in the Magellanic region of Chile (Venegas & Jory 1979).

19 Rock Shag *Phalacrocorax magellanicus*

Detectability: **5 Very High**

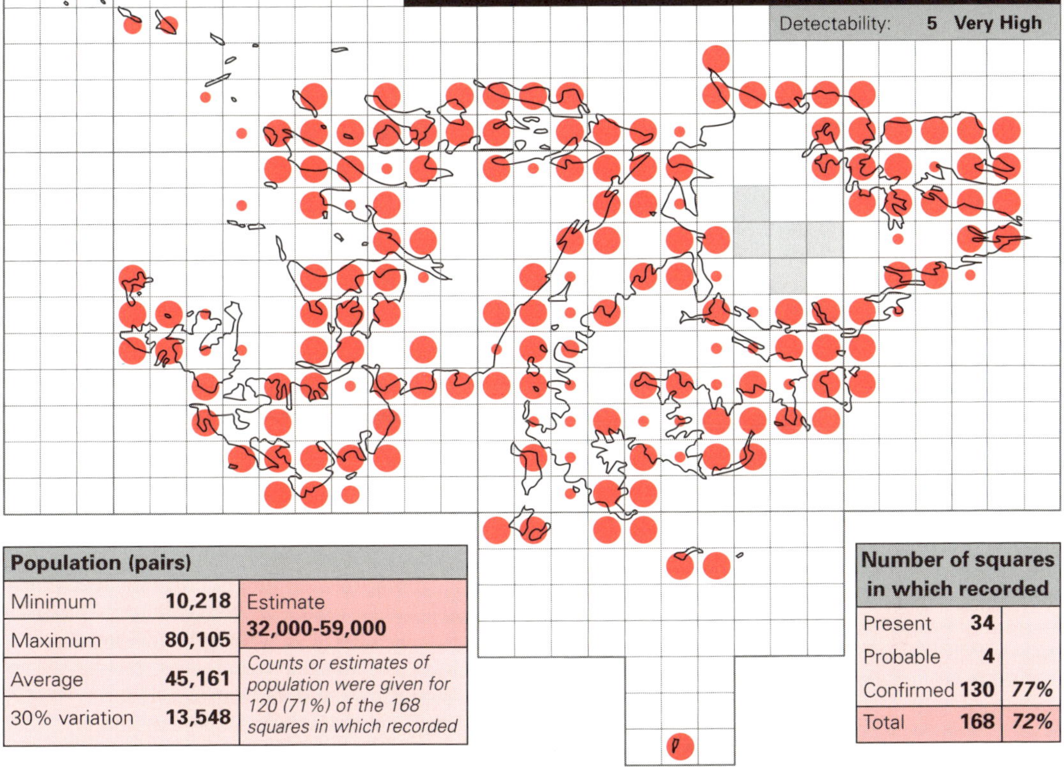

Population (pairs)		
Minimum	**10,218**	Estimate **32,000-59,000**
Maximum	**80,105**	
Average	**45,161**	*Counts or estimates of population were given for 120 (71%) of the 168 squares in which recorded*
30% variation	**13,548**	

Number of squares in which recorded		
Present	**34**	
Probable	**4**	
Confirmed	**130**	***77%***
Total	**168**	***72%***

Comparisons between species records and topography	Dry/Wet: Number of wet tetrads					Inland/Coastal: Amount of coast			Upland/Lowland: Proportion of square > 500′				
	Nil	**1-5**	**6-10**	**11-15**	**16-22**	**None**	**<10%**	**≥10%**	**≥1/2**	**≥1/4**	**<1/4**	**None**	**Total**
Surveyed squares	38	104	52	24	16	17	14	203	7	25	73	129	234
% of surveyed squares	16	45	22	10	7	7	6	87	3	11	31	55	100
Grid squares in which recorded	23	80	40	18	7	0	3	165	1	13	59	95	168
% records in each category	14	47	24	11	4	0	2	98	1	8	35	56	100
Significant variation								***			♦		

Implications for Conservation

The sedentary Falkland Islands population is substantial and widespread with many breeding sites. This distribution pattern means that it is potentially less vulnerable to environmental disasters than those species that breed in a few large colonies and the Falkland population does not appear to be under any specific threat. Little is known about diet in the Falklands, except that it catches small fish in inshore waters, often among kelp beds. A study in Patagonia has shown that it takes small fish, crustaceans and cephalopods, mainly from the sea bed in water from 2-12m deep, with the most frequent prey species being rock cod (Punta *et al.* 1993).

References

TBFI (1975) p97-8, GBFI (1988) p127-8.

20 Imperial/King Shag

Phalacrocorax atriceps albiventer

A. Viguá blanco, Cormorán real
C. Cormorán de las Malvinas

History

Quoy and Gaimard collected the first examples of this species in 1820 in the Falklands, and it was named *Carbo albiventer* from their specimens by Lesson in 1831 (Hellmayr & Conover 1948). Sixty years earlier, Pernety (1771) had noticed what he called 'Divers' that assembled in flocks upon the rocks near the sea coast and said that they were 'tolerably good though rather oily', presumably referring to their edibility. One of his 'three kinds' of 'Divers' appears to be this species. Abbott (1861) stated that this shag was common along the coast of the Falklands, all the year round. He noted that it bred with Rockhopper Penguins and that the 'nests were not placed together, but here and there, all over the rookery, amongst the penguins'. Vallentin (1904) remarked on a 'fairly large rookery' of this species on a rocky hillside at West Point Island and Wace (1921) simply stated that it bred in the Falklands and that earlier observers had noted its presence. Bennett (1926) believed that it was 'exceedingly abundant' in the Falklands. Cawkell & Hamilton (1961) considered that the King Shag was much more common than the Rock Shag and more a species of open cliffs and seas. They note that, 'usually rare in Stanley Harbour, it was attracted in numbers on one occasion by great masses of krill'.

Habitat and Breeding

It usually breeds in closely packed colonies, often in association with Rockhopper Penguins and Black-browed Albatrosses on flat cliff top sites, which are also used as roosts in winter. The nest is a column of mud, tussac grass and algae about 30cm (1 foot) high topped by a shallow cup and lined with grass. Building of new nest pillars or renovation of old ones starts in October and continues until November. Two to 4, usually 3 pale green-blue limy eggs are laid between late October and (mostly) in mid-November. Eggs hatch in December and most juveniles are flying by mid-February.

Breeding Birds Survey: analysis of data

The map shows that the King Shag is a widespread breeding species present around all coasts. All records came from grid squares with a considerable length of coastline. In ten squares, observers recorded more than 1,000 nests and the largest colony of about 2,000 pairs, with two smaller colonies totalling about 500 pairs, was noted by Mike Riddy on the northern coast of Berkeley Sound between Rugged Hill and Eagle Hill in 1987/88. During the concurrent Seabird Monitoring Programme of Falklands Conservation, food samples were collected from over 50 adults breeding at New Island in 1986/87. Analysis showed that these King Shags took mainly crustaceans between the pre-laying and the incubation period, but during chick-rearing, up to 90% of the prey was small fish, mainly nototheniids, individually weighing less than 1·5g. It was therefore concluded that commercially exploited species did not seem to be significant in the diet of the King Shag on New Island during the breeding season. Observations suggest that at times it may travel further offshore for food than the Rock Shag (Thompson 1989).

World Distribution

There has been controversy and confusion for many years about the similarities and differences between the King Shag, which was thought to be restricted to the Falklands, and the Blue-eyed Shag, which occurs in southern South America and at other subantarctic and antarctic islands. Investigations in the past 20 years have shown that the 'Blue-eyed' type co-exists with the 'King Shag' type at colonies in southern South America where there are many mixed pairs and, apparently, hybrid birds (Devillers & Terschuren 1978). Genetic and behavioural studies also support the conclusion that the 'King' and 'Blue-eyed' Shags are two forms of the same species that often occur together and hybridise, though only the 'King' Shag form has been found in the Falklands (Rasmussen 1991).

This species group includes about eight forms which are treated either as geographical races or island species. They occur on southern coasts and inland lakes of Chile and Argentina from about

20 Imperial/King Shag *Phalacrocorax atriceps*

Detectability: **5 Very High**

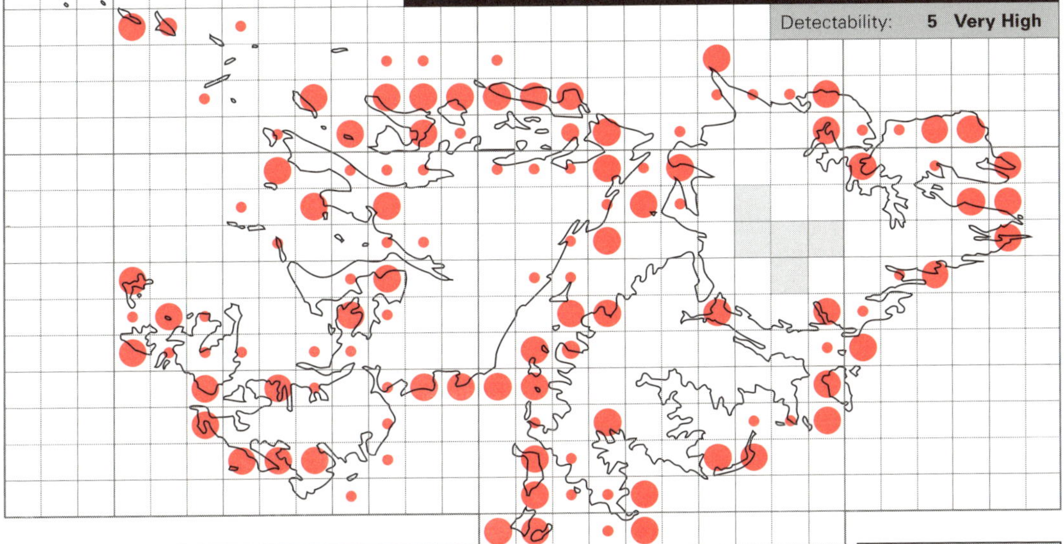

Population (pairs)		
Minimum	**19,313**	Estimate **45,000-84,000**
Maximum	**110,470**	
Average	**64,891**	*Counts or estimates of population were given for 56 (46%) of the 121 squares in which recorded*
30% variation	**19,467**	

Number of squares in which recorded		
Present	**54**	
Probable	**3**	
Confirmed	**64**	***53%***
Total	**121**	***52%***

Comparisons between species records and topography	Dry/Wet Number of wet tetrads					Inland/Coastal Amount of coast			Upland/Lowland Proportion of square > 500′				
	Nil	1-5	6-10	11-15	16-22	None	<10%	≥10%	≥1/2	≥1/4	<1/4	None	Total
Surveyed squares	38	104	52	24	16	17	14	203	7	25	73	129	234
% of surveyed squares	16	45	22	10	7	7	6	87	3	11	31	55	100
Grid squares in which recorded	22	64	21	10	4	0	0	121	1	7	45	68	121
% records in each category	>18	53	>17	>8	>3	0	0	100	1	6	37	56	100
Significant variation								***			♦		

40°S to Tierra del Fuego, at South Georgia, South Shetland and other antarctic islands to about 65°S and at Heard, Crozet and islands in the southern Indian and Pacific Oceans.

Implications for Conservation

There is no evidence of significant competition with human activities, destruction of nesting habitats or predation in the Falklands. From a few studies of the closely related forms at the South Shetlands and South Orkneys, Croxall (1992) noted signs of gradual long term increases in population of 3 to 7% over periods of at least 20 years. He also remarked that assessment of population trends for the *P. atriceps* group is very difficult because there are large variations from year to year in the timing of breeding and in population size.

References

TBFI (1975) p99-103, GBFI (1988) p128-9.

21 Black-crowned Night Heron

Nycticorax nycticorax falklandicus

A. Cuaco, Martinete

C. Huairavo

Local name: Quark

History

Pernety (1771) mentioned a 'small heron' which may be the first record of a Night Heron in the Falklands but Garnot (1826) apparently did not see any during his one month stay in the same area of East Falkland. Abbott (1861) described a visit to a colony of at least 100 pairs near Hope Place in December 1859. Nests had been built of coarse sticks on grass tufts along one side of a freshwater pond; half-grown nestlings were seen in some nests, while in others were clutches of eggs, some freshly laid. Sclater (1861) only listed the species and noted that the English Antarctic Expedition had encountered Night Herons in Berkeley Sound in 1842. Vallentin (1904) stated that it was 'universally distributed throughout these islands, but scarce everywhere. A very beautiful bird, but usually seen alone'. Wace (1921) simply recorded that it 'breeds in the islands'. Bennett (1926) stated that it was fairly common while Cawkell & Hamilton (1961) regarded it as 'well-known, conspicuous and widely-spread'.

Habitat and Breeding

The Night Heron is often seen singly or in pairs on beaches with rock pools, though Cawkell & Hamilton observed that it ascended streams for short distances, presumably in pursuit of small fish. It usually breeds in small colonies, occasionally up to 100 pairs strong, on steep low cliffs with shrubby plants or tussac grass, in rushes or tussac by ponds and on Carcass Island, in dense Monterey Cypress trees. The nest is a substantial flat platform of grass, dead kelp stems, rushes or small sticks in which 2-3, sometimes 4, smooth blue-green eggs are laid from late October. It is possible that it has two broods in a season as eggs have been found in late January.

Breeding Birds Survey: analysis of data

Records indicate that it is well-spread around the archipelago, with few birds seen inland and, in comparison with the analysis of Inland/Coastal squares, a highly significant bias ($p < 0{\cdot}007$) towards occurrence in coastal regions, suggesting that it is more a bird of the tidal zone than freshwater ponds or lakes. However, all of the substantial colonies reported by Mike Riddy were at large ponds or lakes, near coasts. The largest held 50 to 100 nests with more than 100 adults present at a coastal pond in Shag Rincon camp, Walker Creek (UC73) on 9 December 1987. Another with 40 nests was seen at West Lagoons, north of Hill Cove (TC89) on 13 December 1988. Mike found two colonies with about 30 nests at Quark Pond, Port Howard (UC27) on 17 November 1988 and at Red Pond in Hill Gap camp near Shag Cove on 2 December 1988. At Goose Green, M Conroy noted that the *Vicar of Bray* hulk was used as a nest and roost site between December 1982 and April 1983. The colony in trees at the settlement on Carcass Island held about 50 birds when Harro Müller visited in March 1987 and Donna O'Daniel found 25 nests in trees at Weddell Island settlement (TC34) in January 1988.

World Distribution

The Black-crowned Night Heron has a world wide distribution, except in Australasia. The Falkland Islands race *N. n. falklandicus* was separated from those of southern South America in 1914 by Hartert (Hellmayr & Conover 1948) on the grounds that it was smaller and generally darker. However, there is still uncertainty because continental and Falkland birds show considerable variation in plumage.

Implications for Conservation

There is no evidence that this species has suffered from waste disposal in coastal waters of the Falklands but it is potentially vulnerable to pollution of beaches. It has not been studied closely and there is scope for field work on the plumage, morphology and behaviour of this apparently endemic, sedentary race of an otherwise highly dispersive species.

References

TBFI (1975) p105-7, GBFI (1988) p132-3.

21 *Black-crowned Night Heron *Nycticorax nycticorax*

Detectability: 5 **Very High**

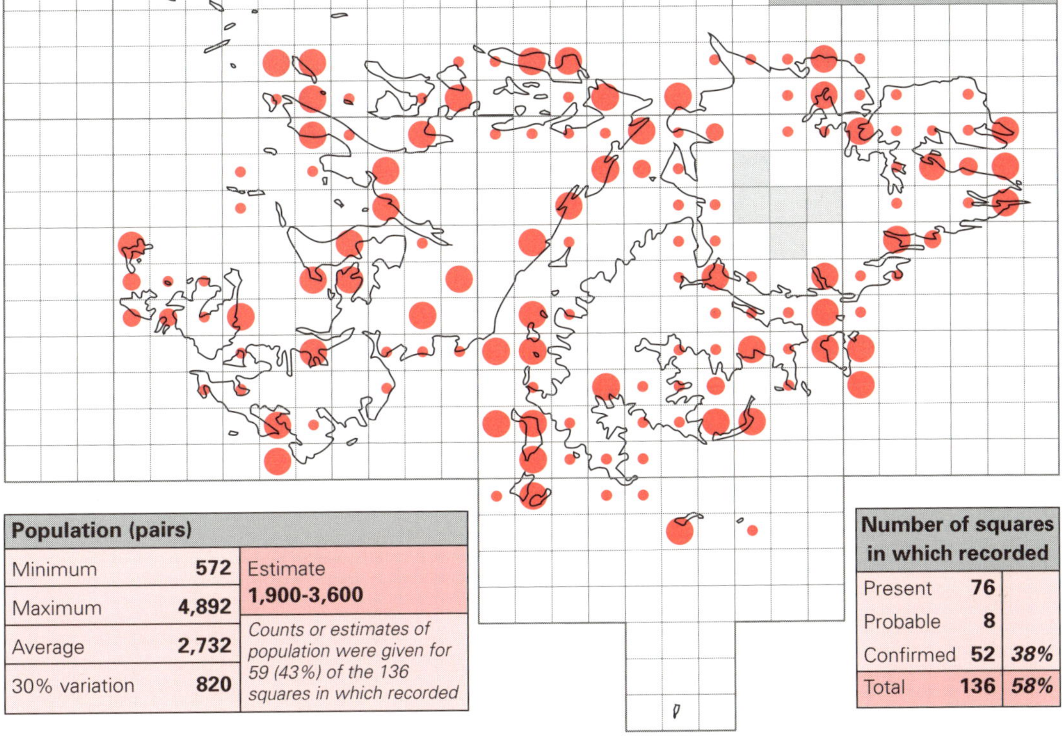

Population (pairs)		
Minimum	**572**	Estimate **1,900-3,600**
Maximum	**4,892**	
Average	**2,732**	*Counts or estimates of population were given for 59 (43%) of the 136 squares in which recorded*
30% variation	**820**	

Number of squares in which recorded		
Present	**76**	
Probable	**8**	
Confirmed	**52**	***38%***
Total	**136**	***58%***

Comparisons between species records and topography	Dry/Wet Number of wet tetrads					Inland/Coastal Amount of coast			Upland/Lowland Proportion of square > 500′				
	Nil	1-5	6-10	11-15	16-22	None	<10%	≥10%	≥1/2	≥1/4	<1/4	None	Total
Surveyed squares	38	104	52	24	16	17	14	203	7	25	73	129	234
% of surveyed squares	16	45	22	10	7	7	6	87	3	11	31	55	100
Grid squares in which recorded	14	69	32	15	6	1	4	131	1	10	47	78	136
% records in each category	10	51	24	11	4	1	3	96	1	7	35	57	100
Significant variation								**					

22 Black-necked Swan

Cygnus melancoryphus

A. & C. Cisne cuello negro

History

This swan was apparently first recognised in the Falklands in 1789 when Gmelin named it *Anas nigricollis* (Hellmayr & Conover 1948). Garnot (1826) looked for this species which Bougainville had mentioned from his visits in the 1760s, but did not see any. Sclater (1860) only listed the Black-necked Swan as having been collected by Captain Pack. Abbott (1861) stated that it was rather scarce, very wild and was found all the year round in East Falkland. He saw 'a number' on the River Murrell in 1859, and noted that most of the group moulted there. He thought that it bred mainly on the adjacent islands, as he had only heard of one nest on the mainland, at the edge of a pond in Mare Harbour. Vallentin (1904) said that it was 'very locally distributed' and that a few still bred in places on West Falkland. He had heard of a pair near Stanley early in November. Bennett (1926) only noted it as a resident breeding species while Cobb (1933) remarked that they were shy and adults usually flew out to sea when disturbed. He believed that there were more on East than on West Falkland and that 'they usually frequent places remote from a settlement'. Cawkell & Hamilton (1961) stated that it no longer bred on West Falkland but mainly in Lafonia, though only in small numbers. Wintering flocks assembled in May at Port Howard (up to 15), Cape Dolphin, Murrell Creek and Swan Inlet (40 or more birds at each). They suspected migration between southern South America and the Falklands, mentioning a few that arrived at Port Purvis (West Falkland) and a pair at Sea Lion Island each spring, which did not stay.

The comments of Vallentin (1904) and Cawkell & Hamilton (1961) both suggest that the Black-necked Swan population of the Falklands was declining during the first half of the 20th century. However, at the start of the survey, it was thought to be numerous in Lafonia and was also found on Pebble Island and on large ponds or lakes in West Falkland. Flocks of wintering and nonbreeding immature birds were known to gather in favoured estuaries, particularly Swan Inlet at Choiseul Sound entrance and the River Murrell estuary near Stanley.

Habitat and Breeding

The Black-necked Swan breeds between early August and mid-September, building a large nest of twigs and grass usually on an islet in a pond or near a shallow lake. The 4-7 cream-coloured eggs need about five weeks incubation before hatching and juveniles are fledged after about another three months (del Hoyo *et al.* 1992).

Breeding Birds Survey: analysis of data

As the map shows, Lafonia seems to be the stronghold with extensions into the Mare Harbour/Mount Pleasant/Fitzroy area south of the Wickham Heights. It also breeds on Speedwell Island and possibly on Lively and Bleaker Islands and was found in the Swan Pond/Elephant Beach area towards Cape Dolphin. On West Falkland the main area was between Chartres and Fox Bay with breeding confirmed in the lowlying wet region known as Hawks Nest ponds (TC95) and also on Pebble Island. The largest count, in Upper Swan Inlet, was 132 on 23 March 1986 by Mike Strickland. A flock of 46, possibly nonbreeding birds, was seen in the same area by Nicholas Beswick in November 1992. At Pebble Island, Andrew Clarke saw 95 including a few cygnets on the ponds east of the settlement in November 1990 and Harro Müller reported 63 from the same group of ponds in March 1987. There were no large numbers reported from other localities but there is no evidence that the population is declining. Analyses of topographical parameters show a highly significant bias ($p = 0{\cdot}0007$) towards the grid squares with more than 11 tetrads containing ponds or lakes. The wider occurrence in Lafonia may be partly explained by the distribution of squares with wet tetrads; West Falkland has only 8 squares with 11 or more wet tetrads while East Falkland, with 32 squares, has four times as many. There was a slight tendency for more records in squares with no land higher than 500 feet.

World Distribution

The Black-necked Swan is common in southern South America from about 30°S to Tierra del

22 Black-necked Swan *Cygnus melancoryphus*

Detectability: **3 Moderate**

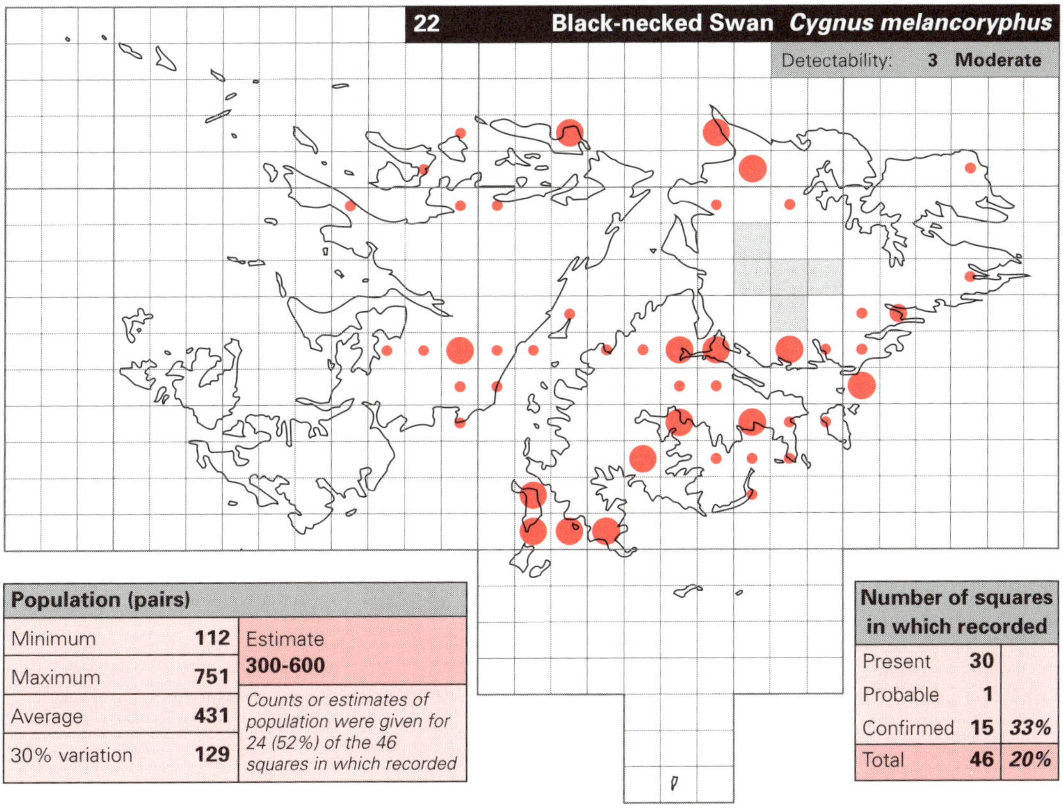

Population (pairs)		
Minimum	**112**	Estimate **300-600**
Maximum	**751**	
Average	**431**	*Counts or estimates of population were given for 24 (52%) of the 46 squares in which recorded*
30% variation	**129**	

Number of squares in which recorded		
Present	**30**	
Probable	**1**	
Confirmed	**15**	***33%***
Total	**46**	***20%***

Comparisons between species records and topography	**Dry/Wet** Number of wet tetrads					**Inland/Coastal** Amount of coast			**Upland/Lowland** Proportion of square > 500′				
	Nil	**1-5**	**6-10**	**11-15**	**16-22**	**None**	**<10%**	**≥10%**	**≥1/2**	**≥1/4**	**<1/4**	**None**	**Total**
Surveyed squares	38	104	52	24	16	17	14	203	7	25	73	129	234
% of surveyed squares	16	45	22	10	7	7	6	87	3	11	31	55	100
Grid squares in which recorded	1	13	13	10	9	3	5	38	0	3	14	29	46
% records in each category	2	28	28	22	20	6	11	83	0	7	30	63	100
Significant variation				***	***								

Fuego, breeding mainly in the La Plata area and in low valleys in the southern Andes (Fjeldså & Krabbe 1990). There is some northward migration after breeding, though Venegas & Jory (1979) and Clark (1986) both state that it is resident in the Magellanic region and Tierra del Fuego. Bennett (1926) noted a solitary half-dead bird captured in the summer of 1916/17, in the Charlotte Channel, Graham Land, 64°30′S. In the austral summer of 1988/89, between November and March, there were about 60 sightings of this species at 16 localities along the coast of the Antarctic Peninsula between 62°S and 68°S (Orgeira & Fogliatto 1991). It is possible that these records concerned no more than ten to 20 individuals (J Croxall pers. comm.). The authors linked these accidental occurrences to drought conditions in Argentina.

Implications for Conservation

This shy swan is an uncommon and distinctive species among the resident waterfowl. It is not currently under threat but would be vulnerable to drainage of the more extensive wetland areas. The Falklands hold a very small fraction of the world population and it is not known whether this is a separate race because there have been no comparative studies of morphology or behaviour.

References

TBFI (1975) p108-9, GBFI (1988) p136-7.

23 Feral Domestic Goose

Anser anser

A. & C. Ganso

History
It is thought that domestic geese were introduced to the Falklands during the first English settlement at Saunders Island, in or soon after 1765. In 1774 Clayton wrote that their 'English Geese all paird in August and September, laid and hatch'd at the Season regular' when he was on Saunders Island between March 1773 and May 1774. Bougainville's expeditions which established the Port Louis settlement from 1764, introduced several domestic animals but apparently not domestic geese (Pernety 1771). Later authors did not mention a domesticated form of the Greylag Goose as naturalised in the Falklands until 1988.

In September 1983, I visited West Point Island after an absence of 20 years and soon noticed that small flocks of domestic geese were prominent along the eastern coast. Nests with eggs had been found and it was obvious that the species was naturalised. In discussions with Roddy and Lily Napier, I learned that a pair of geese from Johnson's Harbour, East Falkland were taken to West Point Island in 1935. They produced apparently infertile eggs annually for nine years. In 1944 a pair of the curly-feathered 'Sebastopol' breed were imported to West Point from Montevideo, Uruguay. These successfully hatched their first eggs in October 1944. Interbreeding with the original pair occurred and by 1955 numbers had increased to about 50. The population remained stable for ten years, then increased to 150-200 by October 1983. During a walk from Sand Paddock to Hope Harbour, following the lower coastal slopes on the eastern side of the island on 1 October 1983, I counted all visible domestic geese. The total of 132 included 57 pure white and 68 brown/white smooth-feathered, three white Sebastopol and four brown/white Sebastopol birds. The smooth-feathered birds appeared to be of the West of England type which was taken to America by the Pilgrim Fathers in 1620 (Soames 1980). Feral domestic geese first appeared on nearby Carcass Island, 9km (5½ miles) to the northeast in the late 1950s, possibly after being blown from West Point. They have survived and in 1983 at least 30 were present. Roddy Napier reported that Weddell Island, Dunbar, Roy Cove, Hill Cove and Chartres settlements also had feral geese.

Habitat and Breeding
On West Point Island it nests from late September and lays up to at least 9 large buff eggs in a grass nest which may be on top of a tussac clump hidden by overhanging leaves, on the ground sheltered in grass, or in dense low gorse. On Carcass in October 1983, four occupied nests were within 2m (6 feet) of each other in a circular gorse patch of about 25sq.m (30sq.yds). The first goslings seen on Carcass Island on 28 October 1983 had down that was either golden-yellow with a brown saddle across the back, or white with dark brown markings and all had orangey-yellow bill and legs.

Breeding Birds Survey: analysis of data
Publicity about feral geese in the *Guide to Birds of the Falkland Islands* (1988) led to records being received from 17 settlements and from seven other localities. These records include about ten birds breeding at Sea Lion Island, 32 seen at Dunbar House, a cluster of records of up to ten birds in the northeastern extremity of East Falkland, east of Salvador and others at Port Stephens, Beaver Island and Cape Meredith. No reports suggest that there are more birds present at any location than were counted at West Point Island in 1983. All records are coastal and there are no significant results from comparison with the topographical analyses.

World Distribution
The Greylag Goose *Anser anser* is the most likely original ancestor for the domestic geese kept over the past 4,000 years. Cross-breeding between Greylag and other large species and artificial selection of characteristics have produced domestic breeds which are found in most parts of the world. Domestic geese were first introduced to North America in 1620 (Soames 1980). Fjeldså & Krabbe (1990) note that domestic geese are found in most of the andean chain.

23 Feral Domestic Goose *Anser anser*

Detectability: 5 Very High

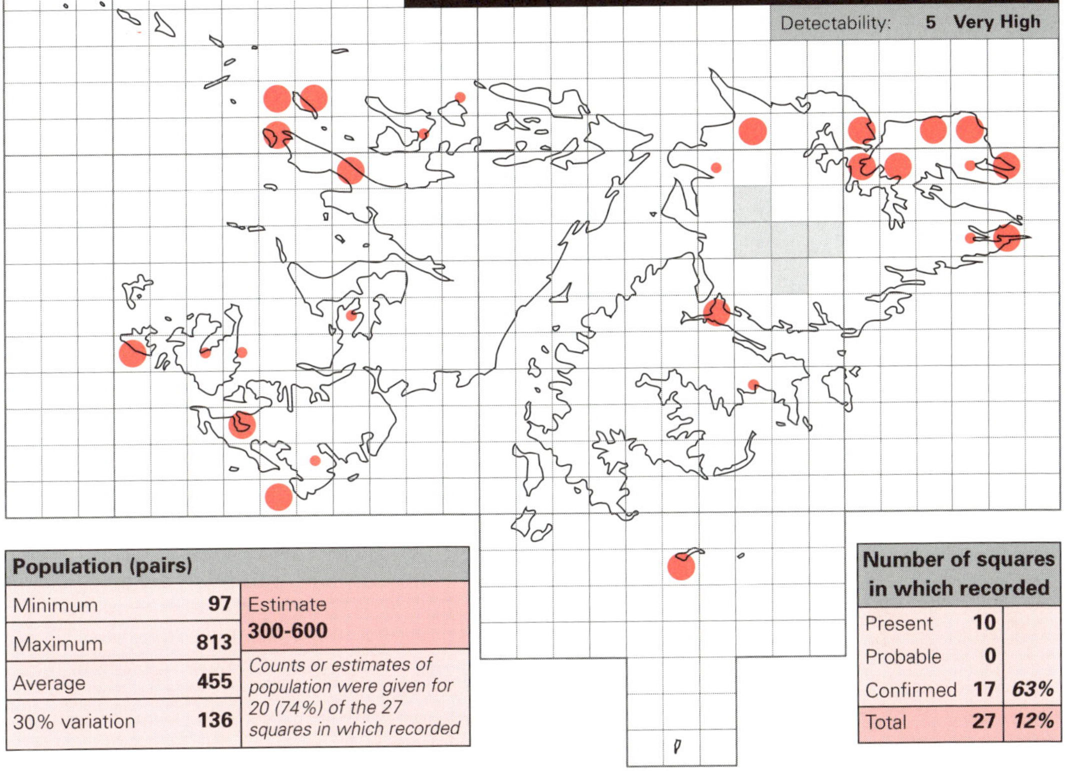

Population (pairs)		
Minimum	97	Estimate **300-600**
Maximum	813	
Average	455	*Counts or estimates of population were given for 20 (74%) of the 27 squares in which recorded*
30% variation	136	

Number of squares in which recorded		
Present	10	
Probable	0	
Confirmed	17	*63%*
Total	27	*12%*

Comparisons between species records and topography	Dry/Wet: Number of wet tetrads Nil	1-5	6-10	11-15	16-22	Inland/Coastal: Amount of coast None	<10%	≥10%	Upland/Lowland: Proportion of square > 500′ ≥1/2	≥1/4	<1/4	None	Total
Surveyed squares	38	104	52	24	16	17	14	203	7	25	73	129	234
% of surveyed squares	16	45	22	10	7	7	6	87	3	11	31	55	100
Grid squares in which recorded	2	14	4	4	3	0	0	27	0	3	12	12	27
% records in each category	7	52	15	15	11	0	0	100	0	>11	>44	>44	100
Significant variation													

Implications for Conservation

Since domestic geese were recognised as naturalised in the Falklands, there has been speculation that they could oust the native Upland and/or the Ruddy-headed Geese and Phil Gregory reported an apparent pairing between a white female Feral Domestic Goose and a male Upland Goose on Carcass Island in August 1989. On West Point Island in 1983, it was noticeable that domestic geese were more wary of humans than Upland Geese and were liable to fly out into coastal bays rather than walk away, which could be a reversion to characteristics of their wild ancestors. However, their powers of flight were variable; smooth-feathered birds could take off easily from land but some of the Sebastopol type could barely get off the ground. On Carcass Island, none seemed able or willing to fly. Considering that these birds have been naturalised on West Point Island for over 50 years and that there was a noticeable increase in numbers of both native geese there between 1963 and 1983, the eventual dominance of domestic geese seems unlikely.

References
GBFI (1988) p141-3.

24 Upland Goose

Chloephaga picta leucoptera

A. Cauquén común
C. Caiquén

History

The male Upland Goose was named *Anas leucoptera* by Gmelin in 1789 based on an earlier description of 'L'Oie des Iles Malouines' by Buffon and an illustration by Brown of a 'White-winged Antarctic Goose' from the Falklands. The strong sexual dimorphism led Gmelin to name the female *Anas magellanica* from a Staten Island specimen (Hellmayr & Conover 1948). The tameness and abundance of the 'wild Geese' or 'Painted Geese' were sufficiently notable to be recorded by contemporary writers since the Falklands were first visited by John Strong in 1690 (Summers & McAdam 1993). Pernety (1771) called them 'Bustards ... exquisite either boiled, roasted or fricasseed' and was clearly amazed that they were present in such numbers around Port Louis that, '150 men in perfect health, with good stomachs, found a quantity sufficient for their subsistence during a stay of more than two months, within a tract of country not exceeding three leagues' (15km or about 9 miles). Clayton (1774) also noted that these geese were 'very good food'. In 1833 and 1834 Darwin found them, 'common ... they lived in pairs or small flocks throughout the interior of the island [East Falkland], rarely or never found on the sea-coast and seldom even near fresh-water lakes'. Darwin said that it nested on small outlying islets, supposedly due to fear of the foxes and that the seamen called them the 'Upland geese' (Gould & Darwin 1838-41). Abbott (1861) described it as abundant 'everywhere in East Falkland'. He said that the Cow Bay area 'where the grass is short and sweet' (near Volunteer Point) was known as 'the Farm-yard' because rabbits, Upland Geese and Magellanic Penguins were so plentiful there. He felt that Darwin had been mistaken because they nested 'all over the country, as well as on the adjoining islets', 'unless the disappearance of the Fox from East Falkland has caused a change in this respect'.

Vallentin (1904) described this species as abundant over the whole archipelago, mentioned the bounties paid for bills and said it was 'estimated that three geese devour as much grass as one sheep'. He felt that in January, young birds were 'delicious eating' having fattened themselves on diddle-dee berries. Cobb (1910) remarked that it was to be seen in all parts of the Falklands and that eggs were eaten by people early in the season, as were goslings when old enough and the adults, which were 'considered in perfection when diddle-dee berries (on which they feed largely) are ripe'. Cobb later (1933) repeated the 'estimate' given by Vallentin in 1904 and elaborated on the domestic potential of the species, including the common use of goose wings as brushes around peat-burning fireplaces or for painting in awkward places. Wace (1917 & 1921) considered that it could be a 'semi migrant' with some birds remaining the whole year. Bennett (1926) stated that it was 'abundant; war is waged on it on account of its grass-eating propensities'. He confirmed that it was excellent eating and remarked, 'Even if they do eat grass, are they not disseminating the seeds ?' Cott (1953) quoted Hamilton as having informed him that possibly up to 10,000 eggs were taken annually. Cawkell & Hamilton (1961) felt that it was still 'very plentiful', could not believe that it could continue to withstand 'ceaseless persecution' and thought that it was only the 'very low human population' that saved it from extinction.

The apparent competition for pasture land with introduced sheep (c807,000 by about 1900) became an issue late in the 19th century. In 1905, Government officially classed them as pests and authorised payment of bounties for goose beaks. In 1924, H Munro, a visiting specialist from New Zealand, recommended the use of grain poisoned with phosphorus to kill geese. Luckily for them and other species, this recommendation was not adopted (Harradine 1976). Another agricultural team visiting in 1969/70 to advise on improving productivity, concluded that the geese did pose a problem, recommended greater efforts to control what they saw as a major pest to sheep farming and felt that a specialist should study the problem (Davies *et al.* 1971). A recommendation in 1972 for ecological studies before implementing any control measures brought John Harradine in 1974 and he was followed by Ron Summers in 1977.

24 *Upland Goose *Chloephaga picta*

Detectability: **5 Very High**

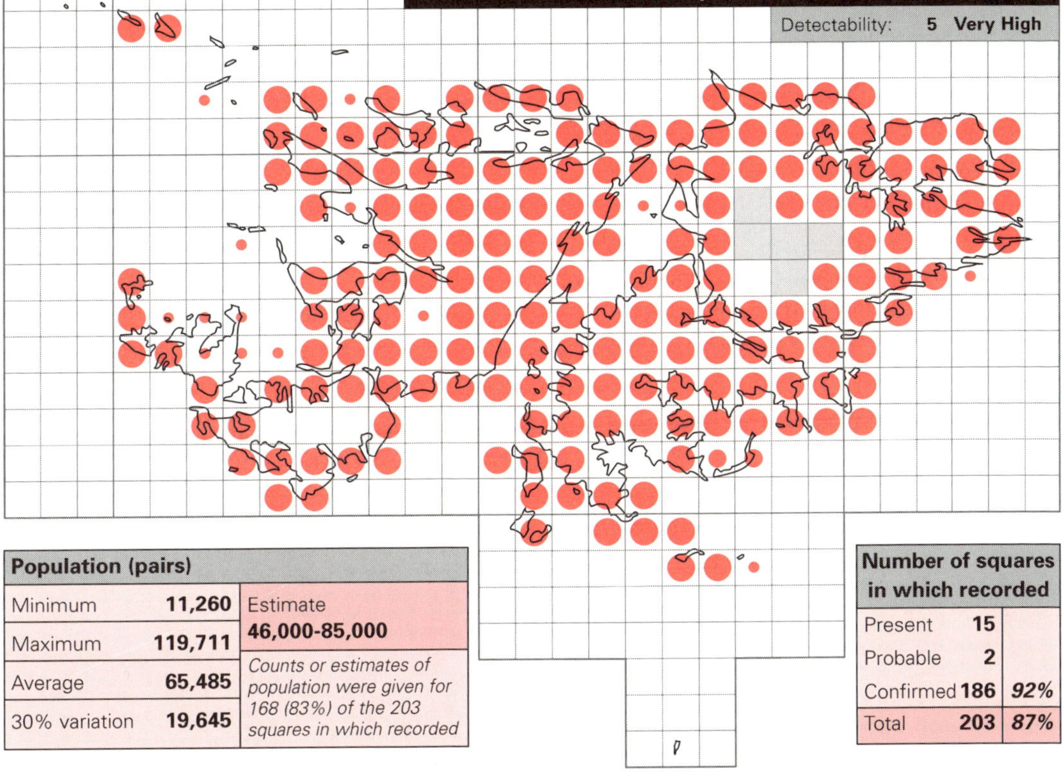

Population (pairs)		
Minimum	**11,260**	Estimate **46,000-85,000**
Maximum	**119,711**	
Average	**65,485**	*Counts or estimates of population were given for 168 (83%) of the 203 squares in which recorded*
30% variation	**19,645**	

Number of squares in which recorded		
Present	**15**	
Probable	**2**	
Confirmed	**186**	***92%***
Total	**203**	***87%***

Comparisons between species records and topography	Dry/Wet Number of wet tetrads					Inland/Coastal Amount of coast			Upland/Lowland Proportion of square > 500′				
	Nil	**1-5**	**6-10**	**11-15**	**16-22**	**None**	**<10%**	**≥10%**	**≥1/2**	**≥1/4**	**<1/4**	**None**	**Total**
Surveyed squares	38	104	52	24	16	17	14	203	7	25	73	129	234
% of surveyed squares	16	45	22	10	7	7	6	87	3	11	31	55	100
Grid squares in which recorded	21	93	49	24	16	15	13	175	6	22	70	105	203
% records in each category	10	46	24	12	8	>7	>6	>86	3	11	34	52	100
Significant variation													

Studies of the effects of geese on reseeded and permanent pastures, the nutritive value of goose droppings, the annual cycle, diet, food intake and populations, showed that geese took only about 1·5% of the total annual organic matter production of herbage. This represented only 7% of what was eaten by all herbivores (Summers & Dunnet 1984). Most of this consumption was concentrated on small areas of particular vegetation types, such as greens and reseeded pastures. It was on the latter areas that geese caused most problems for farmers. They questioned the value of culling and felt that Upland and Ruddy-headed Geese should perhaps be regarded as tolerable nuisances and that farmers should site reseeded grassland areas away from ponds and green valleys, where geese particularly congregated during moult of their flight feathers between late November and early January. Summers (1985) made sample counts between 1977 and 1980 mainly within a 10km radius of Darwin. From the data on several different pastures and the detailed vegetation map produced in 1980 by Jim McAdam, he calculated that there could be 34,317 individual Upland Geese (about 17,000 pairs) on the 124,000ha of the North Arm Farm area. It is not possible to extrapolate accurately to the whole archipelago from this estimate. The controversy over the conflicting interests of geese, sheep and humans in the Falklands continues and is well-documented by Summers and McAdam (1993).

Habitat and Breeding

It is especially common on coastal greens near settlements and ponds and by creeks. Nesting can occur between early August and early January with the main laying season from mid-

September to late October. Five to 8 large cream-coloured eggs are laid in a grass nest amongst ferns, diddle-dee or white grass, or on a tussac grass clump up to 1·5m (5 feet) above ground. The female incubates, for about a month, while the male 'waits off' at a distance. Fledging takes a further 9-10 weeks. Both parents defend their young vigorously but many small goslings are taken by Kelp Gulls and Skuas.

Breeding Birds Survey: analysis of data
The Upland Goose achieved the distinction of being the most widely recorded species. The very high proportion of records where breeding was confirmed is probably due to its size, tameness and abundance. The calculations from estimates and counts suggest that there are fewer than 100,000 pairs, but, given Summers' calculations (above), it is possible that the Upland Goose population of the Falklands may be as high as 150,000 to 200,000 pairs. The topographical analyses do not show any significant differences; there is only a tendency for Upland Geese to be seen more often in squares with at least some standing water rather than in squares without ponds or lakes. These results probably reflect the changes in vegetation through overgrazing since Darwin's visits and the consequent changes in habitats used by Upland Geese, concurrent with increases in their population.

World Distribution
The Falkland race is larger and resident, favouring coastal areas. On the continent, the Upland Goose breeds from southern Tierra del Fuego to about 38°S in Argentina and Chile. Some birds from Tierra del Fuego, where 25,000 to 30,000 were recorded in 1973 (del Hoyo *et al.* 1992), migrate northwards to about 35°S in winter. Harradine (1976) mentions that occasional individuals of the smaller continental race *C. p. picta* reach the Falklands but there are no dated records. Although it has been heavily persecuted by cattle and sheep farmers in southern South America and the Falklands for over two centuries, it is still widely distributed and common.

Implications for Conservation
Upland Geese are a good alternative food to mutton, their fresh eggs are enjoyed by many local people and full-grown geese are a delicacy when they have been feeding on diddle-dee berries in autumn. They have remained numerous due to the low human population, the inefficiency of culling, and in the last thirty years, through the growth of enlightened attitudes towards a bird that can co-exist with sheep. In the extreme winter of 1995, with the ground snow-covered for about three months, it is thought that thousands of Upland Geese died of starvation. Apart from such unpredictable events, this local race is not currently threatened. The benefits of the large goose population should be appreciated and it should be valued as a typical part of the Falkland scene, particularly interesting to the increasing number of tourists.

References
TBFI (1975) p112-4, GBFI (1988) p139-40.

---

25 Kelp Goose

Chloephaga hybrida malvinarum

A. Cauquén caranca
C. Caranca

History
The beautiful and highly ornamental Kelp Goose first came to the notice of naturalists when Molina described a specimen from Chiloe Island in southern Chile in 1782 (Hellmayr & Conover 1948). In the Falklands, the settlers on Saunders Island had already noticed that the 'Sea Goose' was always along the shore, that the ganders were 'quite white' and that 'they are only eatable when made into soop' (Clayton 1774). Garnot (1826) independently named the female *Anser antarcticus* and remarked that their flesh was less good than that of the Upland Goose because it had a little of the flavour of the seaweed from which they gained their nourishment. He thought that the Kelp Goose was not resident because they had hardly caught a glimpse of these geese since their arrival at Port Louis on the *Coquille*. Their

25 * Kelp Goose *Chloephaga hybrida*

Detectability: 5 Very High

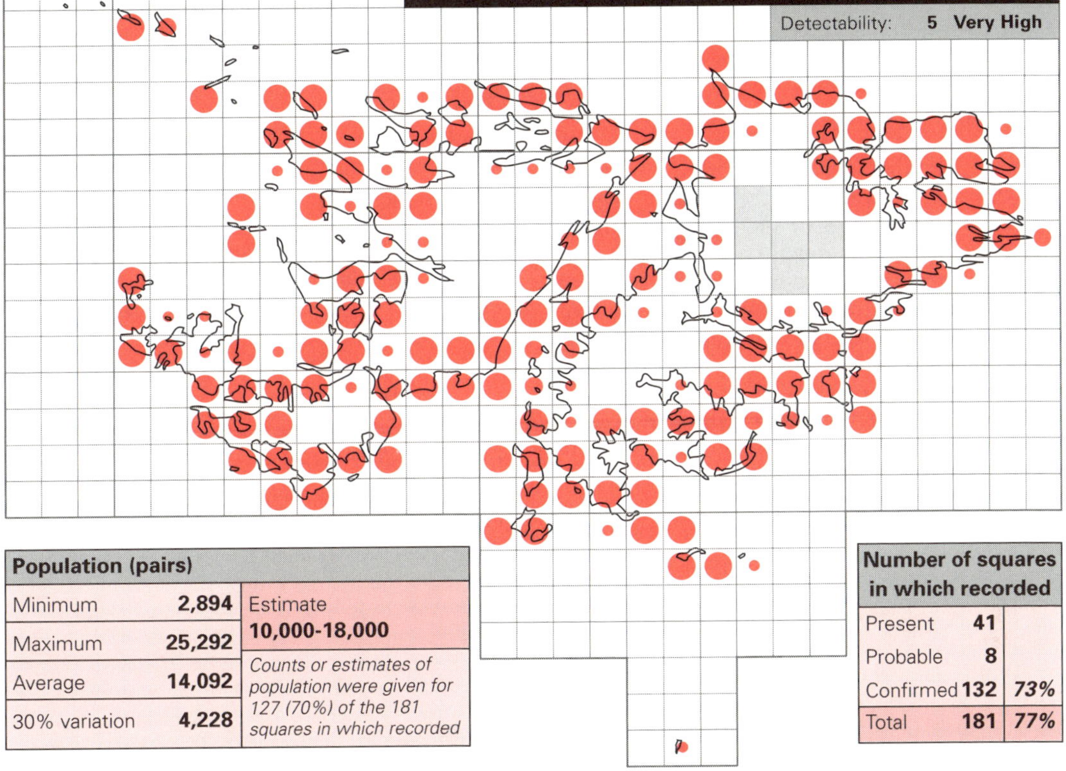

Population (pairs)		
Minimum	2,894	Estimate **10,000-18,000**
Maximum	25,292	
Average	14,092	*Counts or estimates of population were given for 127 (70%) of the 181 squares in which recorded*
30% variation	4,228	

Number of squares in which recorded		
Present	41	
Probable	8	
Confirmed	132	73%
Total	181	77%

Comparisons between species records and topography	Dry/Wet Number of wet tetrads					Inland/Coastal Amount of coast			Upland/Lowland Proportion of square > 500′				
	Nil	1-5	6-10	11-15	16-22	None	<10%	≥10%	≥1/2	≥1/4	<1/4	None	Total
Surveyed squares	38	104	52	24	16	17	14	203	7	25	73	129	234
% of surveyed squares	16	45	22	10	7	7	6	87	3	11	31	55	100
Grid squares in which recorded	23	87	43	19	9	0	5	176	2	13	62	104	181
% records in each category	13	48	24	10	5	0	3	97	1	7	34	58	100
Significant variation								***					

visit of one month coincided with the incubation or early gosling stage of the breeding season and it is possible that most adults were nesting on offshore islands rather than along the coasts.

Darwin (Gould & Darwin, 1838-41) remarked that the Kelp Goose was common and that the sailors called them 'Rock Geese'. Abbott (1861) noted that it was, 'A very common bird along the coast' while Vallentin (1904) considered that this species was, 'universally distributed around the shores of the Falkland isles'. He felt that they paired for life and commented that this species did not eat molluscs, 'as stated by numerous writers' but fed mainly on a common seaweed, *Porphyra vulgaris*. Brooks (1917) described this goose as, 'one of the most handsome in existence', common in the Falklands and very tame and stated that it never ventured more than 50 yards inland, and then only to feed on ripe berries from mid-December. Specimens collected by Brooks were used to distinguish the Kelp Geese of the Falklands as a separate race from that occupying the coasts of Tierra del Fuego (Phillips 1916); the bill, legs, feet and wings are all larger than those of the continental birds. Bennett (1926) simply stated that it was a 'common shore resident' and added, 'though it has beautiful plumage, is quite unfit to eat'. He was still judging this aberrant goose using the standards of settlers over a century earlier.

Although Kelp Geese have not been persecuted as pests or game recently, Brooks (1917) commented that the snowy breasts of the males were sold in considerable numbers in London, until this became illegal early this century. Cobb (1933) understood that the plucked skins of the males were worn round the scalps of native Fuegians ('doctors') who also painted their faces scarlet.

Habitat and Breeding
Kelp Geese are found mainly on rocky coasts where their usual diet of algae is available. They sometimes visit nearby freshwater ponds for bathing and drinking and may graze on short grass or take ripe diddle-dee berries in autumn. Small flocks squabble on the shores in spring and flocks of a hundred or more nonbreeders gather on rocky shores throughout the summer. They nest between mid-October and early November, laying 4-6, sometimes 7 eggs of a light buff colour in a rough grass nest lined with down. Nests are near the shore, sheltered amongst tussac grass or rushes, sometimes on a low cliff ledge. The female incubates while the male stands guard close by, unwittingly revealing the general location of the nest. Immature males pair at one year but probably do not breed until at least two years old. There is evidence of local migrations from offshore islands to more sheltered shores in winter (Woods 1975).

Breeding Birds Survey: analysis of data
The Kelp Goose continues to have a widespread distribution. There are no records from any of the seven inland squares surveyed and comparison with topographical parameters shows a very significant association ($p = 0{\cdot}0003$) with coastal squares. It probably breeds in most of the squares in which it was recorded though breeding was confirmed in only 73%.

World Distribution
The continental race of the Kelp Goose *C. h. hybrida* occurs from about 42°S in Chile through the coasts and islands to southern Tierra del Fuego. It is apparently less common on the coast of Argentina north to about 50°S. In winter there is some migration to about 39°S in coastal Chubut, Argentina and to about 33°S on the coast of Chile. PW Reynolds felt that Falkland Island Kelp Geese visited the east coast of Tierra del Fuego during the winter (Humphrey *et al.* 1970), but there has been no evidence of emigration from Falkland shores.

Implications for Conservation
The Falkland race of this very attractive goose is not currently threatened. However, as it is dependent on the fine green and red algae found on rocky coasts, any large spillage of toxic substances affecting these algae could have serious effects on populations of Falkland Kelp Geese.

References
TBFI (1975) p114-6, GBFI (1988) p140-1.

---

26 Ashy-headed Goose

Chloephaga poliocephala

A. Cauquén cabecigrís

C. Canquén

Local names: Coast/White-breasted Brant

History
The Ashy-headed Goose was first described scientifically by Sclater in 1857 from a Chiloe Island (Chile) specimen and he later (1861) commented that it was an occasional straggler to the Falklands from the coast of South America. In 1861, Abbott stated that it could 'hardly be called a Falkland-Island bird' because during the three years he had been in East Falkland, he had seen only three single birds amongst flocks of Upland Geese. Sclater & Salvin (1876) noted that their collector Lecomte had brought home one skin, obtained between late 1867 and mid-1868. Wace (1917 & 1921) classed it only as an accidental visitor, while Bennett (1922) noted that 'a certain number' reached the Falklands in August 1920 and a large number in the summer of 1920/21. In 1926 Bennett described it as a frequent spring visitor that 'may perhaps breed in the Falklands'. Cawkell & Hamilton (1961) stated that it had bred 'in recent years' at Spring Point and Roy Cove and that one of two pairs at New Island in 1953 had certainly bred. A single bird was seen at a pond in Port Harriet camp with Ruddy-headed Geese in October 1960 by Millie and Len Grant. In late June 1961, Stan Bennett reported that he had shot two near Surf Bay, Stanley. The last breeding record before the survey was of a pair with goslings seen by Richard Cockwell at Port Howard in 1968 (Wolsey 1986).

Habitat and Breeding
In South America the Ashy-headed Goose breeds in open, often lightly wooded country with lakes

26 Ashy-headed Goose *Chloephaga poliocephala*

Detectability: 2 Low

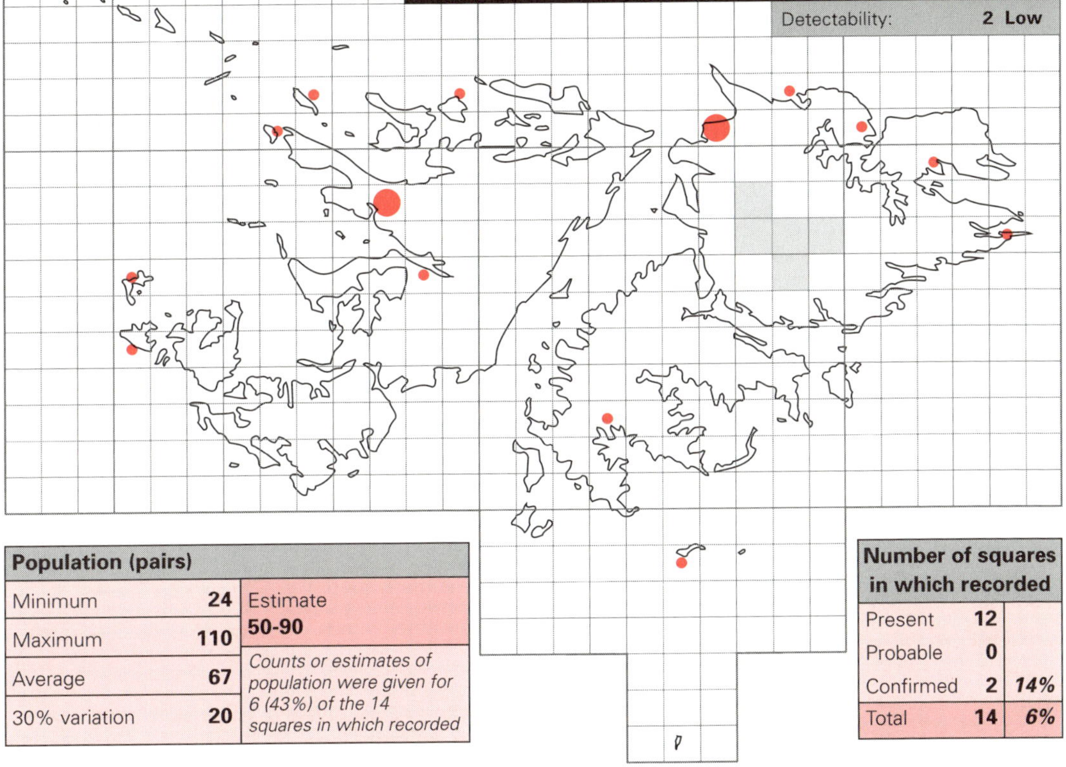

Population (pairs)		
Minimum	24	Estimate 50-90
Maximum	110	
Average	67	*Counts or estimates of population were given for 6 (43%) of the 14 squares in which recorded*
30% variation	20	

Number of squares in which recorded		
Present	12	
Probable	0	
Confirmed	2	***14%***
Total	14	***6%***

and marshes, sometimes perching on low branches of trees and nesting in tree-hollows or among tall grass (Madge & Burn 1988). It winters in natural meadows or pastures (del Hoyo *et al.* 1992). In the Falklands, individuals are usually seen on coastal greens, but there are no records of nesting sites.

Breeding Birds Survey: analysis of data

There have been at least a dozen records of single birds and small parties, usually seen with Ruddy-headed or Upland Geese on short grass. The first confirmed breeding record came from Danny Donnelly at Crooked Inlet (TC78, West Falkland) in 1985. The second was received from Phil Middleton, who visited Smylies Village camp north of Port San Carlos (UD60) in January 1992. Phil checked with some local residents, Jill and Les Harris and Clara MacKay who confirmed the presence of a small flock, which had been seen in this area for several years. Tim Duggleby saw one at New Island in December 1986 (Duggleby 1987). Harro Müller visited many islands during March 1987 and reported one or two on Keppel, Beaver, Carcass and Pebble Islands between 7 and 22 March; this group of records in the same autumn month could represent birds drifted on northward migration from southern South America. A single bird was seen at Sea Lion Island by David Gray in November 1987 and Helen Hannah reported at least one at Carcass Island in 1988. Alan Henry reported one at Penguin Walk, Cape Pembroke in August 1989 and it was later seen by Phil Gregory, M Jones and Digby Munns with nine Upland Geese. Mike Morrison saw one at Port Louis on 15 September 1990 and David Gray had another single bird at Sea Lion Island in early October 1990. Alan Henry and Phil Gregory again saw one at Penguin Walk between 11 and 13 October 1990 and Sue Bonner reported 'several' at Pickthorne Farm, West Falkland before 15 November 1990. The group of records in September/October 1990 could also have been wind blown migrants from the continent.

Overall, these records seem to be randomly scattered around East and West Falkland, again suggesting that occurrence in the Falklands is accidental, apart from the reported breeding group in the Smylies Village area. There were insufficient records to compare distribution with topographical parameters.

World Distribution

The Ashy-headed Goose occurs in Argentina and southern Chile from about 37°S to Tierra del Fuego. Much of the southern population migrates to the pampas of southeastern Buenos Aires province between April and September. According to Madge & Burn (1988) it is not abundant but Fjeldså & Krabbe (1990) state that it is common.

References

TBFI (1975) p109-10, GBFI (1988) p137-8.

Implications for Conservation

The status of the Ashy-headed Goose may not have changed significantly since early human colonisation in the late 18th and early 19th centuries. It seems that individuals are drifted to the Falklands during migration and that pairs occasionally breed. The reported small flock north of Port San Carlos needs further investigation; if confirmed as resident and breeding, it would appear that they have adapted to a new habitat. The species is not considered threatened in southern South America.

27 Ruddy-headed Goose

Chloephaga rubidiceps

A. Cauquén colorado

C. Canquén colorada

Local names: Brent, Brant

History

The Ruddy-headed Goose was named scientifically by Sclater in 1860 from specimens obtained in the Falkland Islands by the Antarctic Expedition of Captain James Clark Ross in 1842. Zoologists in London had previously confused it with the Ashy-headed Goose. Sclater compared specimens of these two smaller species carefully and pointed out the differences in plumage. In the Falklands, British settlers at Saunders Island had distinguished it from the Upland Goose nearly a century earlier. Clayton (1774) quite clearly described the first of three types of geese as the, 'Mountain Goose...a small sort about the size of a Muscovy Duck'. The Ruddy-headed Goose was not recognised by Darwin in 1833 and 1834 but was well-described by Abbott (1861). He felt that it was less common than the Upland and Kelp Geese 'except in some places in the North Camp [of East Falkland] where I have seen very large numbers, probably a hundred, but always in pairs'. Vallentin (in Boyson 1924) stated that it was scarce on East Falkland and only fairly abundant on West Falkland. Bennett (1926) felt that the Ruddy-headed Goose was, 'a coast lover, feeding on grasses near the seashore'. Specimens with white abdomens being 'not uncommon', he speculated that these might be hybrids with the Ashy-headed Goose. Cawkell & Hamilton (1961) said that it 'appears to be commoner on the West Falkland than on the East'. They also felt that collecting beaks of this species for payment of bounties was regrettable because, 'the population is far less than that of the Upland Goose'. They noted that it was usually found with the Upland Goose as they liked similar feeding grounds and agreed with Vallentin that it was shyer.

From his study of Upland and Ruddy-headed Geese in the Darwin and North Arm areas of East Falkland, Summers (1985) estimated a population of about 9,003 (sic) individual Ruddy-headed Geese on 124,000ha of the North Arm farm. It is impossible to extrapolate accurately to the whole of the Falklands from this estimate because the relative abundance of pasture-types is not known outside the North Arm farm area. Other factors such as introduced predators, mountainous areas with unsuitable vegetation and variations in pressure from shooting in the past have all had unknown effects on the Ruddy-headed Goose population. However, if his estimate is multiplied on the basis of area alone, there are possibly up to 42,000 pairs of Ruddy-headed Geese in the Falklands.

Habitat and Breeding

It nests from late September to early November, laying 5-8 slightly shiny creamy-buff eggs in a grass nest lined with down. Nest sites may be well-hidden in long grass or rushes, in an old Magellanic Penguin burrow, on top of a tussac grass clump or quite exposed amongst low

27 # Ruddy-headed Goose *Chloephaga rubidiceps*

Detectability: **4 High**

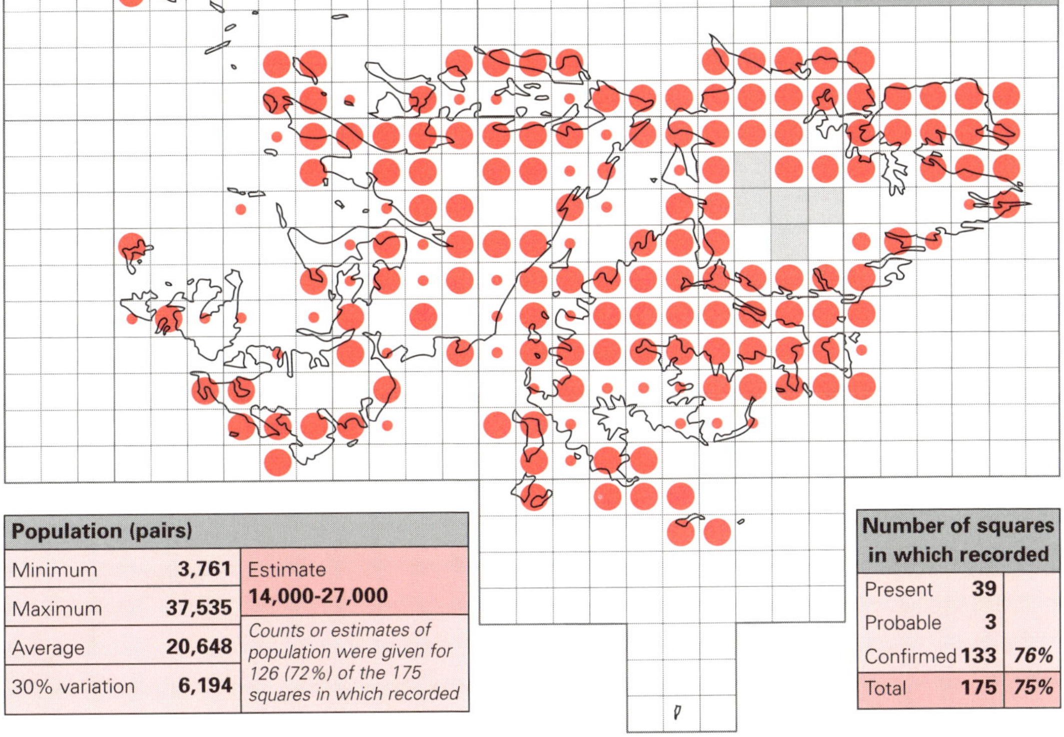

Population (pairs)		
Minimum	**3,761**	Estimate
Maximum	**37,535**	**14,000-27,000**
Average	**20,648**	*Counts or estimates of population were given for 126 (72%) of the 175 squares in which recorded*
30% variation	**6,194**	

Number of squares in which recorded		
Present	**39**	
Probable	**3**	
Confirmed	**133**	***76%***
Total	**175**	***75%***

Comparisons between species records and topography	Dry/Wet Number of wet tetrads					Inland/Coastal Amount of coast			Upland/Lowland Proportion of square > 500′				
	Nil	1-5	6-10	11-15	16-22	None	<10%	≥10%	≥1/2	≥1/4	<1/4	None	Total
Surveyed squares	38	104	52	24	16	17	14	203	7	25	73	129	234
% of surveyed squares	16	45	22	10	7	7	6	87	3	11	31	55	100
Grid squares in which recorded	14	81	44	21	15	10	11	154	3	18	59	95	175
% records in each category	8	46	25	12	9	6	6	88	2	10	34	54	100
Significant variation													

diddle-dee. The female alone incubates while the male 'waits off', often at a considerable distance from the nest. He will return rapidly, however, if the female is flushed. When with small goslings, the male may perform a 'broken-wing' display and call loudly while the female leads the goslings to safety.

Breeding Birds Survey: analysis of data

Records give the impression that this species is widely distributed throughout the Falklands, except perhaps on some of the areas of higher ground, including the Mount Alice region and Hornby Mountains of West Falkland. It also seems to be absent from the Wickham Heights in East Falkland, though five of these squares were not surveyed. Earlier statements about it being fairly common have been confirmed but not the contention that it was more numerous on West Falkland. It is not restricted to coastal areas as much as was thought by Bennett. In the statistical analyses, there were no significant differences between the distribution of records and the parameters studied though there was a slight tendency for more records from squares with lakes and ponds.

World Distribution

In South America the Ruddy-headed Goose formerly occurred in lowland grassland from southern Tierra del Fuego to about 52°S. A spectacular decline in numbers has been reported within the last 35 years (Madge & Burn 1988), apparently caused by foxes *Dusicyon griseus*, which were introduced in the late 1940s to prey on the rabbits introduced earlier. The Ruddy-headed Goose population had apparently reduced to less than 1,000 birds by 1976, only occurring in northern Tierra del Fuego. These birds migrate to southern Buenos Aires province of Argentina between April and September. Because this population has been so seriously reduced, it is now classed as '**Near-threatened**' by Collar *et al.* (1994), meaning that it is considered worthy of monitoring in order to detect any evidence of deteriorating status.

References

TBFI (1975) p110-12, GBFI (1988) p138-9

Implications for Conservation

The Falkland archipelago holds most of the world population of the Ruddy-headed Goose. Although not threatened in the Falkland Islands, this sedentary population has a global importance. The protection given from 1985 when it was removed from the list of pest species (Schedule I of the 1964 Wild Animals and Birds Protection Ordinance) should continue in new conservation legislation. Information on status and the differences between Ruddy-headed and Upland Geese should be given local publicity.

28 Flying Steamer Duck

Tachyeres patachonicus

A. Pato vapor volador

C. Quetru volador

Local name: Canvasback

History

The Flying Steamer Duck was described by King in 1828 from the western parts of the Straits of Magellan (Hellmayr & Conover 1948). Abbott (1861) was apparently the first to recognise it as occurring in the Falklands. He said that it was, 'not uncommon', noted that it bred in the same places as the flightless species, 'but rather frequents the freshwater ponds near the sea, and is a difficult bird to approach from its wariness'. He shot only one specimen and observed that it was much smaller and lighter than the local flightless Steamer Duck (Logger) *T. brachydactyla* and the wings were more developed, though still small for the size of the body. Oustalet (1891) noted that it occurred in the Falklands but this information did not reach Dr. Wace at the Darwin settlement on East Falkland. In his manuscript list (1917) he described the 'Canvas-back Duck' as possibly a variety of the Flightless Steamer Duck and commented that it was a doubtful species, though it 'flies high and well'. Dabbene (in his editorial comments in Wace, 1921) was still not sure that two species existed and put forward the idea that the juveniles of the flightless species could fly, but lost this ability as they matured and became heavier.

Bennett (1926) was certain that two species were present. He stated that it is, 'a very distinct bird, even though its plumage is almost identical with the [Logger]... and it can fly as well as any Duck'. He said that it was found on freshwater and went as far as to say that, 'if you see a Logger-like Duck on a freshwater pond in daylight it is sure to be a Canvas-back'. He also stated that the Canvas-back is 'excellent eating' whereas the Logger 'is quite impossible'. The Canvas-back was much smaller and less robust, showed slightly different white edges on the inner primaries and had noticeable serrations on the bill, which were not visible on the male Logger. Both species of steamer duck were accepted as being present by the time Cawkell & Hamilton's list was published (1961). They felt that this species could be distinguished by its build, proportionate wing-length, behaviour and habitat. However, due to the long standing confusion with the Logger it is only possible to speculate that the Flying Steamer Duck has remained relatively uncommon.

Habitat and Breeding

The Flying Steamer Duck is usually found on freshwater ponds. It nests from October, possibly earlier, laying 5-8 buff eggs, much smaller and more pointed than those of the Logger. Nest sites are apparently similar to those of the Logger though there are few nest records; one was found in a small gorse bush about 3m (10 feet) diameter and another in a large rush patch, up to 350m (400 yds) from the sea by Rodney Hutching on Carcass Island in 1983. Sitting females exploded from both these nests when disturbed by a dog and flew out onto the sea; eggs were chipping in one nest on 5 November. Other records suggest that this species is as territorial as the Logger. Ponds which are in areas suitable for breeding are not likely to have more than one pair unless

28 Flying Steamer Duck *Tachyeres patachonicus*

Detectability: **3 Moderate**

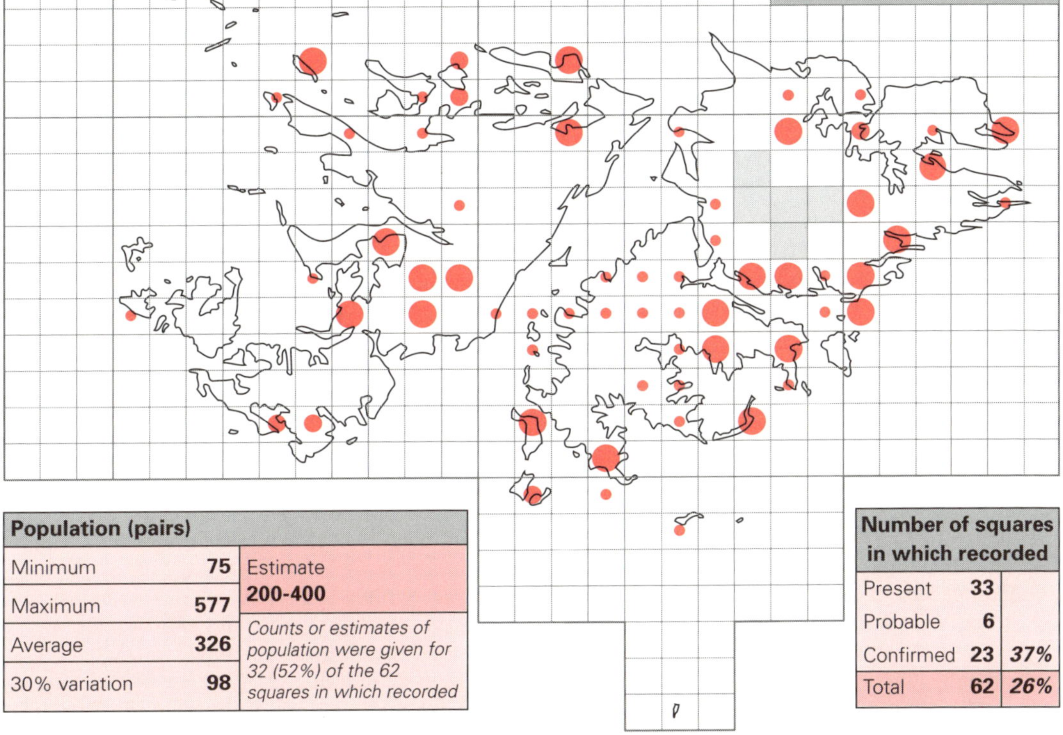

Population (pairs)		
Minimum	**75**	Estimate **200-400**
Maximum	**577**	
Average	**326**	*Counts or estimates of population were given for 32 (52%) of the 62 squares in which recorded*
30% variation	**98**	

Number of squares in which recorded		
Present	**33**	
Probable	**6**	
Confirmed	**23**	***37%***
Total	**62**	***26%***

Comparisons between species records and topography	Dry/Wet: Number of wet tetrads					Inland/Coastal: Amount of coast			Upland/Lowland: Proportion of square > 500′				
	Nil	**1-5**	**6-10**	**11-15**	**16-22**	**None**	**<10%**	**≥10%**	**≥1/2**	**≥1/4**	**<1/4**	**None**	**Total**
Surveyed squares	38	104	52	24	16	17	14	203	7	25	73	129	234
% of surveyed squares	16	45	22	10	7	7	6	87	3	11	31	55	100
Grid squares in which recorded	2	23	13	15	9	5	6	51	0	5	19	38	62
% records in each category	3	37	21	24	15	8	10	82	0	8	31	61	100
Significant variation				**	**		**						

they are particularly rich in aquatic vegetation and invertebrates (Weller 1972).

Breeding Birds Survey: analysis of data

The map shows a wide spread of records; two thirds of the grid squares in which it was recorded being on or around East Falkland. However, the numbers are low, with no more than ten birds being seen at any one locality. The fact that it was recorded in one quarter of the squares surveyed supports the earlier statements about it being widely scattered but the population estimate indicates that it is one of the least common waterfowl. Analyses of topographical parameters show a significant bias towards occurrence in squares with very little coastline, rather than squares with a long coastline or squares that are landlocked ($p = 0{\cdot}007$) and a greater bias towards occurrence in squares with 11 or more 'wet' tetrads ($p = 0{\cdot}002$). This adds weight to Bennett's belief that inland ponds are the places to look for this species.

World Distribution

In southern South America, the Flying Steamer Duck inhabits lakes, rivers and coastal waters from about 37°S in Chile and about 44°S in Argentina to Tierra del Fuego. It also occurs inland in Argentina to about 38°S and is common on several inland lakes in Santa Cruz province (51°S) (del Hoyo *et al.* 1992).

Implications for Conservation

Though it is not numerous, this species is not endangered in the Falklands. The survey has increased knowledge of its distribution but its breeding biology needs further investigation.

References

TBFI (1975) p123, GBFI (1988) p145-6.

29 Falkland Steamer Duck

Tachyeres brachydactyla

A. Pato vapor malvinero

Local names: Loggerhead, Logger

History

The Falkland Steamer Duck was first described in 1789 by Gmelin, apparently from skins of the 'Loggerhead Goose' obtained by Pernety in 1764 at Port Louis (Hellmayr & Conover 1948). Clayton on Saunders Island (1774) noted a wild duck 'which never flies, their wings being very short heads large (thence name Loggerhead Ducks) they only flutter along the water, they keep in great flocks and are easily drove onshore'. Abbott (1861) stated that this duck 'frequents the salt water' and said that 'the harbour of Stanley is full of them, as well as every other part of the coast'. He commented on the strongly territorial behaviour of pairs which drove away any other Loggers. Their eggs were considered a great delicacy and looking for nests was 'a great amusement to all the boys in Stanley'.

Vallentin (1901) was impressed by their extreme tameness. Later he visited West Falkland and commented that the Logger was, 'one of the commonest birds on the Falkland Archipelago, perhaps more abundant in the numerous estuaries than along the shores of the ocean' (Vallentin 1904). Cobb (1933) said that the eggs were good for eating and that 'quantities are collected and pickled for winter use'. Both observers commented on the large flocks of immature birds which are a feature of many inlets. Hamilton wrote in 1949 and 1950 that he felt the eggs, 'though musky are much valued'. He estimated that 'the annual crop was low down in the 1,000-10,000 category, some being used for chickenfeed' (Cott 1953). Bennett (1926) and Cawkell & Hamilton (1961) repeated that it was very common and that it could be found at most beaches and inlets. Cawkell & Hamilton and later, Pettingill (1965), gave detailed information about this unusual endemic duck, which is familiar near every settlement.

Habitat and Breeding

The Logger is common around the coasts of the Falklands, especially in sheltered harbours and creeks and where large offshore kelp beds occur. It is almost restricted to salt water in daylight, though it visits coastal freshwater ponds and streams for bathing and drinking. It ranges regularly to the outer edge of kelp beds and has been seen up to 5km (3 miles) offshore. Pairs defend their territories throughout the year against adjacent pairs and other species. Nests are made in short grass, diddle-dee, dry kelp or tussac grass or in old Magellanic Penguin burrows and may be found from a few metres above high water mark to about 400m (¼ mile) inland. The nest is a depression scantily filled with grasses and lined with down in which 5-8, occasionally up to 12, large cream-coloured eggs are laid usually between mid-September and December, though nests with eggs are known from all months. The female alone incubates while the male waits offshore, patrolling their territory. Mortality among newly hatched ducklings is high due to predation by Kelp Gulls and Falkland Skuas, but full-grown birds have no natural predators except seals.

Breeding Birds Survey: analysis of data

Records confirm the very wide distribution of this endemic species and the Inland/Coastal analysis shows a very strong bias towards occurrence in coastal squares (p = 0·0006) rather than those with little or no coastline. The presence of more or fewer ponds or lakes has no effect on its distribution nor does the general altitude reached in each square. These results contrast strongly with those for the Flying Steamer Duck and confirm the maritime preferences of the Logger. It appears that the status of this species has not changed since the Falklands were settled.

World Distribution

The Falkland Steamer Duck, found only in the Falkland Islands, is now recognised as one of three flightless species of Steamer Duck in southern South America, all of which overlap in distribution with the Flying Steamer Duck in different regions.

29 **Falkland Steamer Duck *Tachyeres brachydactyla*

Detectability: 5 **Very High**

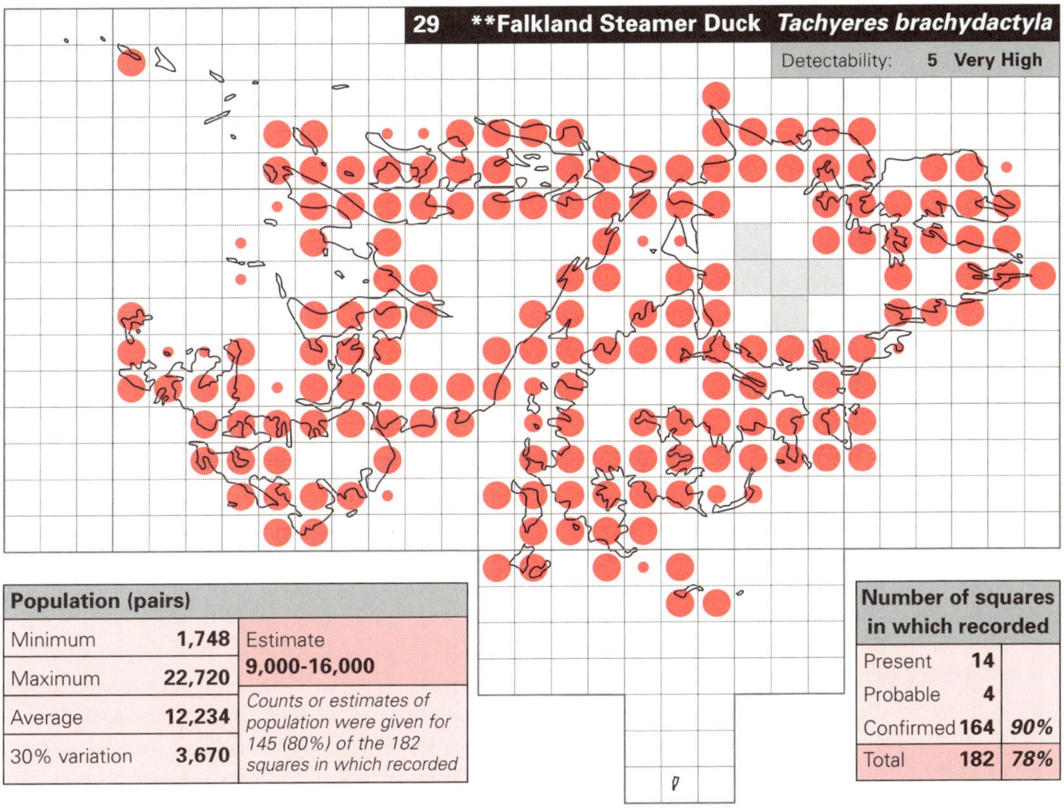

Population (pairs)		
Minimum	**1,748**	Estimate **9,000-16,000**
Maximum	**22,720**	
Average	**12,234**	*Counts or estimates of population were given for 145 (80%) of the 182 squares in which recorded*
30% variation	**3,670**	

Number of squares in which recorded		
Present	**14**	
Probable	**4**	
Confirmed	**164**	***90%***
Total	**182**	***78%***

Comparisons between species records and topography	Dry/Wet Number of wet tetrads					Inland/Coastal Amount of coast			Upland/Lowland Proportion of square > 500′				
	Nil	1-5	6-10	11-15	16-22	None	<10%	≥10%	≥1/2	≥1/4	<1/4	None	Total
Surveyed squares	38	104	52	24	16	17	14	203	7	25	73	129	234
% of surveyed squares	16	45	22	10	7	7	6	87	3	11	31	55	100
Grid squares in which recorded	24	88	40	20	10	0	7	175	2	14	60	106	182
% records in each category	13	>48	22	11	>5	0	4	96	1	8	33	58	100
Significant variation								***					

Implications for Conservation

The population of this species is not at present threatened around Falkland coasts. Some eggs are still taken for human consumption but the quantities are small and would have little effect on the whole population. As it occurs around all coasts with any shelter, it is potentially vulnerable to pollution in the coastal waters where its marine invertebrate food is obtained (Weller 1972). Any future offshore oil exploration or extraction could be a particular threat as accidental spillages would almost inevitably affect some part of the Logger population.

References
TBFI (1975) p118-22, GBFI (1988) p144-5.

30 Chiloe Wigeon

Anas sibilatrix

A. Pato overo C. Pato real

Local names: Black and White Wigeon, Wigeon

History
This species was named in 1829 from Chilean examples and seems to have been first recorded in the Falklands by the Antarctic Expedition of James Clark Ross in 1842. Sclater (1860) notes only that this expedition obtained some specimens while Abbott (1861) felt that it was 'one of the wildest and scarcest birds in East Falkland'. He found ducklings with a parent in a pond at Port Louis in January; on 7 April he shot some near the north shore which had imperfect wings and questioned whether they were young birds or simply moulting. Vallentin (in Boyson 1924) felt that its status was still as described by Abbott, though he had shot at more than one specimen. He found only one nest, with 8 eggs, by Byron Sound some time between 1901-2 or 1909-11. Bennett (1926) noted that it was a resident but not common while Cawkell & Hamilton (1961) stated that it was usually only seen in pairs or very small parties which were very wary, except when tending ducklings. They believed that its stronghold was in Lafonia with some on West Falkland and they noted that three pairs nested on New Island in 1953.

Habitat and Breeding
The Chiloe Wigeon is found on larger ponds, lakes and rivers. It also occurs in coastal kelp patches and often grazes on green pasture near shores or the banks of ponds. It nests from September to December, laying 5-8 white eggs in a well-hidden grass nest often a long way from the nearest water. It may sometimes be double-brooded but little information has been collected on its breeding behaviour.

Breeding Birds Survey: analysis of data
It was recorded in 68 grid squares on East Falkland and the adjacent islands where there are slightly more records from Lafonia, but it is also well-distributed in the lower regions north of the Wickham Heights. On West Falkland it was found in only 21 squares with a noticeable concentration in the wetter areas between Fox Bay and Chartres. The maximum reported during the survey was 60, seen by Harro Müller on Pebble Island in March 1987. In the southern part of East Falkland, Mike Riddy counted 56 birds on one pond in the Bull Roads area of Bull Point (UC30) in December 1987. David Osborn reported a maximum of 33 at East Cove on 4 November 1989 and Bill Bourne saw a flock of 23, including four broods of three, six, three and five juveniles on Bertha's Pond on 8 February 1990. Apart from these flocks, most reports were of groups of less than ten birds. The population estimate suggests that it is much less numerous than the Silver Teal and Speckled Teal. The most striking result of the topographical analyses is the very significant ($p = 0{\cdot}0004$) bias towards presence in the wetter squares, those with 6 to 22 wet tetrads. There is also a non-significant tendency for more records from squares with no land above 500 feet. In comparison with the distribution of Inland/Coastal squares, records show no bias towards any category. In summary, this species is highly dependant on the wetter, lowlying parts of both East and West Falkland, though it also inhabits coastal areas adjoining higher ground. The largest numbers were recorded in regions known to have many large ponds.

World Distribution
The Chiloe Wigeon is common in Chile from 40°S and Argentina from 34°S to Tierra del Fuego. The southern population is partly migratory, some reaching Paraguay, Uruguay and southeastern Brazil. Vagrants have been recorded at South Georgia on six occasions between 1972 and 1985 and on several occasions at the South Shetland and South Orkney Islands since 1966 (Prince & Croxall 1996).

Implications for Conservation
This uncommon duck is considered to be a game species that may be shot during the open season. The numbers killed are not known and it appears to be maintaining its population. However, it seems rare enough to qualify for removal from the list of game birds and the main wetland areas on which it depends should be protected.

References
TBFI (1975) p125-6, GBFI (1988) p148.

30 Chiloe Wigeon *Anas sibilatrix*

Detectability: 4 High

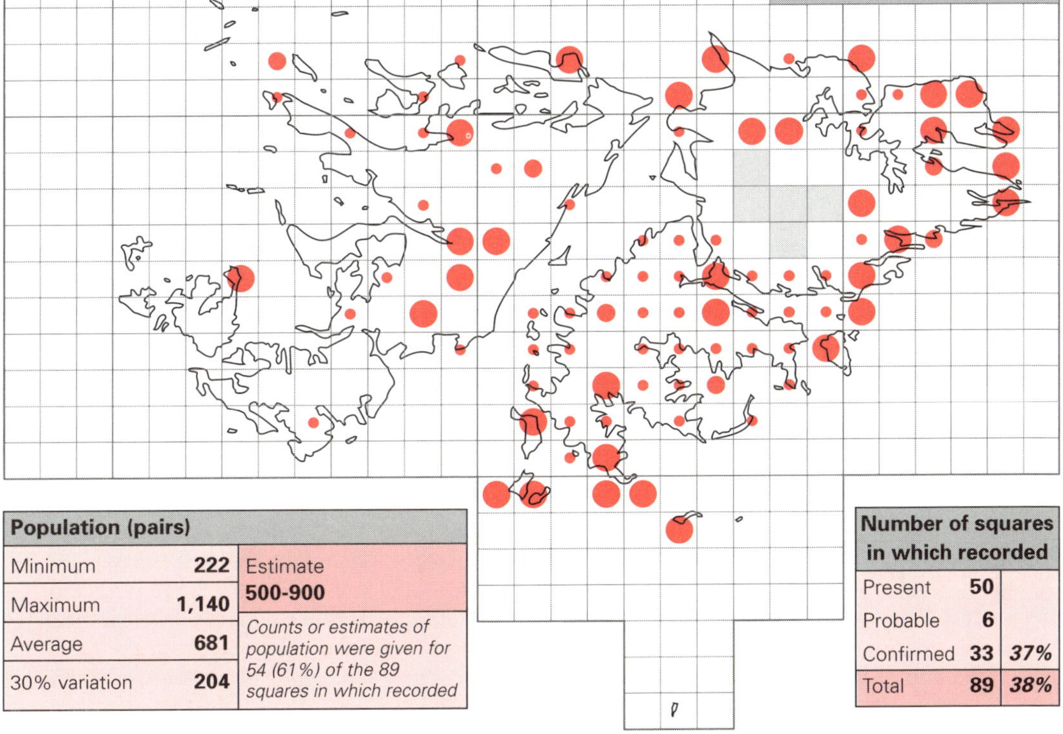

Population (pairs)		
Minimum	**222**	Estimate **500-900**
Maximum	**1,140**	
Average	**681**	*Counts or estimates of population were given for 54 (61%) of the 89 squares in which recorded*
30% variation	**204**	

Number of squares in which recorded		
Present	**50**	
Probable	**6**	
Confirmed	**33**	***37%***
Total	**89**	***38%***

Comparisons between species records and topography	Dry/Wet: Number of wet tetrads					Inland/Coastal: Amount of coast			Upland/Lowland: Proportion of square > 500′				
	Nil	1-5	6-10	11-15	16-22	None	<10%	≥10%	≥1/2	≥1/4	<1/4	None	Total
Surveyed squares	38	104	52	24	16	17	14	203	7	25	73	129	234
% of surveyed squares	16	45	22	10	7	7	6	87	3	11	31	55	100
Grid squares in which recorded	1	34	25	16	13	7	7	75	0	7	20	62	89
% records in each category	1	38	28	18	15	8	8	84	0	8	22	70	100
Significant variation			***	***	***							♦	

31 Speckled Teal

Anas flavirostris flavirostris

A. Pato barcino

C. Pato jergón chico

Local names: Teal, Duck

History
When Pernety (1771) remarked on 'numerous' teal at Port Louis in 1764 and Clayton (1774) at Saunders Island observed that the teal were very good [to eat] they were probably referring to this species. Garnot (1826) mentioned a very small species, 'le canard à bec jaune et noir', a translation of the name previously given by Azara to the Speckled Teal in Argentina. Ross's Antarctic Expedition collected some specimens in Salvador Bay, East Falkland and Captain Pack collected others before 1860. Abbott (1861) found this teal 'more plentiful in the interior than in the neighbourhood of civilization'. He noticed it in large flocks in some of the streams and thought it surprising, when it was so tame, that it often nested in out of the way places. In 1904 Vallentin stated that this teal was abundant in places throughout the Falklands and that it was 'highly esteemed for table purposes'. Brooks (1917) felt it was fairly common along small streams and creeks while Bennett (1926) merely stated that it was a common resident. Cawkell & Hamilton (1961) believed it to be widely spread and that 'it would be more common if not shot so much'. They noted that its favourite wintering lakes, especially on Pebble Island, could attract hundreds of birds.

Habitat and Breeding
It can be found on lakes, ponds and streams during the summer and sometimes on salt water in sheltered bays and creeks in the winter. It nests from mid-August onwards, often rearing two broods and ducklings have been seen as late as April. It lays 5-8 reddish-cream eggs in a well-hidden grass nest which may be near water or up to 1·6km (1 mile) from a pond.

Breeding Birds Survey: analysis of data
The Speckled Teal was found almost throughout the Falklands and breeding was proved on the two most remote sizeable islands, Steeple Jason and Beauchêne. There were no records from the Hornby Mountains and the Mount Alice area in southern West Falkland nor from the Wickham Heights. In comparison with the analysis of Dry/Wet squares, this species was much more likely ($p = 0{\cdot}01$) to occur in squares with 6-22 wet tetrads than in the drier squares. Perhaps it favours areas with many ponds more than was traditionally thought. Comparison of population estimates with other species indicates that it is second to the Logger, and about equal in numbers to the Crested Duck.

World Distribution
Four races of the Speckled Teal *A. flavirostris* have been described from South America. Fjeldså & Krabbe (1990) considered that it is generally common, but abundant only in a few areas; *A. f. altipetens* and *A. f. andium* (sometimes treated as a separate species, del Hoyo *et al.* 1992) occur in the Andes of Venezuela, Colombia and Ecuador. *A. f. oxyptera* is found in andean regions of southern Peru and northern Chile and Argentina to about 30°S. The race occurring in the Falklands, *A. f. flavirostris,* is also widespread in southern South America from 30°S to Tierra del Fuego. From April, most southern birds migrate northwards to Uruguay, Paraguay and southeastern Brazil and return during August; some remain in Tierra del Fuego through the winter. In 1971 a breeding population of 40 to 50 birds was discovered in Cumberland East Bay, South Georgia; they may have been introduced by early whalers (del Hoyo *et al.* 1992). Twelve birds recorded in six years between 1979 and 1989 at Bird Island, northwest South Georgia (Prince & Croxall 1996), could have dispersed from this population.

Implications for Conservation
The Falkland population forms only a small fraction of the world population. It seems to have remained common and widespread over the past two centuries and is still tame in places, despite being a game bird. It is not at present under any particular threat but numbers could be reduced if any large scale development of wetlands occurred.

References
TBFI (1975) p123-5, GBFI (1988) p147-8.

31 Speckled Teal *Anas flavirostris*

Detectability: **5 Very High**

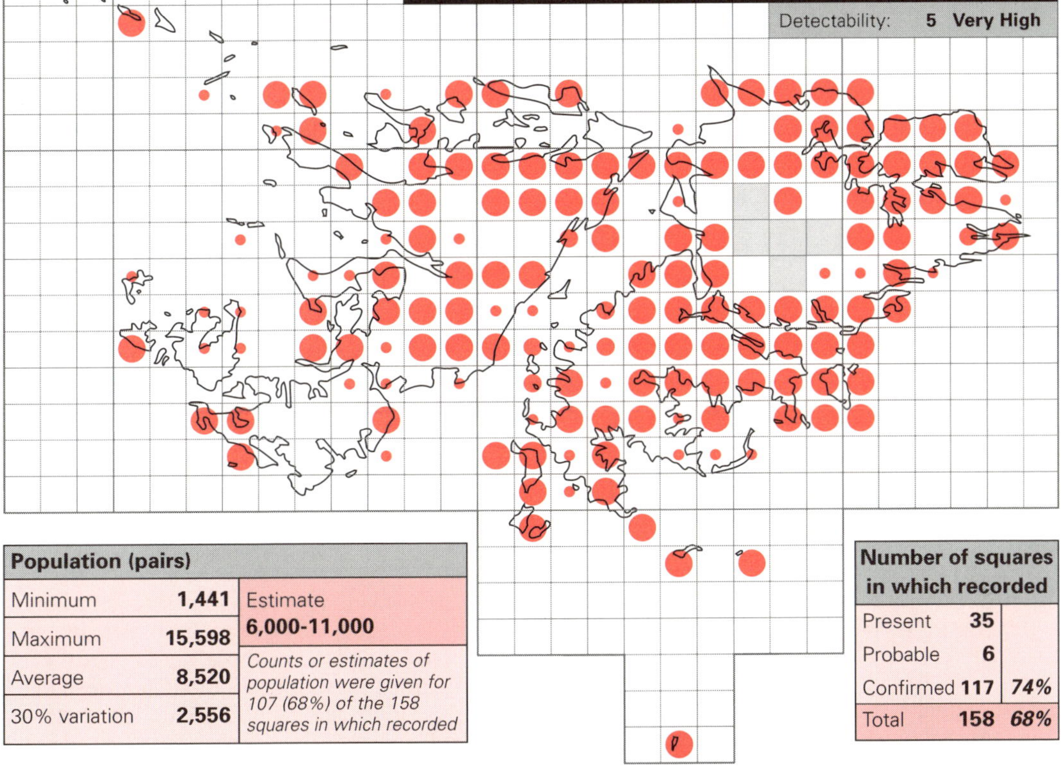

Population (pairs)		
Minimum	**1,441**	Estimate **6,000-11,000**
Maximum	**15,598**	
Average	**8,520**	*Counts or estimates of population were given for 107 (68%) of the 158 squares in which recorded*
30% variation	**2,556**	

Number of squares in which recorded		
Present	**35**	
Probable	**6**	
Confirmed	**117**	***74%***
Total	**158**	***68%***

Comparisons between species records and topography	**Dry/Wet** Number of wet tetrads					**Inland/Coastal** Amount of coast			**Upland/Lowland** Proportion of square > 500′				
	Nil	**1-5**	**6-10**	**11-15**	**16-22**	**None**	**<10%**	**≥10%**	**≥1/2**	**≥1/4**	**<1/4**	**None**	**Total**
Surveyed squares	38	104	52	24	16	17	14	203	7	25	73	129	234
% of surveyed squares	16	45	22	10	7	7	6	87	3	11	31	55	100
Grid squares in which recorded	8	69	43	22	16	12	11	135	3	16	55	84	158
% records in each category	5	44	27	14	10	8	7	85	2	10	35	53	100
Significant variation			**	**	**								

32 Crested Duck

Anas specularioides specularioides

A. Pato crestón

C. Pato juarjual

Local name: Grey Duck

History

The Antarctic Expedition of 1842 and Captain Pack in the 1850s (Sclater 1860) were the first to record this duck in the Falklands. Abbott (1861) felt that it was common everywhere, mainly occurring on salt water but sometimes on freshwater ponds, that pairs were territorial along the coast and that nests were made inland among the grass or on the edges of ponds. Brooks (1917) stated that it was common and that pairs could be found scattered along protected beaches almost anywhere. Vallentin (in Boyson 1924) said that it was one of the commonest and most prolific birds on West Falkland, but 'not nearly so abundant on the Eastern Island as in former times'. He believed that it had two or even three broods annually, laying 7-9 eggs per clutch. Vallentin also considered it made excellent shooting and good eating. Bennett (1926) stated that it was a common resident, nesting from July onwards. Cawkell & Hamilton (1961) felt it was common, sometimes occurring in small flocks in bays and creeks and also on freshwater near the sea, and that it had decreased 'in the past few years' because it was increasingly being shot.

Habitat and Breeding

The Crested Duck frequents sheltered coves and creeks and ponds near the sea. Birds often bathe and drink where streams flow onto beaches. Territories may be held throughout the year; Bob Reid found that 1300m (1400 yds) of coastline on George Island held eight pairs in August 1961 (Woods 1975). The breeding season is prolonged and eggs or very small ducklings have been seen in every month between July and May, but the main breeding season is September to November. It is often double-brooded and may even have three broods in a year. The nest, sometimes close to water but often well inland, is lined with down amongst grass, tussac, diddle-dee or fern and is made of these materials. It usually lays 5-7 cream-coloured eggs, though clutches of 9 and 11 are known.

Breeding Birds Survey: analysis of data

Vallentin (1904) and, nearly sixty years later, Cawkell & Hamilton (1961), felt that the numbers of this species had declined, but the evidence from the survey is that it is widespread, both in coastal waters and on inland freshwater ponds and lakes. There is no evidence that it is more common on either West or East Falkland. Records show no noticeable deviation from the distribution of 'wet' tetrads; it is equally likely to be found in wetter or drier grid squares. Considering the Inland/Coastal analysis, there is a non-significant tendency ($p = 0{\cdot}09$) for more records to come from coastal squares. Even so, Crested Ducks were recorded from five completely inland squares and from eight of the 14 surveyed squares with very little coastline. In the Upland/Lowland analysis, the distribution of records shows no significant difference from that of the four altitude categories though on West Falkland, the large areas where they were not recorded coincide mostly with the regions of high ground.

An unsubstantiated report of a hybrid between this species and a domestic duck at Horseshoe Bay, East Falkland was received in January 1988 from Mike Riddy. There have been no other reports of hybridisation, though the Mallard *A. platyrhynchos*, one of the ancestors of domestic breeds, is notorious for hybridising with many other species of duck in several genera.

World Distribution

This race of the Crested Duck, *A. s. specularioides*, is common in coastal and andean regions of southern South America from Tierra del Fuego to about 35°S in Chile and 38°S in Argentina. Some birds migrate to northeast Argentina (about 35°S) in winter. A somewhat larger race *A. s. alticola* with yellow rather than red eyes, occupies the Andes of Peru, Bolivia and Chile to about 35°S, mostly above 4000m. The two races intergrade around 35°S in the Chilean Andes (Fjeldså & Krabbe 1990).

32 Crested Duck *Anas specularioides*

Detectability: **5 Very High**

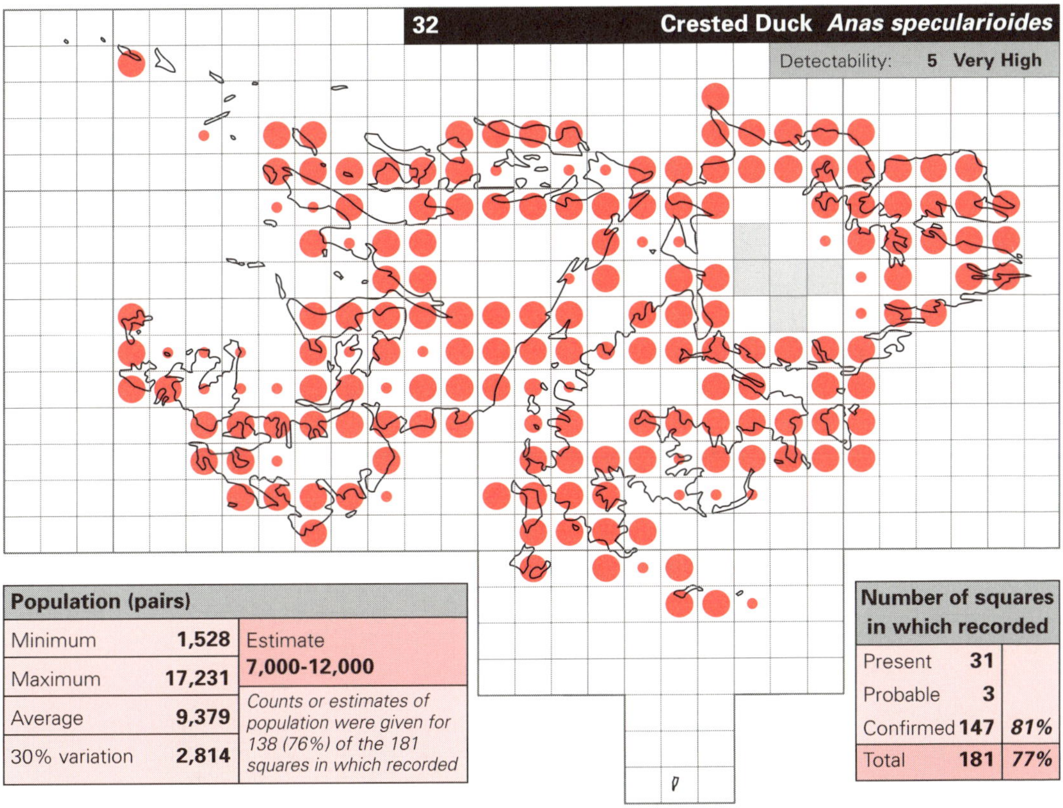

Population (pairs)		
Minimum	**1,528**	Estimate **7,000-12,000**
Maximum	**17,231**	
Average	**9,379**	*Counts or estimates of population were given for 138 (76%) of the 181 squares in which recorded*
30% variation	**2,814**	

Number of squares in which recorded		
Present	**31**	
Probable	**3**	
Confirmed	**147**	***81%***
Total	**181**	***77%***

Comparisons between species records and topography	Dry/Wet: Number of wet tetrads					Inland/Coastal: Amount of coast			Upland/Lowland: Proportion of square > 500′				
	Nil	**1-5**	**6-10**	**11-15**	**16-22**	**None**	**<10%**	**≥10%**	**≥1/2**	**≥1/4**	**<1/4**	**None**	**Total**
Surveyed squares	38	104	52	24	16	17	14	203	7	25	73	129	234
% of surveyed squares	16	45	22	10	7	7	6	87	3	11	31	55	100
Grid squares in which recorded	21	85	42	21	12	5	8	168	3	18	63	97	181
% records in each category	>11	47	>23	>11	>6	3	4	93	2	10	35	53	100
Significant variation								♦					

Implications for Conservation

One of the more numerous and widespread ducks of the Falklands, it does not appear to be under any threat. It is still considered to be a game bird and may be shot during the open season.

References

TBFI (1975) p117-8, GBFI (1988) p143-4.

33 Yellow-billed Pintail

Anas georgica spinicauda

A. Pato maicero

C. Pato jergón grande

Local names: Grey Teal, Coast Teal

History
Sclater (1860) was the first to suggest that the Yellow-billed Pintail occurred in the Falklands, when he reported on a pair in winter plumage that had been collected by Captain Pack. However, Sclater was not sure of the species because their plumage did not agree with that shown in Eyton's monograph on the Anatidae. Abbott (1861) had no doubts about its identity, stating that it occurred 'rather sparingly in the interior of the island on the freshwater ponds, where it is resident all the year round' but Wace (1921) only rated it as a visitor. Bennett (1926) described it as an occasional visitor that 'may sometimes remain to nest' in the Falklands but in a letter of September 1936, now in the Natural History Museum, Tring, B Elliott of North Arm wrote to Hamilton that it was 'not uncommon round here' and that nests had been found at Bull Point. Cawkell & Hamilton (1961) stated that the Yellow-billed (or Brown) Pintail was an uncommon resident, 'but as it has a local name it was presumably at one time less so'.

Habitat and Breeding
This rather dull plumaged duck is found in the breeding season and autumn on freshwater ponds with emergent vegetation. In winter it has also been recorded from kelp beds in sheltered coastal waters. The Yellow-billed Pintail has probably been overlooked frequently because it has a close resemblance to the more common and widespread, though smaller, Speckled Teal. It is possibly migratory as most recorded sightings are from September to April, with few seen in winter. It nests between September and December, laying 7-10 cream-coloured eggs in a grass nest hidden amongst coarse grass or rushes.

Breeding Birds Survey: analysis of data
There were twice as many records from East Falkland than from West and they were unevenly distributed around the archipelago. Analyses of topographical parameters show a very strong bias towards occurrence in squares with 11 or more 'wet' tetrads ($p = 0{\cdot}001$) and there was a non-significant tendency ($p = 0{\cdot}056$) towards more records from squares with no land above 500 feet. Although the population estimate may be low due to under-recording, calculations suggest that it is only present in small numbers and that it is one of the least common of the regularly breeding ducks. The largest counts were made by Mike Riddy when he visited the area between Lion Creek and Bull Roads (UC30) in Lafonia on 16 and 17 December 1987 and found a total of 54 birds on four of the six ponds examined. Between October and December 1989, David Osborn saw up to nine in East Cove. In 1990, Bill Bourne saw between 12 and 22 on the western part of Gull Island Pond between 10 and 15 February and on 12 March, Mike Winter found 23 on the same pond. Phil Gregory confirmed breeding at the Stanley airport pools in November 1990 when he saw distraction display from a female with a juvenile. Nicholas Beswick reported a few in the Mount Pleasant Airport to Bertha's Beach area from September to December 1992, including a pair with five small ducklings on Champina Pond on 13 November, but none was seen in the period from June to August.

World Distribution
The race *spinicauda* of the Yellow-billed Pintail, found in the Falklands, is one of the commonest ducks in Argentina and Chile. It occurs from Tierra del Fuego through Chile to southwestern Colombia and east of the Andes, to Bolivia, Paraguay and southeastern Brazil. It is migratory in the south, arriving in Tierra del Fuego in August/September and leaving in March/April. It has occurred twice as a vagrant in South Georgia (Prince & Croxall 1996). A smaller, darker race *A. g. georgica* is endemic in South Georgia, where the population in 1971 was estimated to be several thousands (Weller 1975). The race *A. g. niceforoi* was rare in the Colombian Andes and is probably now extinct (Fjeldså & Krabbe 1990).

33 Yellow-billed Pintail *Anas georgica*

Detectability: 3 Moderate

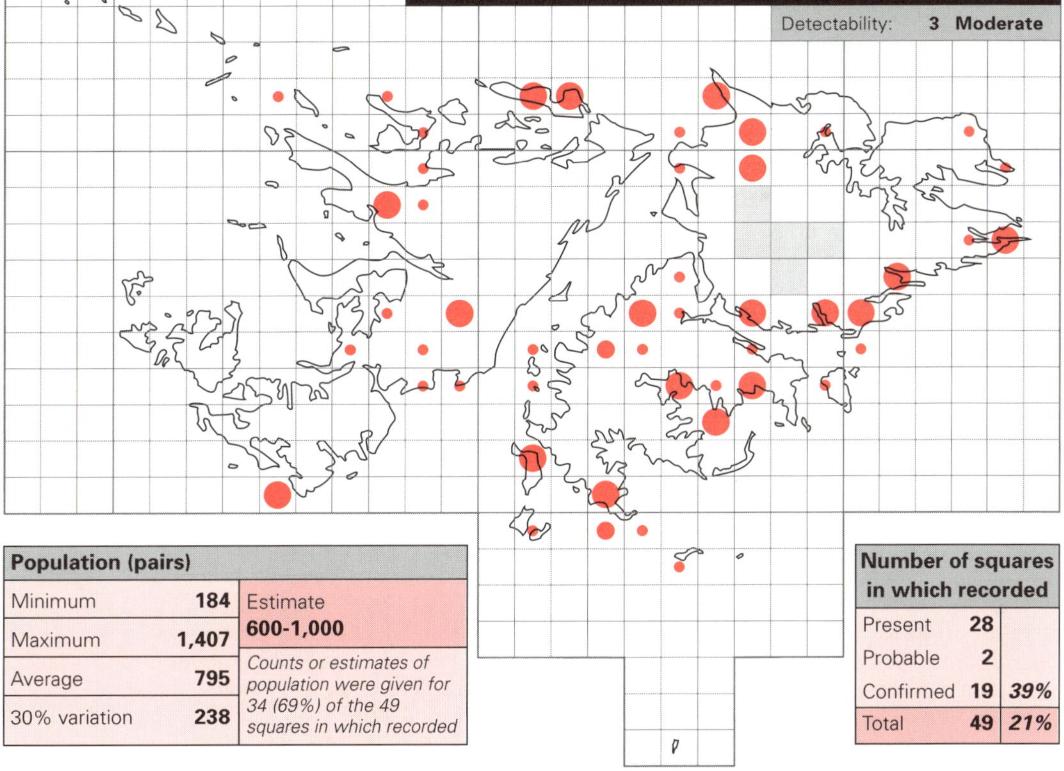

Population (pairs)		
Minimum	184	Estimate **600-1,000**
Maximum	1,407	
Average	795	*Counts or estimates of population were given for 34 (69%) of the 49 squares in which recorded*
30% variation	238	

Number of squares in which recorded		
Present	28	
Probable	2	
Confirmed	19	*39%*
Total	49	*21%*

Comparisons between species records and topography	Dry/Wet Number of wet tetrads					Inland/Coastal Amount of coast			Upland/Lowland Proportion of square > 500′				
	Nil	1-5	6-10	11-15	16-22	None	<10%	≥10%	≥1/2	≥1/4	<1/4	None	Total
Surveyed squares	38	104	52	24	16	17	14	203	7	25	73	129	234
% of surveyed squares	16	45	22	10	7	7	6	87	3	11	31	55	100
Grid squares in which recorded	2	15	11	12	9	2	2	45	1	0	13	35	49
% records in each category	4	31	22	25	18	4	4	92	2	0	27	71	100
Significant variation				**	**							♦	

Implications for Conservation

Although it is legally considered as a game bird, the survey suggests that the population is very small and it should therefore be removed from this category. The most extensive wetland areas are of considerable importance to this duck but a study of its annual cycle, including seasonal changes in habitat, would be necessary to provide reliable information for any conservation measures.

References

TBFI (1975) p126-7, GBFI (1988) p149-50.

34 Silver Teal

Anas versicolor fretensis

A. Pato capuchino

C. Pato capuchino austral

Local name: Pampa Teal

History

The first published record of this species in the Falklands is that it was collected by Captain Pack, probably in the 1850s (Sclater 1860). However, according to Dabbene's version of Wace (1921), the English Antarctic Expedition had collected specimens in 1842. Abbott (1861) felt that the Silver Teal was 'not common in East Falkland, occurring in but few places, but where found is generally seen in numbers'. Vallentin (in Boyson 1924) agreed with Abbott that it was seen 'in numbers' and recorded that he had shot a flock of nine at Port North Lake within ten minutes. He misquoted in stating that Abbott thought it 'not uncommon'. Bennett (1926) remarked that it was a fairly common breeding resident. Cawkell & Hamilton (1961) felt that it was local, but met with in fair numbers in winter in certain places, such as Cow Pond in Fitzroy camp, Swan Pond at Cape Dolphin and on the Pebble Island ponds.

Habitat and Breeding

The Silver Teal occurs on larger freshwater ponds with emergent vegetation. There is little information on breeding in the Falklands, though laying seems to start in October and can occur to at least the end of December. At least 5 eggs, maybe as many as 10, are laid in a grass nest, well-hidden in thick vegetation, and ducklings have been seen between mid-November and late January.

Breeding Birds Survey: analysis of data

The map shows a concentration of records in the lowlying Lafonia region of East Falkland with other scattered records, mainly in the east of West and East Falkland. Comparison with the analysis of standing freshwater shows a very significant bias ($p = 0{\cdot}0001$); 70% of the squares in which it was recorded were in the three wettest categories containing 6-22 tetrads with ponds or lakes. The bias towards Lafonia is related to the significantly high occurrence of freshwater ponds and lakes in this area. It was also recorded in the other noted wetland areas north of Fox Bay on West Falkland and on Pebble Island. All confirmed breeding records were of adults with ducklings or larger fledged juveniles and no nests with eggs were reported. Surprisingly close to Stanley, Tim Cowley saw adults with juveniles on the pool behind Surf Bay some time between July and December 1985. The largest number counted in any square was 26 pairs on five ponds between Lion Creek and Bull Roads, south of North Arm settlement recorded by Mike Riddy in December 1987. Several records from the MPA/Bertha's Beach area included seven on small ponds in the dunes in early summer 1989 noted by David Osborn and a maximum count by Mike Winter of 28 on East Cove Pond South on 5 May 1990. In the spring of 1992, Nicholas Beswick saw two pairs on Champina Pond with four small ducklings in mid-November.

World Distribution

Two races of the Silver Teal are recognised. *A. v. fretensis* occurs in the Falklands and from central Chile and central Argentina (about 40°S) to Tierra del Fuego. A smaller race *A. v. versicolor*, occurs from southern Bolivia, central Chile and southeastern Brazil to northern Argentina. The most southerly populations of both races are migratory in South America, occurring north to southern Brazil in winter. Fjeldså & Krabbe (1990) describe it as very numerous in the pampas marshes from eastern Argentina to southeastern Brazil.

Implications for Conservation

Only a very small part of the world population is found in the Falklands but as it appears to be sedentary here and therefore isolated from those on the continent, it is possible that it could be a distinct island race. Investigation of its behaviour, breeding habits and taxonomy would be valuable. The population estimate is low and indicates that it should not be classed as a game bird. Those regions of the Falklands with the most standing freshwater are particularly important to this species.

References

TBFI (1975) p127-8, GBFI (1988) p150.

34 Silver Teal *Anas versicolor*

Detectability: **4 High**

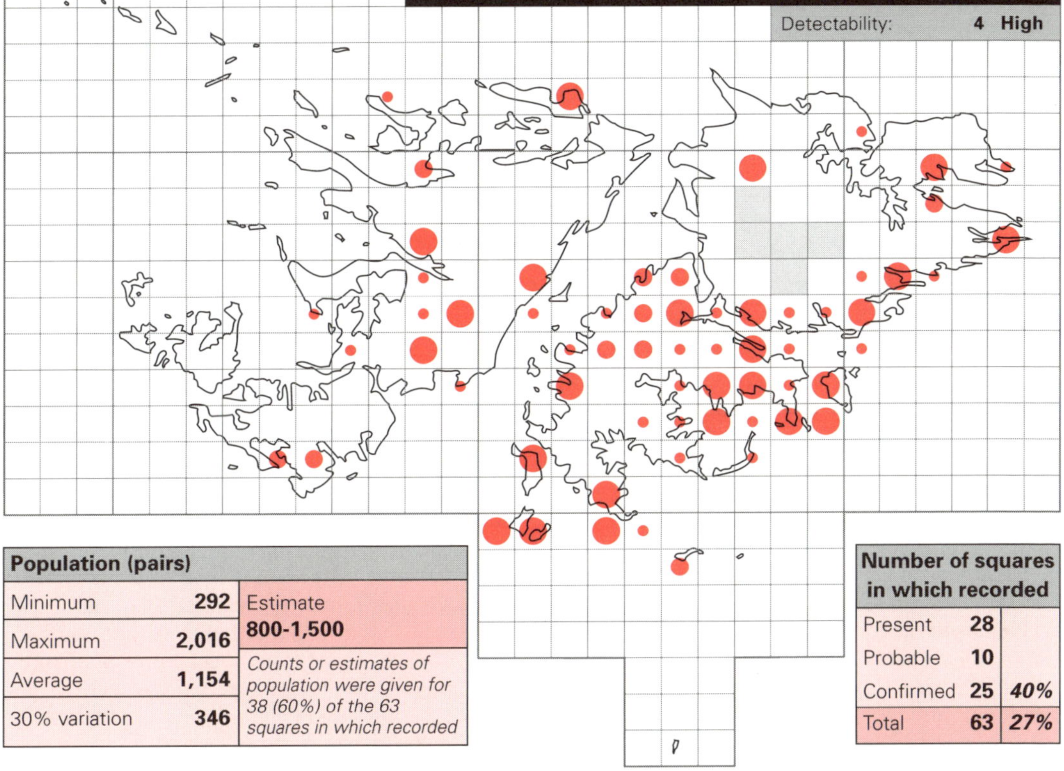

Population (pairs)		
Minimum	**292**	Estimate **800-1,500**
Maximum	**2,016**	
Average	**1,154**	*Counts or estimates of population were given for 38 (60%) of the 63 squares in which recorded*
30% variation	**346**	

Number of squares in which recorded		
Present	**28**	
Probable	**10**	
Confirmed	**25**	***40%***
Total	**63**	***27%***

Comparisons between species records and topography	**Dry/Wet** Number of wet tetrads					**Inland/Coastal** Amount of coast			**Upland/Lowland** Proportion of square > 500′				
	Nil	**1-5**	**6-10**	**11-15**	**16-22**	**None**	**<10%**	**≥10%**	**≥1/2**	**≥1/4**	**<1/4**	**None**	**Total**
Surveyed squares	38	104	52	24	16	17	14	203	7	25	73	129	234
% of surveyed squares	16	45	22	10	7	7	6	87	3	11	31	55	100
Grid squares in which recorded	1	18	20	14	10	4	5	54	1	5	12	45	63
% records in each category	2	28	32	22	16	6	8	86	2	8	19	71	100
Significant variation			***	***	***								

35 Cinnamon Teal

Anas cyanoptera cyanoptera

A. & C. Pato colorado

Local name: Red Teal

History

The first published record is a note that it had been reported in the Falklands by Captain Pack, a resident of several years (Sclater 1860). Abbott (1861) stated that he once killed seven in a day at Mare Harbour, that it was generally very wild and 'far from common'. He had no doubt that it bred, having seen pairs in the summer months, but had not found a nest. He expressed the opinion that a person might go out in East Falkland for a month and not shoot and even perhaps not see a Red Teal. Wace (1921) only said that it bred in the islands and there were specimens in the local collection while Bennett (1926) stated that it bred in the Falklands, but was very scarce. Cawkell & Hamilton (1961) felt that it was 'extremely rare' and that a very few pairs bred in Lafonia and in one place on West Falkland. They also said, 'Now that the shepherd at Bull Point (Lafonia) who took their eggs year after year is dead it may increase a little'. There were no further records until November 1979, when a pair was seen on Stanley Common and in 1981 two males wintered near Surf Bay.

Habitat and Breeding

There is little information on preferred habitats in the Falklands and there are no records of nests and eggs. In South America it is found on shallow lakes with emergent, marginal and abundant floating vegetation in open country, between sea level and about 5000m (Fjeldså & Krabbe 1990).

Breeding Birds Survey: analysis of data

There were at least 11 sightings from eight localities during the survey, mostly in the years 1988-1991. However, the two confirmed breeding records were both in the 1985-86 season. Clara and Bill MacKay reported young birds seen in the Ponds Flat camp (UC79), east of Port San Carlos and Josie Larsen reported adults showing distraction display on Speedwell Island. John Cliffe recorded the largest group of about 20 birds on Big Pond, Saunders Island in October 1984, where local people said that it bred. Shane Wolsey saw ten birds on Saunders Island in April 1987, suggesting that this is a favoured locality. Single birds and up to three pairs were reported from Betts Pond on Pebble Island in December 1988 and November 1990 by Andrew Clarke, Phil Gregory and John Reid. Phil also had two sightings on 9 September 1989 and 9 September 1990 at the northwestern end of Carcass Island (Gregory 1994). He commented that two females were 'very wild, almost certainly newly arrived'. A single male and a single female were seen on at least three dates between mid-October and late December 1989 in the Surf Bay and Cape Pembroke area near Stanley by Phil Gregory and David Osborn. One male was seen on Champina pond in the Mount Pleasant area in August 1991 by David Lee. Apart from Speedwell Island and the Ponds Flat area near Port San Carlos, Saunders Island and Pebble Island appear to be probable breeding localities.

World Distribution

Five races of the Cinnamon Teal occur from western North America to southern South America. The North American breeding race migrates south to Central America and as far as Ecuador. The nominate race *A. c. cyanoptera* which occurs in the Falklands, is widespread in southern South America from southern Peru through central Chile, Bolivia, Paraguay and southeastern Brazil to Argentina (Fjeldså & Krabbe 1990). It is most abundant in the lowlands of eastern Argentina and is an uncommon visitor to Tierra del Fuego. The southern continental breeding population disperses northwards in winter, reaching coastal Peru and central Brazil (Madge & Burn 1988).

Implications for Conservation

The Falkland population of Cinnamon Teal appears to be a very small group, probably originating from continental South America and it may be migratory. In the 1953 Bird Protection Ordinance and the 1964 Wild Animals and Birds Protection Ordinance it was listed as a protected species and it seems likely to retain its status as a very uncommon duck.

References

TBFI (1975) p128-9, GBFI (1988) p150-1

35 Cinnamon Teal *Anas cyanoptera*

Detectability: **3 Moderate**

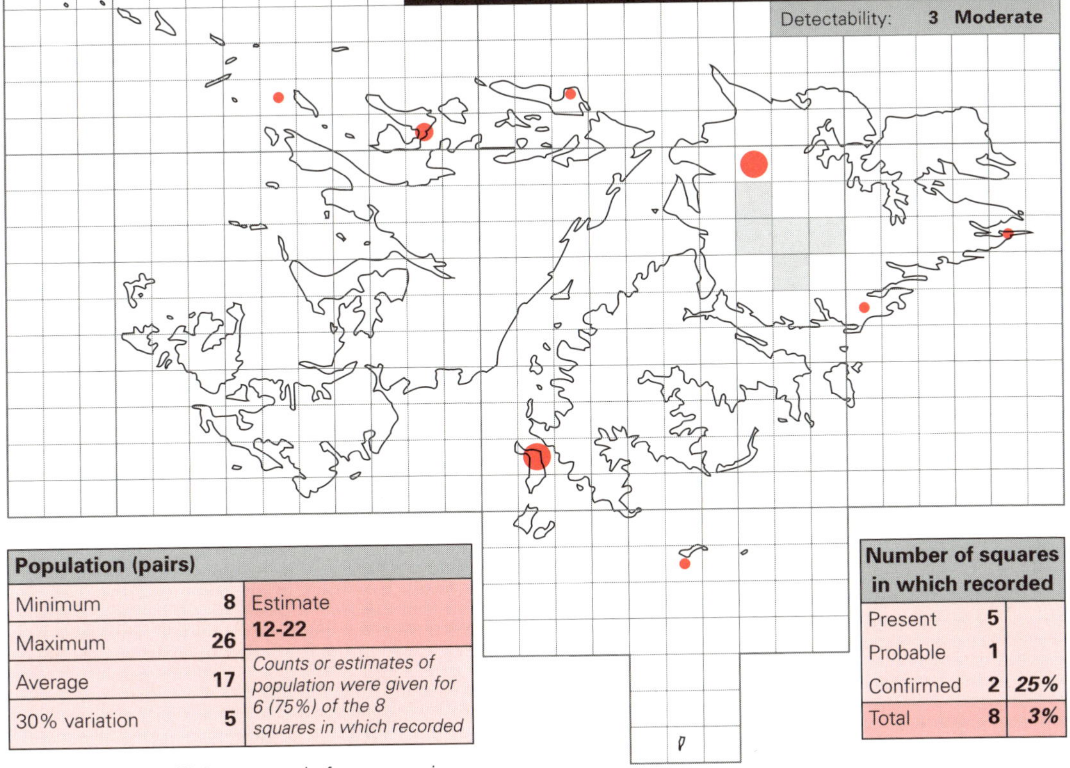

Population (pairs)		
Minimum	**8**	Estimate **12-22**
Maximum	**26**	
Average	**17**	*Counts or estimates of population were given for 6 (75%) of the 8 squares in which recorded*
30% variation	**5**	

Number of squares in which recorded		
Present	**5**	
Probable	**1**	
Confirmed	**2**	***25%***
Total	**8**	***3%***

There were insufficient records for comparison with the analyses of topographical parameters.

36 Red Shoveler

Anas platalea

A. & C. Pato cuchara

History
According to Dabbene in his spanish version of Wace's list (Wace 1921), Adolphe Lecomte indicated that he had seen the Red Shoveler when he was collecting birds and seals in the Falklands for the Zoological Society of London between August 1867 and June 1868. Both Oustalet (1891) and Salvadori (1895) noted that it occurred in the Falklands and Wace (1917 and 1921) described it as an occasional visitor. Bennett (1926) felt that it was only an accidental visitor, as did Cawkell & Hamilton (1961), though Hamilton had an undated record (probably in the 1930s) of females or immatures on Swan Pond, Cape Dolphin. There were no further records of this species before 1983.

Breeding Birds Survey: analysis of data
Between 1985 and 1992, there were at least ten records of Red Shoveler, though none has included breeding behaviour and no ducklings were reported. Seven of the eight grid squares in which it was recorded are on or around the eastern coasts of East Falkland. One was seen on Swan Pond, Seal Bay, East Falkland on 20 and 24 October 1985 by Michael Morrison. A group of seven birds including two males was seen by Mike Riddy on a pond by Low Bay House in Lafonia on 9 December 1987. On Pebble Island, at least two including one female were seen between mid-November and December 1988 by John Reid and Phil Gregory. David Osborn saw a male on 14 November 1989 on Sea Lion Island. During the same period, Graham Bound and Phil Gregory saw a male and a female on the Stanley airfield pool between 15 and 19 December 1989. One female was seen near Volunteer Point by Kate Thompson between December 1989 and January 1990 and up to 12 were seen on five dates between 18 March and 5 May 1990 on East Cove Pond South, near Bertha's Beach by Mike Winter and CPO Calderwood. Single females were seen on Long Pond, Pebble Island by Dann Hale and Phil Gregory on 25 and 26 December 1990 and on Champina pond near Mount Pleasant Airport between 9 and 11 November 1992 by Nicholas Beswick. Except for those near East Cove, which could have been overwintering, all birds were seen in the breeding season. There were insufficient records to compare distribution with the analyses of topographical parameters. Given the apparent tendency for some of these records to group at Pebble Island, on ponds around the eastern end of Choiseul Sound and either side of Berkeley Sound, there is a chance that this species is breeding in small numbers.

World Distribution
The Red Shoveler is common in southern South America, particularly in lowland and coastal Argentina, where it is resident on shallow, often brackish lagoons, but it is much less common in the Magellanic region. The southern continental population migrates northward in winter as far as southern Peru and southeastern Brazil; large numbers use upland plateau lakes of inland Santa Cruz province of Argentina as moulting and staging points (Fjeldså & Krabbe 1990).

Implications for Conservation
In South America it is not apparently under threat from hunting or loss of habitat (Madge & Burn 1988). In the Falklands, its status is still uncertain and needs investigation. It is only likely to occur on ponds or lakes that are rich in vegetation and invertebrates.

References
TBFI (1975) p217-8, GBFI (1988) p151.

36 Red Shoveler *Anas platalea*

Detectability: **2 Low**

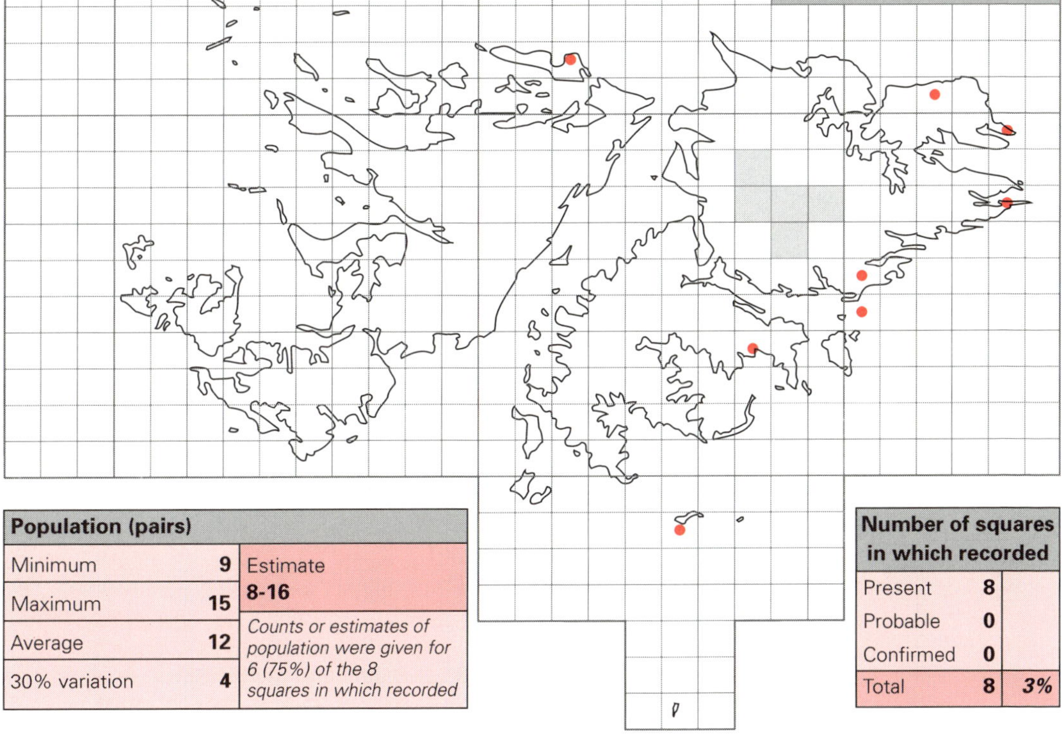

Population (pairs)		
Minimum	9	Estimate **8-16**
Maximum	15	
Average	12	*Counts or estimates of population were given for 6 (75%) of the 8 squares in which recorded*
30% variation	4	

Number of squares in which recorded		
Present	8	
Probable	0	
Confirmed	0	
Total	8	*3%*

37 Turkey Vulture

Cathartes aura falklandicus

A. Jote cabecirrojo

C. Jote de cabeza colorada

Local names: Turkey Buzzard, Turkey

History

Pernety (1771) described a 'kind of eagle' which sounds much like a Turkey Vulture. In 1833 and 1834, Darwin observed that it was tolerably common and usually shy and wondered whether it was migratory. Abbott (1861) stated that it was very common in East Falkland and resident throughout the year. Vallentin's observations (in Boyson 1924) were intermittent between 1897 and 1911. He felt that the Turkey Vulture was rapidly becoming extinct, 'owing to the constant warfare waged on it by farmers and others'. By 1911, however, he described it as 'still fairly common on all the West Falklands', but greatly decreased at Stanley since his previous visit in 1902. It was most easily seen, 'during the killing-season, just before the beginning of winter, when the useless sheep are destroyed for their hides and tallow'. A reward was being offered for beaks, 'owing to their attacking and destroying lambs, and sheep when in difficulties'. In 1915-16, Brooks stayed at Stanley, San Carlos and Port Stephens. He remarked (1917) that the Turkey Vulture was common in many places though, owing to a recently formed habit of eating the eyes and tongue of fallen sheep, it was killed whenever possible and four pence paid per bill. Bennett (1926), after he had been in the Falklands for over ten years, described it as a very common bird, blamed for killing sheep, especially at lambing time. He felt this was hardly fair, because each shepherd could have 5,000 or more ewes under his charge, scattered over a large area, and if sheep were found dead, something had to be blamed. He believed that the benefit of cleaning up carcasses which would otherwise 'pollute the air', outweighed the damage. Cawkell & Hamilton (1961) stated that it was 'common in spite of continuous persecution which makes the bird more wary'. They felt that it was primarily a carrion eater and that seal colonies had probably always been its main source of food.

Habitat and Breeding

The Turkey Vulture is commonly seen in the air in most parts of the Falklands and occurs wherever carrion is available. On Carcass Island, West Point Island and probably elsewhere, it uses the small plantations of introduced trees for roosting. It usually nests in concealed sites on the ground beneath dense tussac grass, in small caves or deserted buildings. Two, occasionally 3, white eggs with variable red-brown spots and lines are laid in a scrape nest between early September and late November, with the main season between mid-September and late October. Flying juveniles are seen from mid-January.

Breeding Birds Survey: analysis of data

Records show that it is still widespread in the Falklands, despite being persecuted for over a century. In comparisons with the analyses of topographical parameters, this species was found in squares with many ponds or none, in coastal squares or inland and in upland and lowland squares. However, most of the confirmed breeding records came from squares with much coastline, perhaps indicating a continuing preference for nesting in coastal regions where tussac grass and breeding colonies of seals are available. The population of this large scavenging bird appears to be maintaining its numbers, against the predictions of naturalists in the early years of the 20th century. Peter Prince and John Croxall suggest that its numbers have increased since the 1970s in the Stanley to Mount Pleasant area, perhaps due to reduced persecution.

World Distribution

The Turkey Vulture is widespread between southern Canada and Tierra del Fuego. Four races have been described though there is still uncertainty about their relationships and some may deserve specific status (del Hoyo *et al.* 1994). It occurs in almost all habitats except dense forest and heavily urbanised areas. The North American race *C. a. meridionalis* is migratory, moving south in large numbers as far as Paraguay. Other races appear to be sedentary, though the southern *C. a. falklandicus* wanders to at least 3500m in southern Peru (Fjeldså & Krabbe 1990). Although there is no evidence that it is migratory in the Falklands, Prince & Croxall (1996) note that individuals have been seen in three of four years since 1991 at South Georgia.

37 Turkey Vulture *Cathartes aura*

Detectability: **5 Very High**

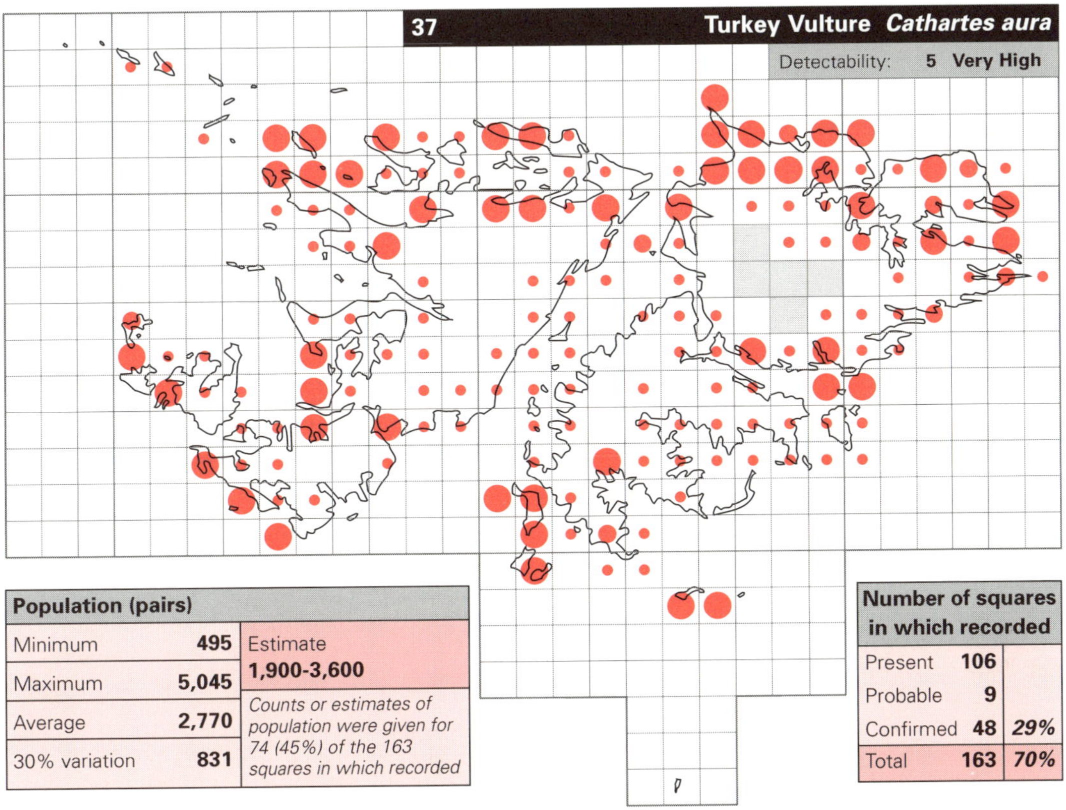

Population (pairs)		
Minimum	**495**	Estimate **1,900-3,600**
Maximum	**5,045**	
Average	**2,770**	*Counts or estimates of population were given for 74 (45%) of the 163 squares in which recorded*
30% variation	**831**	

Number of squares in which recorded		
Present	**106**	
Probable	**9**	
Confirmed	**48**	***29%***
Total	**163**	***70%***

Comparisons between species records and topography	Dry/Wet Number of wet tetrads					Inland/Coastal Amount of coast			Upland/Lowland Proportion of square > 500'				
	Nil	**1-5**	**6-10**	**11-15**	**16-22**	**None**	**<10%**	**≥10%**	**≥1/2**	**≥1/4**	**<1/4**	**None**	**Total**
Surveyed squares	38	104	52	24	16	17	14	203	7	25	73	129	234
% of surveyed squares	16	45	22	10	7	7	6	87	3	11	31	55	100
Grid squares in which recorded	17	79	36	19	12	6	8	149	4	14	57	88	163
% records in each category	10	49	22	12	7	4	5	91	>2	<9	35	54	100
Significant variation													

Implications for Conservation

If sheep farming for wool continues to play a large part in the Falklands economy, the annual supply of culled sheep will support scavengers like the Turkey Vulture. Attitudes vary, and with environmental education and over 80 individual holdings since subdivision of the large farms, it is likely that there is less persecution by farmers than there was even twenty years ago. Consideration should be given to revising its classification as a pest species 'that can be killed by any means' (Wild Animals & Birds Protection Ordinance 1964) in any future wildlife legislation.

References

TBFI (1975) p129-31, GBFI (1988) p154.

38 Red-backed Hawk

Buteo polyosoma polyosoma

A. & C. Aguilucho común
Local names: Hawk, Blue Hawk

History
MM. Quoy and Gaimard in 1820 were the first to record this hawk in the Falklands when they described a melanistic variety as *Falco polyosoma* (in Freycinet 1824). Since then, several ornithologists have named different colour phases as different species, both in the Falklands and South America. A specimen apparently from the Straits of Magellan, later discovered to be a female, was named as *B. erythronotus* by King in 1827. Abbott (1861) considered the red-backed form more common and that birds with dark underparts ('*polyosoma*') always nested near the sea. Wace (1921) continued to separate the two main forms as two species though in his first draft (1917), when referring to the 'Falkland Island Buzzard' *B. poliosomus* and the 'Red Backed Buzzard' *B. erythronotus*, commented, 'a series of skins and an expert opinion required for these'. Bennett (1926) also believed there were two resident species and described the Red-backed as 'common' and the 'Blue Hawk' simply as a resident. Swann (1924-30) described several colour phases of one species and Hellmayr & Conover (1949) stated that it 'occurs in various mutational, strictly alternative plumages'. Cawkell & Hamilton (1961) noted that Hamilton had seen birds of each of the main colour phases paired at a nest, confirming the existence of only one species. They reported a partial census in 1951/52, which suggested that there were only 30 to 40 pairs in the Falklands, though it should be noted that, 'no reports were received from Lafonia and other large areas on the East'. They commented that it had been put on the list of protected species in 1951 and that one nest site was known to have been used annually from at least 1919 to 1951.

Habitat and Breeding
The Red-backed Hawk may be seen in most parts of the Falklands. Nests are usually built on ledges of high crags and are renovated annually with diddle-dee or other twigs, which leads to large accumulations of material. It lays 2-3 almost elliptical whitish eggs, variably marked with red-brown, usually during October though sometimes in late September.

Breeding Birds Survey: analysis of data
The survey and the comparison with the distribution of topographical parameters indicate that this widespread species is found in coastal or inland regions and those with or without standing freshwater. Only the comparison with altitude shows a non-significant bias ($p = 0{\cdot}09$) towards more records from squares with some land over 500 feet. The records were almost equally distributed between West and East Falkland, with a tendency to be absent from smaller outlying islands. There was a slight tendency for more records to come from the northern, higher part of East Falkland, though it was also well-spread over Lafonia. In the 1985/86 season, Clara and Bill MacKay reported that they had seen a pair nesting in old discarded rolls of fencing wire on the ground in the New House, Douglas Station area (UC79) and Pat Short at San Carlos (UC68) had also noticed this habit. In December 1987, Mike Riddy reported a nest 'on fencing near the house' in Ruggles camp near Wreck House (UC23) and in January 1988 he found a nest on the ground in Outside Chata camp (UC89). Nicholas Beswick reported that a pair nested on a mast with a radar array at the east end of the Mount Pleasant Airport runway for more than one year up to the 1991/92 season. Because the adults became notoriously aggressive to intruders and it was very difficult to service the radar, a tower was built specially for the hawks. The nest was physically moved to this new tower in August 1992 and the pair bred there successfully in the 1992/93 season. Since the survey, a study of several pairs nesting in the Mount Pleasant area has been started by Fred Clark which may yield much useful information. The population estimate is about twenty times higher than that obtained in 1951/52 by Cawkell & Hamilton (1961).

World Distribution
The Red-backed Hawk is widely distributed in southern South America, particularly in andean regions, from Colombia to Tierra del Fuego. It is fairly common over open country and partly cultivated areas through much of Argentina from about 32°S and, in winter, occurs north into

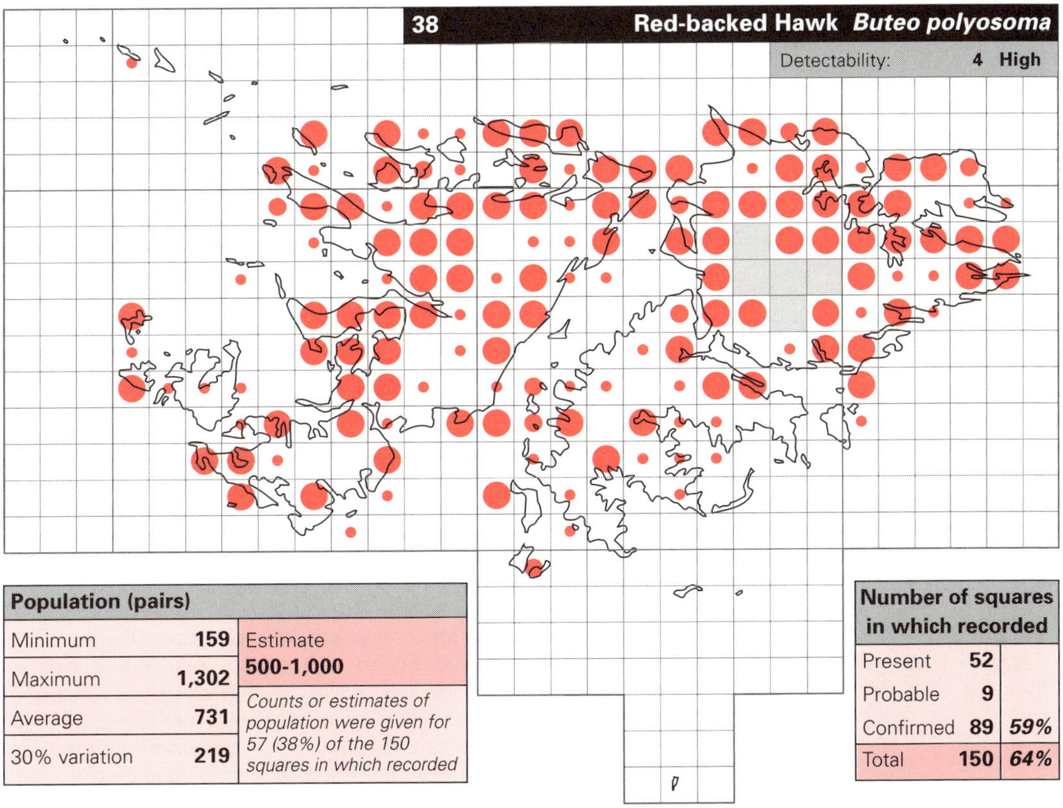

Population (pairs)		
Minimum	159	Estimate **500-1,000**
Maximum	1,302	
Average	731	*Counts or estimates of population were given for 57 (38%) of the 150 squares in which recorded*
30% variation	219	

Number of squares in which recorded		
Present	52	
Probable	9	
Confirmed	89	*59%*
Total	150	*64%*

Comparisons between species records and topography	Dry/Wet Number of wet tetrads					Inland/Coastal Amount of coast			Upland/Lowland Proportion of square > 500′				
	Nil	1-5	6-10	11-15	16-22	None	<10%	≥10%	≥1/2	≥1/4	<1/4	None	Total
Surveyed squares	38	104	52	24	16	17	14	203	7	25	73	129	234
% of surveyed squares	16	45	22	10	7	7	6	87	3	11	31	55	100
Grid squares in which recorded	12	69	39	16	14	13	12	125	6	22	59	63	150
% records in each category	8	46	26	11	9	9	8	83	4	15	39	42	100
Significant variation													

Uruguay and southeastern Brazil. A larger and darker race *B. p. exsul* is endemic on Mas Afuera in the Juan Fernandez Islands to the west of Chile (Goodall *et al.* 1951).

Implications for Conservation

The Falklands hold a very small and apparently sedentary proportion of the world population. The widespread occurrence of introduced mice and rats, with rabbits and hares in some places, and the general abundance of geese will probably ensure the survival of this protected species, provided that it is not seen as a threat to domestic fowls and unlawfully shot. Studies of general breeding biology and diet would be valuable and it would also be interesting to know the proportions of each colour phase present in the Falklands.

References

TBFI (1975) p132-4, GBFI (1988) p155-6.

39 Striated Caracara

Phalcoboenus australis

A. Matamico estriado or Matamico grande

C. Carancho negro

Local names: Johnny Rook or Jack Rook

History

Pernety (1771) noticed what he called 'small eagles' or 'brown hawks' at the first settlement of Port Louis in 1764 and his description, though not detailed, could be of this species. From Saunders Island, Clayton's two accounts are somewhat contradictory because he names three kinds of 'hawks' (1774) and later gives four names, two of them different (1776). In all, he used five names saying that 'hawks are numerous' (1776), but although it was probably present, it is not possible to recognise the Striated Caracara from his writings. Garnot at Port Louis in 1822 was impressed by the daring of a 'variety of caracara' though he may have been describing the behaviour of a Cinereous Harrier rather than a Striated Caracara. Darwin in 1833 and 1834 reported that this species was 'exceedingly numerous', 'extraordinarily tame and fearless' 'very mischievous and inquisitive', 'quarrelsome and passionate', and that it 'constantly haunted the neighbourhood of houses to pick up all kinds of offal' (Gould & Darwin 1838-41). In 1861, Abbott said it was 'one of the commonest birds in East Falkland'. He described finding at least 15 nests along the cliffs of the north shore in the North Camp of East Falkland in December 1860 and thought that Darwin was wrong in saying that they only bred on the adjoining islets.

Vallentin (1904, and in Boyson 1924) appears to confuse this species with *Caracara plancus*, the Crested Caracara and in a 1901 paper, he perhaps refers to the Turkey Vulture, when describing 'carrion hawks' as *Milvago australis.* Ramsay (1913) of the Scottish Antarctic Expedition, reported that 'the hawk known as the Johnny Rook' was logged as seen during the expedition's first visit in January 1903, and described it as 'quite a familiar bird in the vicinity of Port Stanley'. Cobb (1910) remarked that it had 'decreased in numbers in the Falkland Islands of late years', and that it was 'on the list of birds for whose destruction a reward is offered, on account of damage done to sheep and lambs' [Ordinance for the Destruction of Birds of Prey, 1908]. Wace (1917) stated that it was 'being rapidly exterminated, though harmless'.

In April 1922, 14 years after the Ordinance had been enacted, JE Hamilton wrote a memorandum to the FI Government on this species, pointing out that because the Johnny Rook had been subjected to 'a remorseless process of extermination' it occurred in very small numbers on East and West Falkland where he had seen just two adults in two and a half years. It was still numerous only on outlying islands. On those where sheep were kept, he felt that efforts to exterminate it would almost certainly be successful because it had not learned that man is dangerous. Some islands without stock formed refuges where it acted as a scavenger, feeding on eggs and young birds. He remarked that 'the persistent plentifulness of the Upland Goose' could be attributed at least in part to the diminution of the Johnny Rook and the senseless killing of hawks. Hamilton suggested, as Government Naturalist since 1919, that payments for Johnny Rook beaks should cease. He knew that this would not stop people killing them where they felt it necessary, but casual visitors would not be encouraged to kill them. Finally, Hamilton suggested that stopping payments 'would remove from the Government the liability to reproach from scientists that encouragement was given to the extermination of one of the ornaments of the local avifauna'. Hamilton's arguments were accepted by Government and the Striated Caracara was removed from the list in the 1920s.

Bennett (1926) remarked that it was very common on the outer islands, where it bred and he was assured that it and the Crested Caracara did kill sheep, 'displaying considerable cunning in the application of their nefarious deeds'. Attitudes have changed very slowly and by the early 1960s, it was extinct on East Falkland and nearly extinct on West Falkland, apparently only breeding on a few very isolated islands without permanent settlements. Cawkell & Hamilton (1961) remarked that it was still moderately common on the Jason Islands and that Sedge Island had lost its Johnny Rooks through the ill-

39 #Striated Caracara *Phalcoboenus australis*

Detectability: 5 Very High

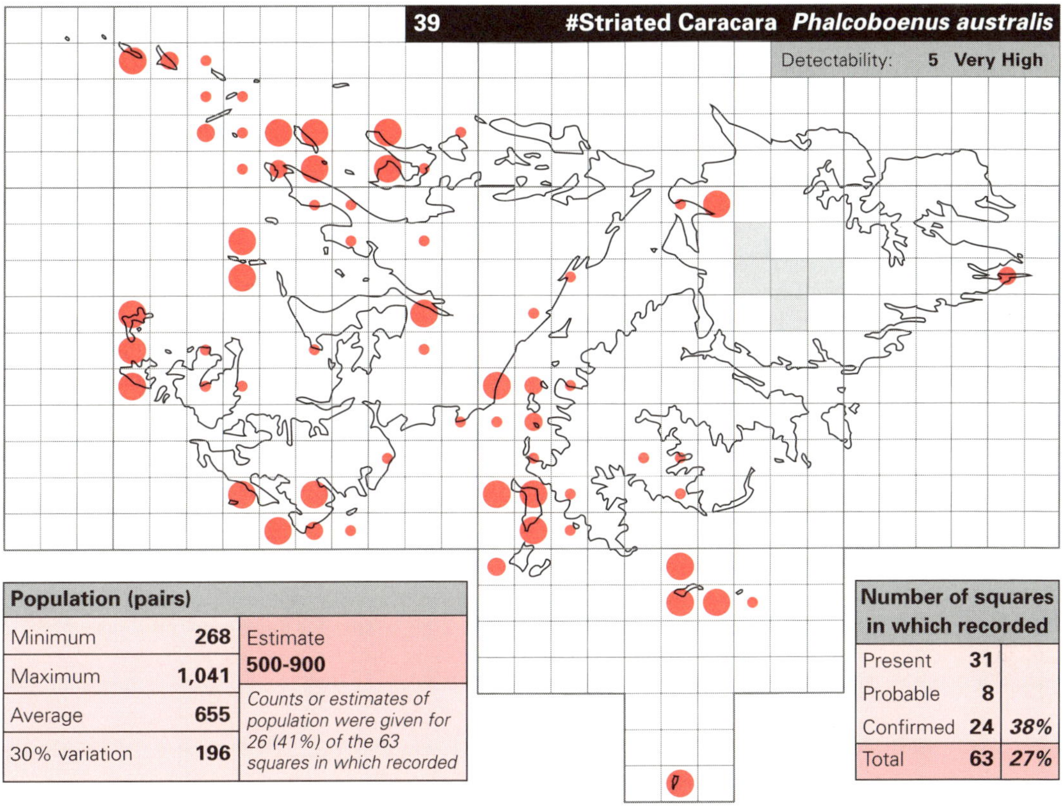

Population (pairs)		
Minimum	268	Estimate **500-900**
Maximum	1,041	
Average	655	*Counts or estimates of population were given for 26 (41%) of the 63 squares in which recorded*
30% variation	196	

Number of squares in which recorded		
Present	31	
Probable	8	
Confirmed	24	*38%*
Total	63	*27%*

Comparisons between species records and topography	Dry/Wet Number of wet tetrads					Inland/Coastal Amount of coast			Upland/Lowland Proportion of square > 500′				
	Nil	1-5	6-10	11-15	16-22	None	<10%	≥10%	≥1/2	≥1/4	<1/4	None	Total
Surveyed squares	38	104	52	24	16	17	14	203	7	25	73	129	234
% of surveyed squares	16	45	22	10	7	7	6	87	3	11	31	55	100
Grid squares in which recorded	14	33	14	2	0	1	1	61	2	8	25	28	63
% records in each category	>22	>52	>22	3	0	<2	<2	>96	3	13	40	44	100
Significant variation	♦	♦						♦					

advised liberation of foxes and skunks. In January 1959 R Johnson (pers.comm.) reported 'hundreds' on Steeple Jason. Three years later, 75 full-grown birds were trapped and ringed there in March-April 1962 by Eric Goss (pers.comm.).

Habitat and Breeding

The Striated Caracara is a coastal scavenger, particularly dependent on albatross, shag and penguin colonies. The nest is built of twigs or grass on cliff ledges, under large rock slabs or on the top of a tussac grass clump, often close to the seabird colonies on which it preys. It lays 2-3 cream-coloured eggs, heavily marked with dark red, in late October or early November. In winter, small numbers tend to congregate at settlements, particularly those on the coast of West Falkland.

Breeding Birds Survey: analysis of data

The distribution of records shows a concentration on the western and southern coasts of West Falkland and its offshore islands, with another group from Great Island in Falkland Sound through Speedwell and Sea Lion Islands to Beauchêne. The recorded presence of this species in 63 grid squares during ten breeding seasons suggests that it has become more common since the 1960s. Comparison with the distribution of topographical parameters shows a non-significant tendency ($p = 0{\cdot}06$) for more records from squares with little or no standing water and a smaller tendency for more records from coastal squares ($p = 0{\cdot}08$). The largest count was of 50+ pairs at Steeple Jason and there were two estimates of 11-100 pairs at Beauchêne and Sea Lion Island. Among several interesting records of behaviour from Dilys Breeze was a filmed sequence from Sea Lion Island in November 1988, showing an adult bird suddenly attacking an adult King Shag on the ground, bowling it over, killing and eating it. Sadly, there was a report that in September 1989 about ten Johnny Rooks were killed at Hill Cove (Munns 1989). In the winter of 1990 a census was conducted by Phil Gregory and members of the Wildlife Club at Stanley Junior School. On 'Johnny Rook Day' (29 July), reports came from most mainland settlements but not from Carcass Island, New Island, Beauchêne and the Jasons, known to be strongholds. On Sea Lion Island, 52 were recorded but the only other reports from East Falkland were of single immature birds at North Arm and Stanley. On West Falkland a total of 241 were counted at 21 settlements, with about 40 to 55 being seen at Port Stephens, Chartres and Boundary Farm and eight farms stating that they thought it had become more common in recent years. The census gave a total of 295 individuals and it was speculated that the total population of Johnny Rooks in the Falklands may not exceed 1,000 individuals. This is within the limits suggested by the survey of breeding birds.

World Distribution

The Falkland Islands almost certainly hold most of the world population of the Striated Caracara. It also breeds, probably in small numbers, on islands south and southwest of Tierra del Fuego, including Isla Noir (54°S, Clark 1988), Ildefonso, Isla Hornos and Navarino and on Staten Island to the east. Some birds wander north to Isla Grande in winter.

Implications for Conservation

The Striated Caracara is considered to be '**Near-threatened**' globally, having only a small population with a restricted range (Collar *et al.* 1994). Scientific studies of all aspects of this bird's biology would be invaluable, including its interaction with sheep farming. Its behaviour is very unusual for a bird of prey and it is a candidate for protected status in any new conservation legislation. As Hellmayr & Conover (1949) remarked, 'it is a very peculiar species'. It depends on colonial nesting seabirds for most of its food and their future will influence its numbers.

References

TBFI (1975) p134-6, GBFI (1988) p158-60.

---

40 Crested Caracara

Caracara plancus plancus

A. Carancho

C. Traro

Local name: Carancho

History

Early writers on Falkland birds (Pernety 1771, Clayton 1776, Darwin who visited the islands in 1833 and 1834, and Abbott 1861) did not mention the Crested Caracara. If it was present, it is difficult to see how it could have been missed because of its large size and distinctive appearance. Garnot (1826) may have seen this species in 1822 when he used the name *Falco brasiliensis*, but it is impossible to be sure because the behaviour described is more typical of the Striated Caracara (Johnny Rook) and this scientific name has also been used for the Cinereous Harrier. It is possible that the Crested Caracara colonised the Falklands after sheep farming became established in the second half of the 19th century. Cobb (unpublished) in an undated note wrote 'Caranchos quite new to Falklands, first pair started at Port Stephens' and he describes one shot by a Mr Allan at Burnside (near Goose Green) on 23 April 1908, which may be the specimen (dated April 1908) held in the Natural History Museum at Tring. Brooks (1917) published the first record of the species, noting that it was 'now uncommon as it is shot at every opportunity'. In 1908 it had been officially classified as vermin with a price on its beak, because of attacks on fallen sheep and newly born lambs. Wace (1917) remarked that 'probably the young bird of [Striated Caracara] which has a dark head, has often been mistaken for this'. Bennett (1926) remarked that it was a resident but not common and Cawkell & Hamilton (1961) also stated that it was not common and 'wary, doubtless through persecution'.

40 Crested Caracara *Caracara plancus*

Detectability: **5 Very High**

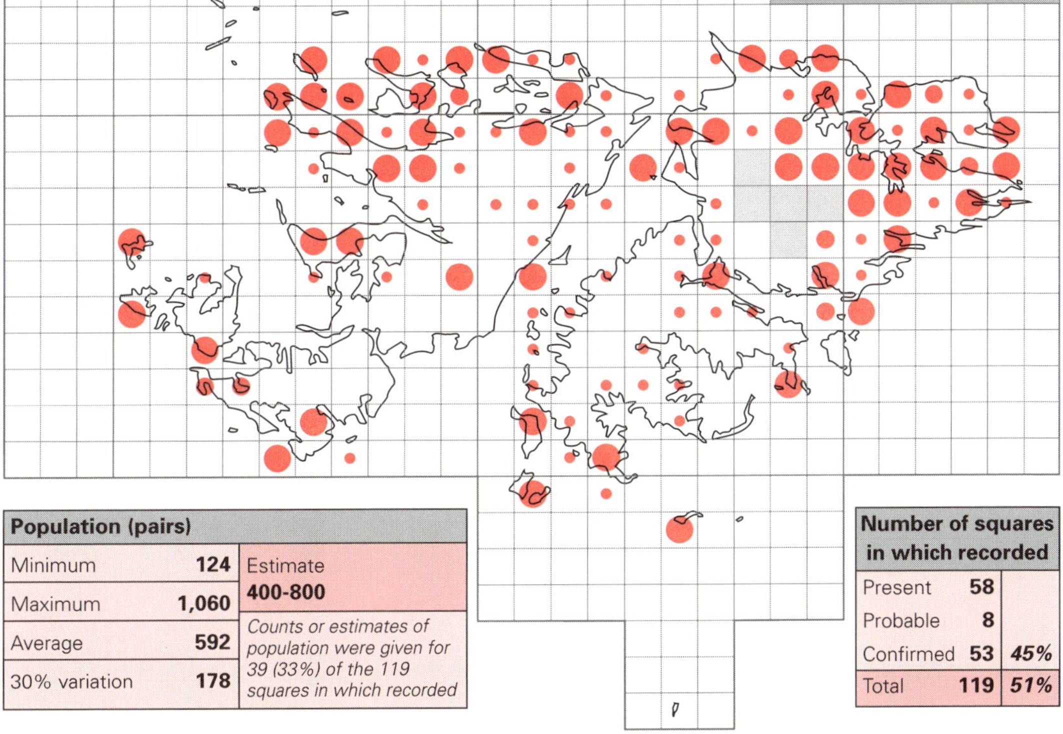

Population (pairs)		
Minimum	**124**	Estimate **400-800**
Maximum	**1,060**	
Average	**592**	*Counts or estimates of population were given for 39 (33%) of the 119 squares in which recorded*
30% variation	**178**	

Number of squares in which recorded		
Present	**58**	
Probable	**8**	
Confirmed	**53**	***45%***
Total	**119**	***51%***

Comparisons between species records and topography	**Dry/Wet** Number of wet tetrads					**Inland/Coastal** Amount of coast			**Upland/Lowland** Proportion of square > 500′				
	Nil	**1-5**	**6-10**	**11-15**	**16-22**	**None**	**<10%**	**≥10%**	**≥1/2**	**≥1/4**	**<1/4**	**None**	**Total**
Surveyed squares	38	104	52	24	16	17	14	203	7	25	73	129	234
% of surveyed squares	16	45	22	10	7	7	6	87	3	11	31	55	100
Grid squares in which recorded	15	51	27	16	10	9	7	103	6	15	43	55	119
% records in each category	13	43	23	13	8	8	6	86	5	13	36	46	100
Significant variation													

Habitat and Breeding

This species can occur almost anywhere in the Falklands and is usually seen singly; in winter, however, small groups of immature birds gather where food is available near settlements. It builds a large nest of diddle-dee twigs, bones and grass, lined with wool, on a coastal or inland cliff ledge and adds to the structure annually. It lays 2 or 3 brownish-red or cream blotched with dark red eggs, between late August and October.

Breeding Birds Survey: analysis of data

The Carancho was found in half of the grid squares surveyed, nearly twice as many as the Johnny Rook (Striated Caracara). However, the estimated population is almost the same for both these birds of prey, indicating that the Carancho is more thinly distributed over a much wider area. Almost all records were of single pairs. Mike Riddy reported in 1989, that some farmers on West Falkland apparently liked having a pair nesting because they kept other Caranchos away. The largest gathering noted was a roost of up to 17, mainly immature birds, at the Stanley abattoir during the 1986/87 season, reported by Tim Duggleby. There were more confirmed breeding records in the northern parts of East and West Falkland with a scatter in other areas. Comparisons of the distribution of records with the analyses of topographical parameters show no significant differences.

World Distribution

The Crested Caracara is widespread from Central America through South America to Tierra del Fuego, mainly in open scrubland and semi-wooded country. Three races have been described. The largest and darkest race *C. p.*

plancus, found in the Falklands, occurs from south of the Amazon in Brazil through most of Chile and Argentina to Tierra del Fuego, where Darwin and others noted its presence and wide distribution before 1840. The other two races *C. p. audubonii* and *C. p. cheriway* are found in southern North America to Panama and in northern South America respectively.

References

TBFI (1975) p136-9, GBFI (1988) p160-1.

Implications for Conservation

The Falkland population forms only a small portion of the world population of this species but it is an impressive large bird of prey and of particular interest to tourists. It is not numerous in the Falklands and may have been a colonist within the past 100 to 150 years, increasing as sheep farming developed to its peak. It has been treated as a pest species for most of this century, though it is primarily an opportunist scavenger. It seems likely to survive while sheep farming remains an important industry but it should be protected in any future wildlife legislation.

41 Peregrine Falcon

Falco peregrinus cassini

A. & C. Halcón peregrino

Local names: Sparrow Hawk, Black Hawk

History

Sharpe of the Natural History Museum in London, first named this form of the Peregrine Falcon as a species *F. cassini* in 1873, using specimens from the Straits of Magellan (Hellmayr & Conover 1949). Vallentin, who was resident for several years between 1898 and 1911, did not see the Peregrine and was 'assured' by a shepherd who gave him an egg for his collection, that 'this species is almost, if not quite, extinct' (in Boyson 1924). Brooks (1917) was the first to describe the Peregrine in the Falklands; it was rare and was 'killed whenever possible, owing to daring and successful raids on poultry'. Wace (1921) listed it as a Falkland species and Bennett (1926) noted that it was an uncommon breeding resident. Cobb (1933) felt that it was comparatively scarce and was sometimes shot at; if only slightly wounded, such birds could be brought home and kept as pets in shepherds' gardens, 'where they do good service by scaring birds from the crops'. Cobb particularly referred to Peregrine Falcons acting as 'tattie-bogles' to keep Long-tailed Meadowlarks away from potatoes. Cawkell & Hamilton (1961) reported that in a partial census of hawks in 1951/52, reports were received of 28 to 29 pairs; West Point Island and New Island were said to have six and four eyries respectively but there were no records from Lafonia and other large areas of East Falkland. They said that the only complaints against it were the killing of domestic pigeons, of which there were few and, in one place, chickens.

Habitat and Breeding

The Peregrine Falcon may be seen singly anywhere over the Falklands. Near their nesting sites on coastal cliffs, pairs can be noisy and sometimes aggressive towards intruders. It lays 2-4 heavily marked red-brown eggs, between late September and late October in a nest scrape or large hollow, usually on an inaccessible ledge.

Breeding Birds Survey: analysis of data

Reports were almost equally distributed between East and West Falkland, with the majority of confirmed breeding records from grid squares on the oceanic coasts of the two main islands, rather than those bordering Falkland Sound. In 1986/87 Tim Duggleby reported at least six pairs breeding on Sea Lion Island, the largest number at any location, but David Osborn noted that only two pairs bred there in 1989/90. Other interesting reports from John Cliffe, Tim Duggleby, Simon Strickland and Andrew Clarke were of Peregrine Falcons attacking gulls, Magellanic Oystercatcher, Silvery Grebe, Rock Shag and Logger, all unsuccessfully, and taking a straying gosling while two male Upland Geese fought. In the analyses, there was a strong but non-significant tendency ($p = 0{\cdot}07$) for more records to come from coastal squares and a weaker non-significant tendency ($p = 0{\cdot}11$) for more records from squares with less than one quarter of the land over 500 feet. Comparison with the data for the Thin-billed Prion shows close agreement. It is possible that

41 Peregrine Falcon *Falco peregrinus*

Detectability: **3 Moderate**

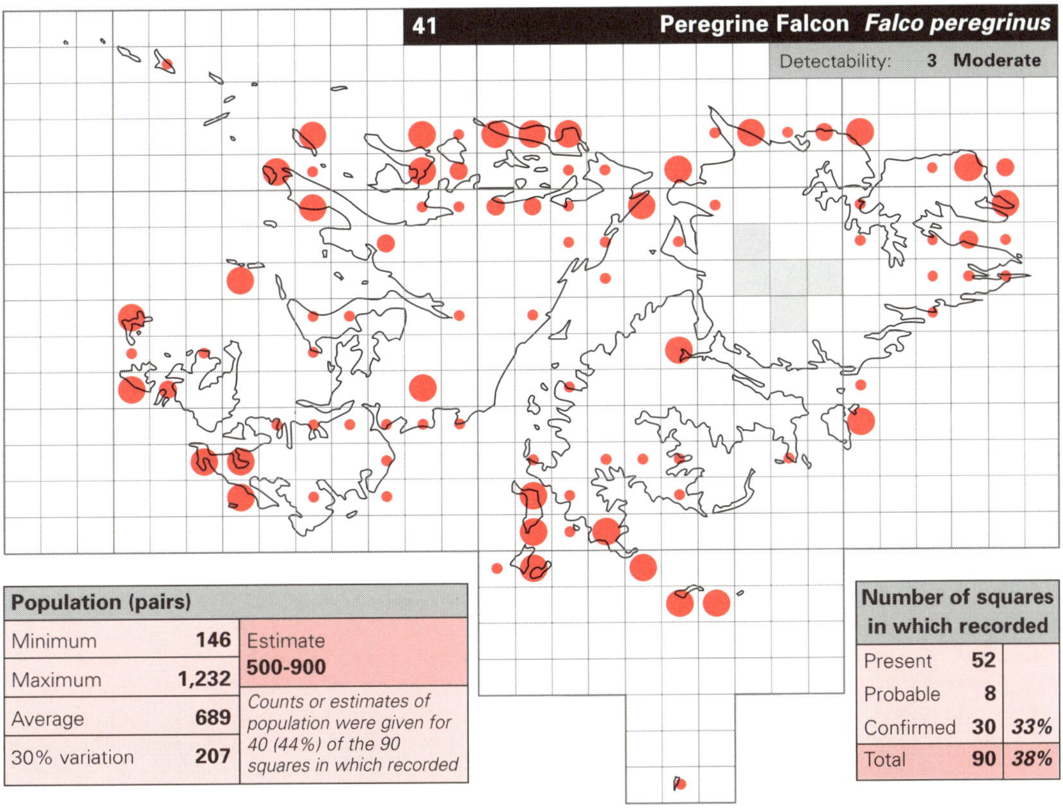

Population (pairs)		
Minimum	**146**	Estimate **500-900**
Maximum	**1,232**	
Average	**689**	*Counts or estimates of population were given for 40 (44%) of the 90 squares in which recorded*
30% variation	**207**	

Number of squares in which recorded		
Present	**52**	
Probable	**8**	
Confirmed	**30**	***33%***
Total	**90**	***38%***

Comparisons between species records and topography	Dry/Wet: Number of wet tetrads					Inland/Coastal: Amount of coast			Upland/Lowland: Proportion of square > 500′				
	Nil	**1-5**	**6-10**	**11-15**	**16-22**	**None**	**<10%**	**≥10%**	**≥1/2**	**≥1/4**	**<1/4**	**None**	**Total**
Surveyed squares	38	104	52	24	16	17	14	203	7	25	73	129	234
% of surveyed squares	16	45	22	10	7	7	6	87	3	11	31	55	100
Grid squares in which recorded	10	49	21	7	3	1	4	85	1	6	40	43	90
% records in each category	11	>54	>23	8	3	1	>4	>94	1	7	44	48	100
Significant variation								♦					

this prion is an important source of food for the Peregrine Falcon in summer. In recent years, there have been at least seven records of Peregrines killing prions or storm-petrels up to 56km (35 miles) offshore during the autumn and then using ships as feeding stations. Bourne & Curtis (1987) noted that all these reports were in the period 16 March to 18 April, suggesting that the Peregrines were on migration. It is also possible that these were immature falcons, dispersing from nest sites and following available prey, while taking advantage of the high perches provided on the masts of ships.

World Distribution

The Peregrine Falcon occurs in most parts of the world and about 18 races have been described. The Falkland race *F. p. cassini* also inhabits southern South America from about 28°S in Chile and 42°S in Argentina to Tierra del Fuego. In winter, continental populations range as far north as southern Colombia. Dispersal may sometimes be aided by shipping; a peregrine flew ashore at South Georgia on 30 April 1986, after joining a ship approaching the island from the west on the previous day (Bourne & Curtis 1986). One was seen chasing a Kelp Gull at Bird Island, South Georgia on 7 August 1991 (Prince & Croxall 1996).

Implications for Conservation

The Falkland population is probably only a small fraction of that occurring in the whole of southern South America and it has not been separated as an island race. Little is known about its diet in winter or to what extent it is migratory. There has been no evidence of a decline due to pesticide contamination of prey species and, despite some persecution in the past, there appears to be no current threat to Peregrines in the Falklands.

References

TBFI (1975) p139-40, GBFI (1988) p161-2.

42 Magellanic Oystercatcher

Haematopus leucopodus

A. Ostrero del Sur

C. Pilpilén austral

Local names: Black and White Curlew, Pied Oystercatcher

History

The Magellanic Oystercatcher was first recorded by Dom Pernety, the Benedictine monk who acted as naturalist with de Bouganville's expedition in 1764, during the first French settlement at Port Louis (Pernety 1771). The species was described scientifically by Garnot (1826) who was the surgeon and naturalist on the corvette *La Coquille*, which was at the Falklands in November and December 1822. Abbott (1861) felt that it was common along the coast and Vallentin (1904) that it was 'fairly numerous' along all Falkland shores. Bennett (1926) said that it was a common breeding species. Cobb (1933) made no comment on abundance but remarked on their noisy displays, which often revealed the nest site instead of distracting a potential predator, and on their habit of flocking after the breeding season. Cawkell & Hamilton (1961) also said that it was a common resident on beaches all round the Falklands and noted the territorial displays and aggressive defence of nests and young.

Habitat and Breeding

It occurs mainly on sand beaches which are backed by low grassy slopes. Parties also feed in short turfed areas and visit freshwater ponds near beaches. The main laying season is from late September to late October, with early clutches from the second week of September and late clutches through to mid-December. The 2 dark olive-brown, buff or greenish eggs, heavily marked with black blotches may be found in a scrape on sand, short turf or dead kelp. Nests may be partly sheltered amongst diddle-dee or sea cabbage plants, from high water mark to at least 91m (100 yards) inland.

Breeding Birds Survey: analysis of data

Several observers have commented that this species appears to be more numerous than the Blackish Oystercatcher and there were estimates of 100-1000 pairs in seven grid squares. This may be due partly to the conspicuous pied colouring and behaviour pattern. However, population estimates confirm this traditional belief and suggest that the Magellanic Oystercatcher is about 1·7 times more numerous. The habit of noisily defending the nesting territory, notably against skuas, caracaras and people, has probably helped to confirm breeding in 73% of the squares in which it was recorded. The largest single count recorded during the survey was an autumn flock of 146 birds seen at Carcass Island by Harro Müller in mid-March 1987. Andrew Clarke counted 60+ feeding in a grass paddock and later gathering on a sand beach at Pebble Island in mid-November 1990. Comparison between the distribution of records and the analyses of topographical parameters indicates a very highly significant bias ($p = 0{\cdot}0005$) towards occurrence in squares with a considerable length of coastline. There are no other significant findings.

World Distribution

The Magellanic Oystercatcher is common on coasts and the shores of lakes and rivers from Tierra del Fuego to about 40°S in Chile and Argentina and nests to at least 1000m in the andean foothills of Santa Cruz province, Argentina. There is some northward migration of the most southerly populations in autumn and though many winter around 46°S (Rio Deseado estuary) on the coast of Argentina, others remain in the south throughout the year (Fjeldså & Krabbe 1990).

Implications for Conservation

The population in the Falklands is probably only a small proportion of the world population and comparative studies would be necessary to determine whether these apparently sedentary birds are separable from those on the mainland. It is still regarded as 'common', suggesting that sand beaches and other feeding areas are providing sufficient animal food, though it is potentially at risk through contamination of the shoreline.

References

TBFI (1975) p142-4, GBFI (1988) p167-8.

42 Magellanic Oystercatcher *Haematopus leucopodus*

Detectability: 5 **Very High**

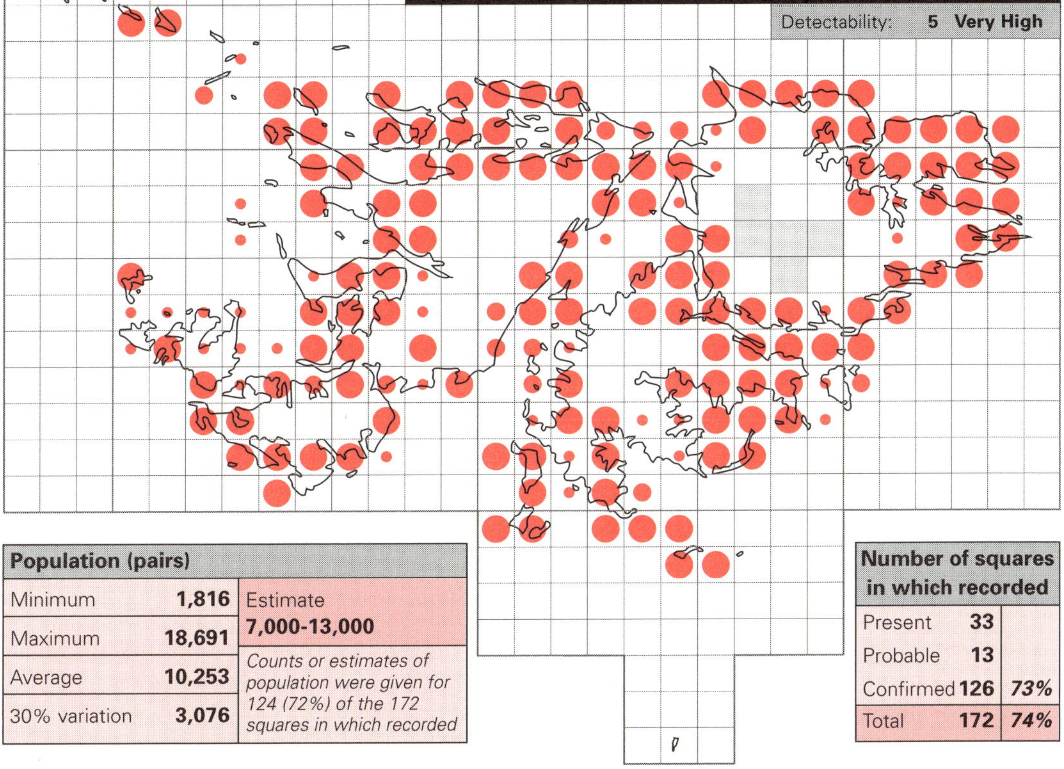

Population (pairs)		
Minimum	**1,816**	Estimate **7,000-13,000**
Maximum	**18,691**	
Average	**10,253**	*Counts or estimates of population were given for 124 (72%) of the 172 squares in which recorded*
30% variation	**3,076**	

Number of squares in which recorded		
Present	**33**	
Probable	**13**	
Confirmed	**126**	***73%***
Total	**172**	***74%***

Comparisons between species records and topography	Dry/Wet: Number of wet tetrads					Inland/Coastal: Amount of coast			Upland/Lowland: Proportion of square > 500′				
	Nil	**1-5**	**6-10**	**11-15**	**16-22**	**None**	**<10%**	**≥10%**	**≥1/2**	**≥1/4**	**<1/4**	**None**	**Total**
Surveyed squares	38	104	52	24	16	17	14	203	7	25	73	129	234
% of surveyed squares	16	45	22	10	7	7	6	87	3	11	31	55	100
Grid squares in which recorded	21	83	40	20	8	1	3	168	2	14	59	97	172
% records in each category	12	48	23	12	5	<1	<2	>97	>1	>8	>34	>56	100
Significant variation								***					

43 Blackish Oystercatcher

Haematopus ater

A. Ostrero negro

C. Pilpilén negro

Local name: Black Curlew

History

This species was first named and described for science from Falkland Islands specimens collected in the vicinity of Port Louis by MM. Quoy and Gaimard, naturalists on the *Uranie* which was wrecked in Berkeley Sound in 1820 (Freycinet 1824). Garnot (1826) implied that he had seen many but Abbott (1861) made no mention of its abundance though he remarked that it was resident throughout the year on East Falkland. Vallentin (in Boyson 1924) was sure that there were more Blackish Oystercatchers in the northward corner of West Falkland than there were Magellanic Oystercatchers. Brooks (1917) noted that this was 'the less common' of the two oystercatcher species and that it was a quiet bird generally. Bennett (1926) stated that it was a common breeding species while Cobb (1933) described it as less numerous than the Magellanic Oystercatcher and quieter in its habits; 'they would often remain unnoticed while standing on the shore among stones and kelp were it not for their coral-coloured bills and pink legs and feet'. Cawkell & Hamilton (1961) described it as a well-known resident, 'which occurs all over the Islands but in smaller numbers than *H. leucopodus*' and they noted that it was 'more common on the eastern side of the group'.

Habitat and Breeding

It is found most frequently on rocky beaches though it also occurs on sandy beaches and in bays where mussels abound. Pairs seem to remain in or near their territories throughout the year and small parties of mostly immature birds have been seen in sheltered creeks during the winter. Blackish Oystercatchers tend to be very quiet when breeding unless their nests are threatened by intruders when their trills can become extremely loud and explosive. They will also chase away larger birds, apparently seen as potential predators. Their nests are often difficult to find, particularly as the sitting bird may remain motionless unless forced to leave, when it will probably walk slowly away without calling. Nests are sited on or near the shore and have been found on low cliffs or rock stacks up to 10m (35 feet) above sea level and on a shingle bank over 91m (100 yds) from a sandy shore. Generally 2 grey-buff or buff eggs, sometimes only one, spotted and streaked below and on the surface with yellow-brown and purple-brown, are laid from late October. The nest may be a shallow scrape on sand or shingle or in a slight depression in bedrock, often containing a few small shells or fragments of dead kelp. Nests with eggs have been found as late as 27 January.

Breeding Birds Survey: analysis of data

There is no significant difference between the number of records for grid squares in East (84) and West (78) Falkland nor is there any indication that it is more numerous on the eastern side of the Falklands. Distribution is similar to that of the Magellanic Oystercatcher though breeding was confirmed in only half as many squares, probably partly due to its drab plumage and quieter behaviour when incubating. Of the topographical analyses, the Inland/Coastal comparison shows a very highly significant bias ($p = 0{\cdot}0003$) towards squares with much coastline; only three records came from squares with less than 10% of coastline. Calculations indicate that the population of this species is roughly 40% less than that of the Magellanic Oystercatcher.

World Distribution

The Blackish Oystercatcher successfully occupies coastal habitats from subantarctic Cape Horn to the subtropical coast of Peru (7°S) on the Pacific and from Tierra del Fuego to the mouth of the River Plate (35°S) on Atlantic shores. No geographical variation in the species has been found (Marchant *et al.* 1986). It is mostly sedentary and usually occurs in small groups, with local migrations to northern and eastern coasts noted in the Magellanic region (Venegas & Jory 1979).

43 Blackish Oystercatcher *Haematopus ater*

Detectability: **4 High**

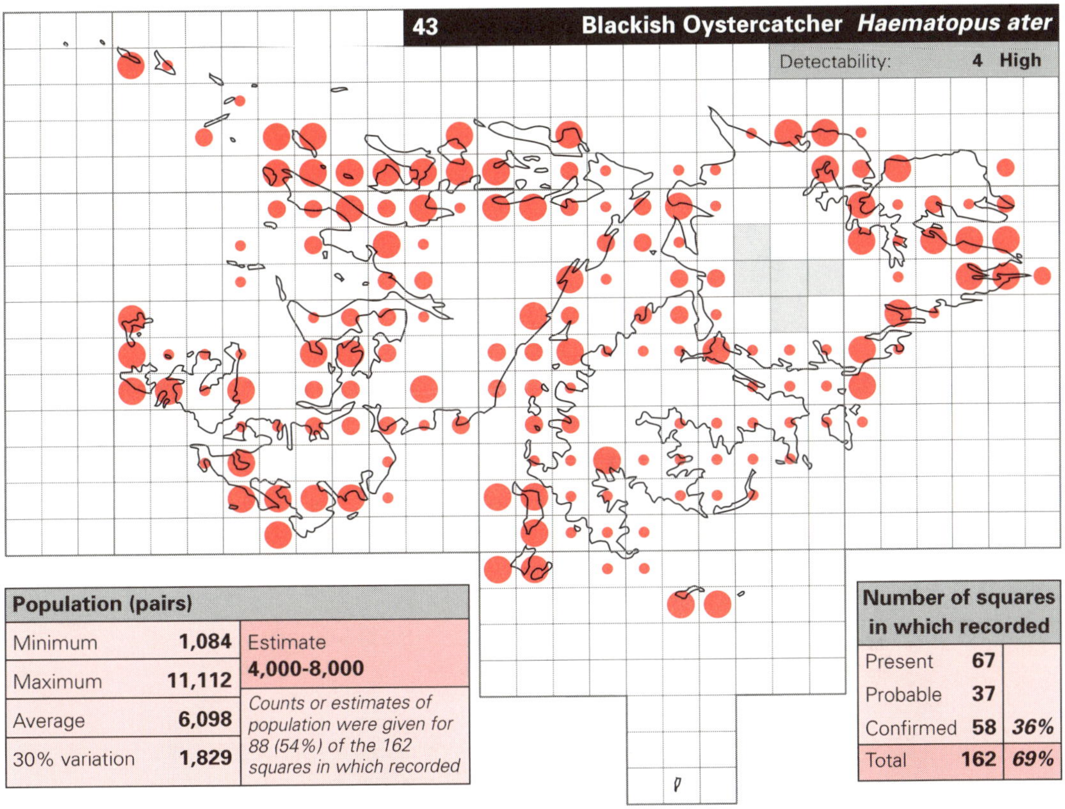

Population (pairs)		
Minimum	**1,084**	Estimate **4,000-8,000**
Maximum	**11,112**	
Average	**6,098**	*Counts or estimates of population were given for 88 (54%) of the 162 squares in which recorded*
30% variation	**1,829**	

Number of squares in which recorded		
Present	**67**	
Probable	**37**	
Confirmed	**58**	***36%***
Total	**162**	***69%***

Comparisons between species records and topography	Dry/Wet: Number of wet tetrads					Inland/Coastal: Amount of coast			Upland/Lowland: Proportion of square > 500′				
	Nil	**1-5**	**6-10**	**11-15**	**16-22**	**None**	**<10%**	**≥10%**	**≥1/2**	**≥1/4**	**<1/4**	**None**	**Total**
Surveyed squares	38	104	52	24	16	17	14	203	7	25	73	129	234
% of surveyed squares	16	45	22	10	7	7	6	87	3	11	31	55	100
Grid squares in which recorded	19	81	38	18	6	0	3	159	2	12	56	92	162
% records in each category	12	50	23	11	4	0	2	98	1	7	35	57	100
Significant variation								***					

Implications for Conservation

The sedentary Falkland population must be a very small proportion of the numbers in South America and studies of behaviour, plumage or proportions in bill, legs and wings would be necessary to determine whether Falkland birds are distinct from those on the continent. There is no evidence of current threats, but any contamination of its coastal habitat could have serious effects on the population of this species in the Falklands.

References

TBFI (1975) p144-7, GBFI (1988) p166-7.

44 Two-banded Plover

Charadrius falklandicus

A. Chorlitejo doble collar

C. Chorlo de doble collar

Local names: Falkland Island Plover, Two-barred Plover, Beach Lark

History

Latham named this species *Charadrius falklandicus* in 1790 from a specimen illlustrated in Portlock's Voyage round the World and described as the 'Rusty-crowned Plover'. It had been obtained at the original British settlement, Port Egmont on Saunders Island. Garnot (1826) named the 'pretty little species of plover' found in the Berkeley Sound area *Charadrius pyrocephalus*. He was apparently unaware of Latham's publication and of Lichtenstein, who in 1823 had named examples of the same species from Montevideo, Uruguay as *Charadrius trifasciatus* (Hellmayr & Conover 1948). Abbott (1861) described it as, 'a spring visitor, arriving about the beginning of September and breeding shortly afterwards', but he made no comments about its abundance. Vallentin (1904) stated that it was 'common on the "camp" throughout the Falklands', Bennett (1926) called it 'a common resident breeding species' and Cawkell & Hamilton (1961) noted that it was a common inhabitant of the outer beaches but preferred creeks and deeply inset bays when breeding. They said that it was fairly approachable and that flocks of shy birds seen in winter could be immigrants. There has been no confirmation of migration to or from the Falklands, but there is evidence of sedentary behaviour. A small scale colour ringing study of 28 birds on the Cape Pembroke peninsula and at Lake Point south of Stanley between 1957 and 1959, showed close attachment to one area throughout the year. The longest movement recorded was of 2km (1½ miles) by three of nine ringed individuals (Woods 1975). One adult male was retrapped once and seen 12 times in the following two years, three times where ringed and nine times within 1·5 km (1 mile).

Habitat and Breeding

The Falkland Island Plover inhabits coastal regions and occurs on sand beaches and mudflats where it is often accompanied by White-rumped Sandpiper and Rufous-chested Dotterel, particularly during the late summer and autumn. In the spring and summer, it may be found up to 1·5 km (1 mile) from the shore, on dry slopes of diddle-dee and short grass. It nests between mid-September and mid-January, with later clutches after November probably being replacements. The nest is a scrape, sometimes lined with wisps of grass, usually in diddle-dee, grass or ferns and often on a small hummock, though some pairs nest on beaches between dry heaps of kelp. It lays 2-3, occasionally 4 greenish or buff eggs, blotched and spotted with brown mostly at the larger end.

Breeding Birds Survey: analysis of data

The largest number recorded was a flock of about 200 seen in March 1987 by Harro Müller and a similar number seen by Phil Gregory on 25 December 1990, on the shores of Elephant Bay, Pebble Island. It was recorded from a noticeably greater number of grid squares on East Falkland (88) than on West Falkland (60) and was apparently more widespread in Lafonia, where hardly any land is above 60m (200 feet). Comparison with the altitude analysis, however, shows no significant bias. There is a strong but statistically non-significant, tendency ($p = 0{\cdot}07$) for more frequent occurrence in squares with considerable coastline.

World Distribution

The Two-banded Plover occurs widely in coastal southern South America and near inland lakes and marshes from Tierra del Fuego to about 25°S and to at least 1200m in the andean foothills of Santa Cruz, Argentina (Fjeldså & Krabbe 1990). It breeds in the southern part of this range to about 37°S and though it is highly migratory some birds remain in the Magellanic region through the winter.

References

TBFI (1975) p148-9, GBFI (1988) p170-1.

44 ?* Two-banded Plover *Charadrius falklandicus*

Detectability: **4 High**

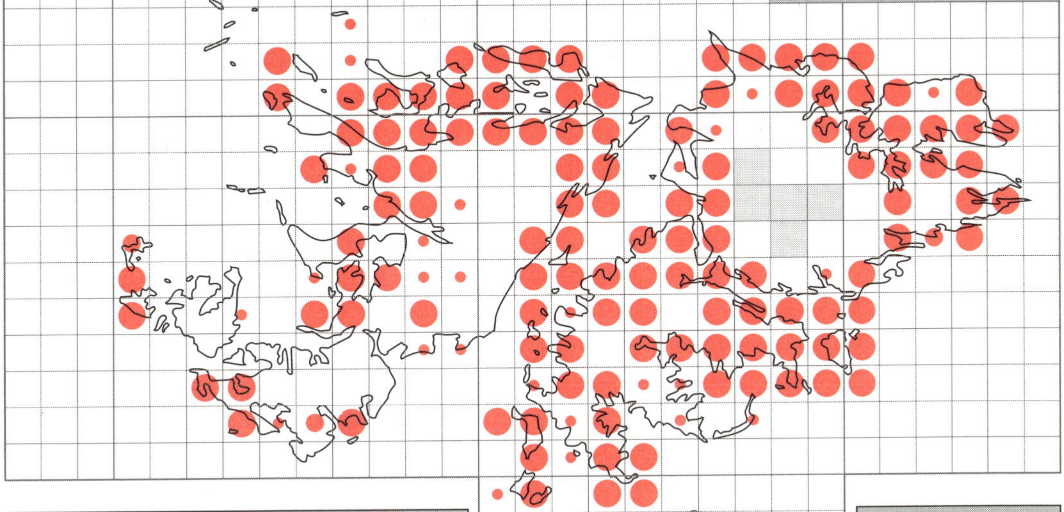

Population (pairs)		
Minimum	**1,397**	Estimate **7,000-13,000**
Maximum	**18,413**	
Average	**9,905**	*Counts or estimates of population were given for 114 (77%) of the 149 squares in which recorded*
30% variation	**2,971**	

Number of squares in which recorded		
Present	**27**	
Probable	**3**	
Confirmed	**119**	***80%***
Total	**149**	***64%***

Comparisons between species records and topography	Dry/Wet Number of wet tetrads					Inland/Coastal Amount of coast			Upland/Lowland Proportion of square > 500′				
	Nil	**1-5**	**6-10**	**11-15**	**16-22**	**None**	**<10%**	**≥10%**	**≥1/2**	**≥1/4**	**<1/4**	**None**	**Total**
Surveyed squares	38	104	52	24	16	17	14	203	7	25	73	129	234
% of surveyed squares	16	45	22	10	7	7	6	87	3	11	31	55	100
Grid squares in which recorded	12	69	38	19	11	4	5	140	2	12	48	87	149
% records in each category	8	46	26	13	7	3	3	94	>1	8	>32	>59	100
Significant variation								♦					

Implications for Conservation

It has been noted that many individuals in the Falklands have an incomplete upper breastband during the breeding season (Marchant *et al.* 1986). Being apparently sedentary, it is possible that the Falkland population is a distinct geographic race but further comparative studies are needed. If separable, it would be described as *Ch. f. falklandicus* and the mainland population as *Ch. f. fasciatus* (Blake 1977). It does not seem to be under any particular threat at present. Extensive coastal habitats provide a choice of feeding areas and grazing by sheep may keep inland grassy slopes in a state which is beneficial to this plover. It could, however, be seriously affected if shores became contaminated by chemical wastes.

45 Rufous-chested Dotterel

Charadrius modestus

A. Chorlo pechicolorado

C. Chorlo chileno

Local name: Dotterel

History
Pernety apparently did not see this beautiful plover in 1764 but Garnot obtained specimens near Port Louis in 1822 and decided to name it *Tringa urvillii* in 1826, after his 'estimable colleague' M. d'Urville, one of the naturalists and deputy leader of the expedition. Unfortunately, Garnot was not aware that Lichtenstein had already named it as *Charadrius modestus* in 1823, from a specimen obtained at Montevideo, Uruguay (Hellmayr & Conover 1948). Darwin obtained specimens on 'the upland marshes' in 1833 or 1834 but did not comment on its abundance. Abbott (1861) concluded that it was migratory and noted that the first birds appeared at the beginning of September; in February they started gathering in flocks along the coast and had disappeared entirely by the end of April. Although Vallentin (in Boyson 1924) confused names and described it under *Ch. falklandicus*, he said that it was one of the earliest immigrants and mentioned hearing the first birds calling on the evening of 29 August 1910 and finding numbers of them a few days later. He stated that it was quite common on the uplands. Brooks (1917) felt that it was 'abundant everywhere in summer frequenting dry "camp" and diddle-dee bushes, even to hill tops; at this season never on beaches'. Wace (1921) described it as a semi-migrant with 'some remaining the whole year' and Bennett (1926) believed it was common and agreed with this description of its status. Cawkell & Hamilton (1961) described migratory behaviour in detail but agreed that some birds remained in the Falklands through the winter. They noted that it was still regarded by some people as game.

Since 1861, there has been controversy about migration. Several observers have produced evidence of semi-migratory behaviour, the latest being Carstairs (1989). Flocks gather in autumn and adults in breeding plumage appear suddenly in late August but records indicate that between April and July, small parties occur on grassland near coasts. They are quiet, in dull plumage and inconspicuous, in sharp contrast to their behaviour during the breeding season.

Habitat and Breeding
The Dotterel occurs in a variety of open habitats from lowland coast and flooded grassland to at least 200m (700 feet) above sea level. It often associates with Two-banded Plover on mudflats in spring and autumn. Between August and November, birds often display in noisy aerial chases and after laying, while one bird incubates, the mate remains on guard, standing on a hummock in the vicinity of the nest and calling. It breeds on the drier slopes where white grass, diddle-dee and rushes are abundant, nesting between mid- September and early February, with the main laying period in October; clutches found in December or later may be replacements. The nest site is merely a scrape sheltered by overhanging low vegetation. It lays only 2 eggs, usually olive-brown heavily blotched with black, occasionally reddish.

Breeding Birds Survey: analysis of data
It was recorded in all parts of the Falklands, including several of the offshore islands. The largest autumn gathering on record was seen by Harro Müller who found a flock of about 350 near Beaver Pond, Sea Lion Island on 27 March 1987. Comparison with the analyses of topographical parameters shows little variation in distribution within the three categories; it occurs as frequently in coastal, lowland squares as in inland squares with much of the land over 500 feet and the amount of standing freshwater appears to be irrelevant.

World Distribution
In southern South America, the Rufous-chested Dotterel is widespread in the Magellanic region and Tierra del Fuego and breeds less commonly to about 42°S. Many birds migrate north as far as 27°S but others remain in coastal parts of the Magellanic region throughout the winter. Recent information from Pablo Canevari (1995 pers. comm.) of Buenos Aires indicates a big reduction in the size of wintering flocks on the Pampas of Argentina.

45 Rufous-chested Dotterel *Charadrius modestus*

Detectability: 5 **Very High**

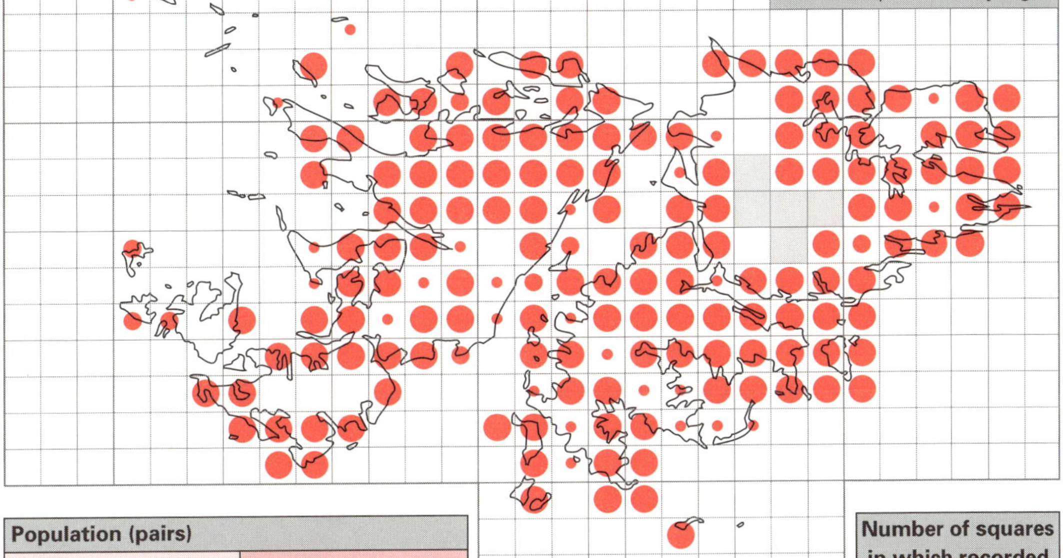

Population (pairs)		
Minimum	2,668	Estimate **11,000-21,000**
Maximum	29,523	
Average	16,095	*Counts or estimates of population were given for 123 (70%) of the 175 squares in which recorded*
30% variation	4,829	

Number of squares in which recorded		
Present	26	
Probable	8	
Confirmed	141	*81%*
Total	175	*75%*

Comparisons between species records and topography	Dry/Wet Number of wet tetrads					Inland/Coastal Amount of coast			Upland/Lowland Proportion of square > 500′				Total
	Nil	1-5	6-10	11-15	16-22	None	<10%	≥10%	≥1/2	≥1/4	<1/4	None	Total
Surveyed squares	38	104	52	24	16	17	14	203	7	25	73	129	234
% of surveyed squares	16	45	22	10	7	7	6	87	3	11	31	55	100
Grid squares in which recorded	15	74	49	22	15	15	12	148	6	20	57	92	175
% records in each category	<9	42	28	<13	<9	8	7	85	3	11	33	53	100
Significant variation													

Implications for Conservation

The Dotterels breeding in the Falklands are only a small proportion of the world population. There has been little Falkland field work on this species but Marchant *et al.* (1986) note that a sample, presumably of museum skins, had noticeably longer wings and legs than mainland birds. Further investigation of the population size and studies of its taxonomy, breeding behaviour and migratory habits would be valuable, especially if the mainland population is declining. It was removed from the list of birds which could be killed outside the close season by the Wild Animals and Birds Protection Ordinance of 1964. There are no known threats in the Falklands now.

References
TBFI (1975) p150-1, GBFI (1988) p171-2

46 Magellanic Snipe

Gallinago magellanica magellanica

A. Agachadiza común

C. Becacina

Local name: Snipe

History
Pernety noted the presence of snipe around Port Louis in 1764 as did Garnot in 1822, but neither gave any indication of numbers. The Antarctic Expedition of James Ross collected four specimens on or around 20 April 1842 in the Berkeley Sound area. Abbott (1861) who was resident between 1858 and 1860 wrote that the snipe generally appeared in East Falkland about mid-August and laid early in September; most birds left in March though a few remained through the winter. Vallentin (in Boyson 1924) felt that owing to the marshy nature of the land, it was 'far more abundant' on East than on West Falkland. He also noted that 'quite big bags of snipe have been made from time to time' on Lively Island. Cobb (unpublished) recorded 'great numbers on Lively Island where there were no rats, mice or cats' and noted that between 26 and 28 April 1910, Mr Pack's party of six guns shot 300 snipe on that island.

Vallentin commented that in the early days of colonisation, the snipe 'abounded in the whole archipelago; but of recent years [up to 1911] they have greatly decreased in numbers owing to the abundance of domestic cats which have taken to a wild existence'. He noted that snipe arrived early in the spring and he found a nest on 2 September 1910. Bennett (1926) felt that it was common and that 'some remain through the winter'. Cawkell & Hamilton (1961) thought that its numbers were 'now much reduced' but it was still 'fairly common' and that some remained throughout the year.

Habitat and Breeding
The Magellanic Snipe occurs in a variety of habitats, including wet moorland with rank grass or rushes, dry ground with diddle-dee cover, eroded and replanted slopes up to at least 300m (1,000 feet) above sea level, open tussac grass paddocks and beaches with rotted kelp heaps. It lays 2, sometimes 3, pear shaped olive-green eggs, spotted and blotched with black in a slight grass lined nest amongst grasses or low diddle-dee. Eggs or young have been found in all months between July and February though the main breeding season is probably August to October. There is no recent evidence of migration; some birds are present through the winter when they rarely call and are very inconspicuous. The population has probably been reduced in the past through the practice of burning white grass in early spring and by shooting in the autumn.

Breeding Birds Survey: analysis of data
It was recorded in 90 East Falkland grid squares and 68 on the West. A 'high density' was noted by Kate Thompson on Steeple Jason in 1987/88 but the maximum count at any location was about 30 pairs at Sea Lion Island in December 1989 by David Osborn. The proportion of squares in which breeding was confirmed is rather low, probably due to difficulty in locating nests or chicks, combined with the somewhat ventriloquial character of the alarm calls of adults. Comparison of the distribution of records in relation to the analyses of topographical parameters shows no statistically significant differences. Even the amount of standing freshwater appears to be irrelevant.

World Distribution
There has long been uncertainty about the classification of the smaller South American snipes. Fjeldså and Krabbe (1990) include *Gallinago paraguaiae* in the species *G. magellanica,* while Marchant *et al.* (1986) give Paraguayan Snipe *G. paraguaiae* as an alternative name for this species. Opinions vary partly because there is a great deal of individual variation between snipe at any one locality. The Magellanic Snipe is widespread and common in South America from Tierra del Fuego to Colombia (Fjeldså & Krabbe 1990). Most populations are generally sedentary but those breeding in Tierra del Fuego and Magallanes are migratory, departing north as far as Uruguay by late April and returning from the beginning of August (Venegas & Jory 1979).

46 ?* Magellanic Snipe *Gallinago magellanica*

Detectability: 4 High

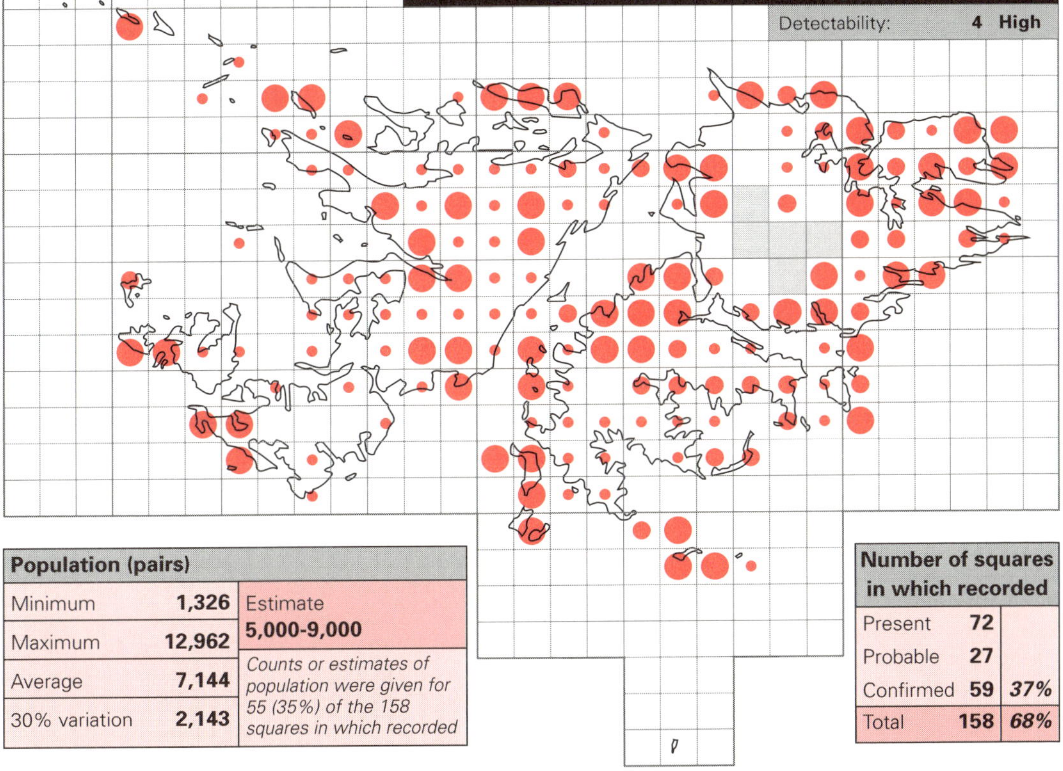

Population (pairs)		
Minimum	1,326	Estimate **5,000-9,000**
Maximum	12,962	
Average	7,144	*Counts or estimates of population were given for 55 (35%) of the 158 squares in which recorded*
30% variation	2,143	

Number of squares in which recorded		
Present	72	
Probable	27	
Confirmed	59	*37%*
Total	158	*68%*

Comparisons between species records and topography	Dry/Wet Number of wet tetrads					Inland/Coastal Amount of coast			Upland/Lowland Proportion of square > 500′				
	Nil	1-5	6-10	11-15	16-22	None	<10%	≥10%	≥1/2	≥1/4	<1/4	None	Total
Surveyed squares	38	104	52	24	16	17	14	203	7	25	73	129	234
% of surveyed squares	16	45	22	10	7	7	6	87	3	11	31	55	100
Grid squares in which recorded	13	65	45	20	15	14	11	133	5	19	53	81	158
% records in each category	8	41	29	13	9	9	7	84	3	12	34	51	100
Significant variation													

Implications for Conservation

The Magellanic Snipe appears to be widely distributed and fairly common but feral cats almost certainly reduce its numbers. It is classed in the 1964 Wild Animals and Birds Protection Ordinance as a game bird that can be shot between 1 March and 31 July. There is no recent evidence that it is being hunted and its listing as a game bird is an anachronism. Given the perceived decline in numbers, it should be protected in any future wildlife legislation.

References

TBFI (1975) p155-7, GBFI (1988) p178-80

47 Falkland Skua

Catharacta antarctica

A. Págalo grande

C. Salteador común

Local name: Sea Hen

History

Pernety (1771) was apparently the first person to write about skuas in the Falklands. He described them as being the size of a common hen and called them 'Grey gulls'. He commented on their predatory habits and said that 'nobody thought it worth while to shoot them, concluding they would be very bad eating'. At Saunders Island, Clayton (1774) used the name 'Sea Hen' and implied that settlers ate the eggs when they were available. He observed that skuas forced cormorants to drop food, but did not comment on their numbers. Quoy and Gaimard (in Freycinet 1824) were the first to name this species scientifically (as *Lestris catarractes*) and said that it was called 'Port Egmont Hen' by 'les matelots de Cook', a local name also recognised by Cobb (1933). Garnot (1826) remarked that it was very common and pursued gulls, making them drop fish which it then seized. Abbott (1861) noted that it was a summer visitor which bred from early December in communities generally near a penguin colony, from which they took eggs and he remarked on their aggressive response to human intrusion. Brooks (1917) stated that it was common though it did not breed at the places he visited. For a museum collector, his vindictive attitude towards a bird whose territory he was invading is surprising; his head was touched only once by a divebombing skua, but once was enough and 'thereafter I would kill any bird taking liberties with me in this manner'. Cobb (1933) described their aggressive behaviour in defence of eggs or young, when they terrified dogs and attacked humans, sometimes striking with the feet. Cawkell & Hamilton (1961) felt that it was 'an unpopular and fairly common breeding visitor' regarded with some merit by farmers because it ate eggs and young of the Upland Goose.

About 200 birds were ringed between 1961 and 1963, mainly at Dunbar Island by Roddy and Lily Napier and at Carcass Island by David Galloway, but no foreign recoveries have been reported. A pair colour ringed at Kidney Island in 1961 returned to the same site the following year. Philip Hutton has reported on two adults individually marked with chicken rings in 1981 at North Arm settlement by two children who had seen a video about ringing birds. By December 1995, it was known that 'Ringo Right' had been visiting North Arm for 15 years though 'Ringo Left', had not been seen since 1992.

Habitat and Breeding

The Falkland Skua, which is a summer resident between mid-October and late April to mid-May, occurs over coastal waters, mainland coasts and on offshore islands and is rarely seen more than 1·5km (1 mile) inland. It usually nests in small colonies on mainland points or on open slopes of smaller islands, up to 1km from the sea, though isolated nests are sometimes found. The nest scrape is scantily lined with grass or diddle-dee and 1-3, usually 2, eggs of variable colour, greyish, pale green or dark olive-brown spotted and blotched with dark brown, are laid between late November and mid-December.

Breeding Birds Survey: analysis of data

Records confirm that this species has a wide distribution around Falkland coasts and in the comparison with the Inland/Coastal analysis, there is a very significant bias ($p = 0{\cdot}002$) towards more records from coastal grid squares. Other comparisons show no significant differences, though there is a slight tendency for more records of presence in squares with no land over 500 feet. The largest counts of about 40 pairs were made at colonies on Steeple Jason Island by Shane Wolsey and in the Cape Orford region of Port Stephens by Mark Jones and Tui de Roy, in the 1986/87 season. There were a few grid square estimates in the range 100-1,000 pairs.

World Distribution

The Falkland Skua was thought to breed only in the Falklands until Devillers (1978) described some typical *C. antarctica* individuals at colonies on the coast of Patagonia between 45°S and 47°S. At Puerto Deseado (47°S) they breed alongside Chilean Skuas and hybrids were described by Devillers. This taxon is thought to be closely related to the Chilean Skua and to the Brown

47 Falkland Skua *Catharacta antarctica*

Detectability: 5 Very High

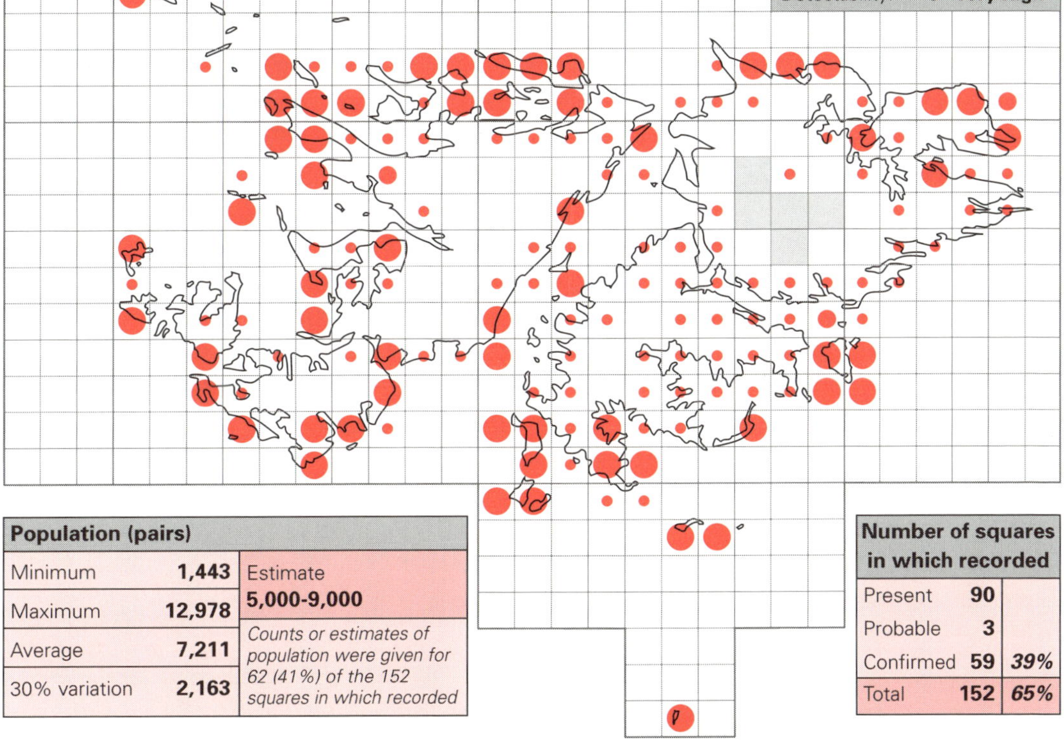

Population (pairs)		
Minimum	1,443	Estimate **5,000-9,000**
Maximum	12,978	
Average	7,211	*Counts or estimates of population were given for 62 (41%) of the 152 squares in which recorded*
30% variation	2,163	

Number of squares in which recorded		
Present	90	
Probable	3	
Confirmed	59	*39%*
Total	152	*65%*

Comparisons between species records and topography	Dry/Wet: Number of wet tetrads					Inland/Coastal: Amount of coast			Upland/Lowland: Proportion of square > 500′				
	Nil	1-5	6-10	11-15	16-22	None	<10%	≥10%	≥1/2	≥1/4	<1/4	None	Total
Surveyed squares	38	104	52	24	16	17	14	203	7	25	73	129	234
% of surveyed squares	16	45	22	10	7	7	6	87	3	11	31	55	100
Grid squares in which recorded	14	74	39	16	9	1	3	148	1	12	48	91	152
% records in each category	9	49	26	10	6	1	2	97	<1	8	32	60	100
Significant variation								**					

(Subantarctic) Skua *C. loennbergi* which has a circumpolar distribution on subantarctic islands and the Antarctic Peninsular (J Croxall pers. comm.). In winter, the Falkland Skua is pelagic, apparently ranging north to about 20°S in southern oceans.

References

TBFI (1975) 160-2, GBFI (1988) 187-8.

Implications for Conservation

The Falkland Skua does not appear to be threatened by environmental changes, though the large but gradual decline in Rockhopper Penguins over the past 60 years may be having some effect (Bingham 1995). Its ability to use a wide range of food, obtained by attacks on seabirds or by scavenging suggests that it can maintain its numbers. It was probably originally labelled as a pest after exaggerated reports by human invaders of their territories. It is included in Schedule I of the 1964 Wild Animals and Birds Protection Ordinance as a bird that may be killed at any time. There is no sound economic reason for this persecution. The Falklands hold most of the world population of this taxon and it should be legally protected.

48 Dolphin Gull

Larus scoresbii

A. & C. Gaviota austral

Local name: Red-billed Gull

History

The beautifully coloured Dolphin Gull was first described scientifically by Traill and named after William Scoresby junior, the arctic whaling captain and historian, from a specimen said to have come from the South Shetlands in 1823 (Hellmayr & Conover 1948). Garnot (1826) mentioned three species of gull at Port Louis and there is a specimen from the English Antarctic Expedition of 1842 in the Natural History Museum at Tring but little else was recorded during the first half of the 19th century. Abbott (1861) observed that it frequently nested among communities of the Kelp Gull and called it the Red-billed Gull, as did Cameron, another collector in 1881. Vallentin (1904) noted that it was fairly common near Stanley and Port Louis where it was attracted to the slaughter houses and at Roy Cove. Cobb (1910) described it as a pretty bird that had no respect for other's property, hovering in the neighbourhood of shag or penguin rookeries on the chance of taking eggs. He said they became very tame in winter at some settlements, coming to feed with poultry and taking more than a fair share of the food 'if they are not sharply looked after'. Vallentin (in Boyson 1924) believed that it was uniformly distributed over the islands but by no means common, whereas Bennett (1926) described it as a very common breeding species. In 1933 Cobb said that 'there must be a goodly number' and they were a familiar sight all over the islands. Cawkell & Hamilton (1961) noted that it was extremely bold and frequented sea lion colonies to consume faeces. They felt that it was not as plentiful as the Kelp Gull and that numbers were decreasing, though no reason was known.

In field work between June 1959 and January 1963, 239 birds were trapped and ringed at seven sites, mostly at Stanley, the Lagoon near Bluff Cove and on Kidney Island. Returns and sight recoveries showed that juveniles from the Lagoon dispersed to the northeast and southwest along the southeastern coast of East Falkland. Colour ringed adults moved between colonies at Kidney Island or the Lagoon and wintering areas in Stanley. One fledgling ringed at the Lagoon in February was seen in Stanley between June and December and on Kidney Island the following April. The greatest longevity shown was nine years.

Habitat and Breeding

The Dolphin Gull is found around all coasts and is particularly attracted to seal, penguin or shag colonies. In winter, flocks may gather where sheep or cattle have been killed. It breeds in small colonies, often closely packed and usually associated with Kelp Gulls, South American Terns or Brown-hooded Gulls on sand or shingle beaches and headlands, by ponds in grass or on rocky ridges with low tussac grass. The largest colony recorded was on a shingle bank at the Lagoon in January 1961. It contained about 56 nests in three small groups of 17, five and four nests only 5 to 15m from each other and a group of about 30 nests over 25m (27 yards) away. The previous year (January 1960) only one close group of 26 nests was found there. Nesting starts in December and 2-3 olive-buff or grey-green eggs, well-marked with dark brown blotches, are laid. The nest is a hollow, lined with grass and dead kelp pieces, often decorated with green plant material. Chicks form crèches from early February and juveniles are usually flying from early March.

Breeding Birds Survey: analysis of data

Breeding was confirmed in less than one third of the grid squares where Dolphin Gulls were present. This probably reflects the relatively small number of places where it breeds and its widespread occurrence where food is available, rather than a failure of observers to find evidence of breeding. Comparison between these records and the Inland/Coastal squares analysis shows a highly significant bias ($p = 0{\cdot}008$) towards more records from squares with considerable lengths of coastline, confirming that it is essentially a coastal species. Other comparisons show no significant bias. There is no apparent difference in distribution between West and East Falkland. Several observers gave additional interesting

48 Dolphin Gull *Larus scoresbii*

Detectability: 5 **Very High**

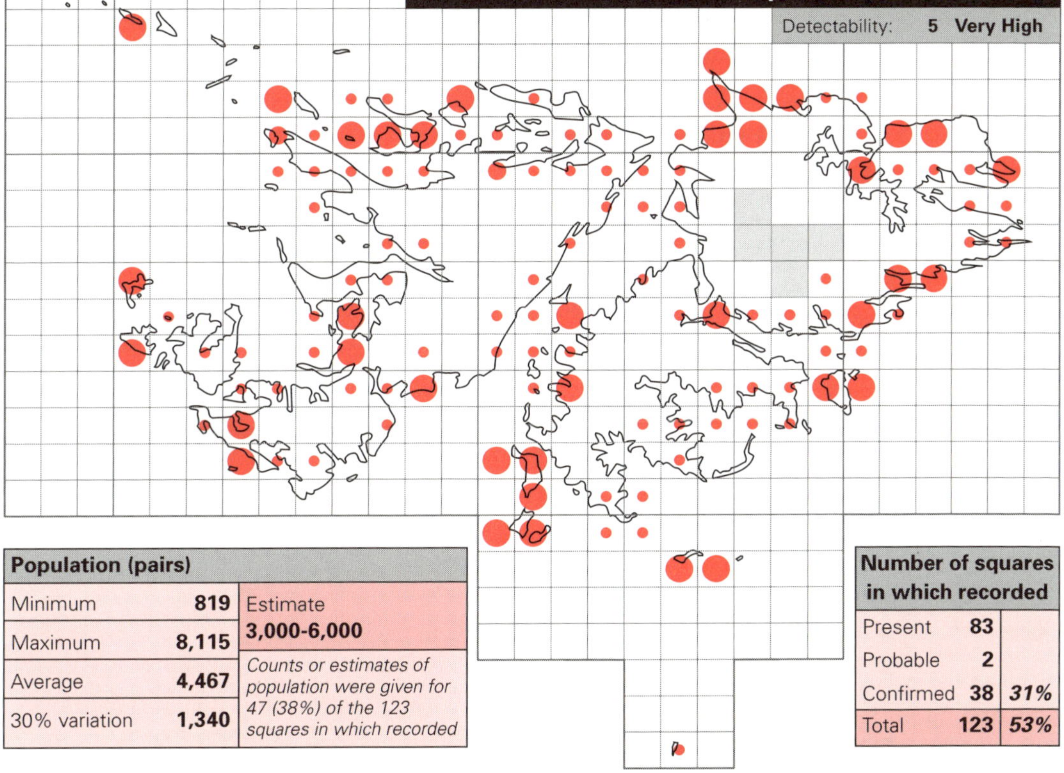

Population (pairs)		
Minimum	**819**	Estimate
Maximum	**8,115**	**3,000-6,000**
Average	**4,467**	*Counts or estimates of population were given for 47 (38%) of the 123 squares in which recorded*
30% variation	**1,340**	

Number of squares in which recorded		
Present	**83**	
Probable	**2**	
Confirmed	**38**	***31%***
Total	**123**	***53%***

Comparisons between species records and topography	**Dry/Wet** Number of wet tetrads					**Inland/Coastal** Amount of coast			**Upland/Lowland** Proportion of square > 500'				
	Nil	1-5	6-10	11-15	16-22	None	<10%	≥10%	≥1/2	≥1/4	<1/4	None	Total
Surveyed squares	38	104	52	24	16	17	14	203	7	25	73	129	234
% of surveyed squares	16	45	22	10	7	7	6	87	3	11	31	55	100
Grid squares in which recorded	13	64	27	14	5	1	3	119	2	7	47	67	123
% records in each category	11	52	22	11	4	1	2	97	2	6	38	54	100
Significant variation								**					

information. When Andy Douse visited the Lagoon in December 1985 he found a colony with 31 nests, in the same area where 26 were seen in 1960 and about 56 in 1961. A group of about 50 pairs was reported as breeding near Port Stephens settlement by Mike Clarke in 1987. Harro Müller saw 62 birds at Sea Lion Island on 26 March 1987 and Nicholas Beswick reported the largest group during the survey, a roost with 150 birds present on 12 November 1992 at East Cove Peninsula (VC021471). Between July and December 1985 Tim Cowley noted that Dolphin Gulls commonly occurred around the coastels (temporary floating hostels installed for military personnel in 1982) moored at the eastern end of Stanley harbour. Later, between December 1986 and March 1987, JY McCallum also reported this gull as common at the coastels, 'where it will feed from the hand (when educated)'.

World Distribution

The Dolphin Gull breeds in a small geographic area, including the Falkland Islands and the coasts of southern South America from about 44°S in Argentina (Punta Tombo) and Chile to islands south of Tierra del Fuego. Census work since 1992 by Adrian Schiavini and Pablo Yorio (in press) has indicated that only about 700 pairs breed between Punta Tombo and the islands belonging to Argentina in the Beagle Channel, with a few more in southern Chile. In winter it occurs north to about 37°S. No Patagonian ringed birds have been seen in the Falklands and no birds ringed in the Falklands were reported in South America. Vagrants have been recorded at South Georgia, one in April 1985 and at least three in February and March 1986 (Bourne & Curtis 1985, 1986).

References

TBFI (1975) 162-4, GBFI (1988) 190-2.

Implications for Conservation

Although it seems that Dolphin Gulls were widespread in the past it was rare to see more than one hundred together. There are no reliable counts before 1960 but numbers do not seem to have changed significantly since. The large decline in sea lion and penguin populations (Thompson & Duck 1995, Bingham 1995) and increase in pollution of the coasts could seriously affect the Dolphin Gull but the smaller farms developed in the past few years may provide more opportunities for scavenging, leading to an increase in numbers. The current population estimate suggests that the Falkland Islands support the majority of the world's Dolphin Gulls and field work on behaviour and breeding biology would be valuable, given the importance of the Falkland environment to this species.

49 Kelp Gull

Larus dominicanus

A.& C. Gaviota dominicana

Local names: Big, White or Saddle-backed Gull

History

Pernety (1771) saw 'White Gulls' which were probably this species, at Port Louis between February and April 1764. Garnot (1826), also at Port Louis, noted that 'Black-backed Gulls' were present but Darwin who remarked on their presence in South America, did not mention them in the Falklands (Gould & Darwin 1838-41). Abbott (1861) described the 'Saddle-backed Gull' as a common resident but believed that many left in the winter because he saw few then and all old birds; large numbers appeared in September and many were immature. Forty years later, when the sheep population had increased to over 700,000 and the human population to about 2,000, Vallentin commented (1904) that the 'Black-backed Gull' was the 'commonest gull in the whole archipelago and universally distributed'. In Boyson (1924), he also stated that it was present all the year round and 'fairly distributed over the whole archipelago'. Brooks (1917) stated that the Kelp Gull was an abundant scavenger and he saw large flocks near slaughter houses in Stanley. According to Bennett (1926), it was 'numerous everywhere'. Cobb (1933) described the 'Big Gull' as the worst enemy among birds, of the sheep farmer; it was 'an excellent scavenger, but when he attacks live sheep and lambs, one forgets his good points,' and he commented on, 'their vast numbers'. More recently, Cawkell & Hamilton (1961) described it as 'one of the commonest birds ... especially near settlements'. They felt that it was a useful scavenger in a country 'where some 80,000 sheep a year die from natural causes'. There is little doubt that the Kelp Gull has benefited greatly from the sheep farming industry.

Over 1,300 full-grown birds were ringed between 1959 and 1964. Local recoveries showed movements across the Falklands of up to 190km (120 miles), regular use of winter feeding areas and the probability that pairs remain together throughout the year. The lack of foreign recoveries suggests that it is resident. Maximum longevity recorded was between 20 and 24 years (Woods 1988).

Habitat and Breeding

The Kelp Gull occurs on lowlying coasts with offshore kelp beds, around settlements and inland. It nests during the first half of December on low coastal ground in colonies of up to several hundred pairs, though isolated nests may be found. Nests are bulky, untidy collections of grass, diddle-dee and dry kelp, irregularly spaced and rarely less than 2m (6 feet) apart. It lays 2-3 eggs, very variable in colour from olive-green to blue-grey or rich buff heavily spotted and blotched with dark brown. Colonies disperse in autumn (early March) and flocks gather wherever food is plentiful in winter. It occurs regularly offshore, particularly around fishing fleets.

Breeding Birds Survey: analysis of data

The high number of grid squares with only presence recorded and the low proportion where breeding was confirmed illustrate its wide distribution and a tendency to gather in large groups for breeding. Comparison between the

49 Kelp Gull *Larus dominicanus*

Detectability: **5 Very High**

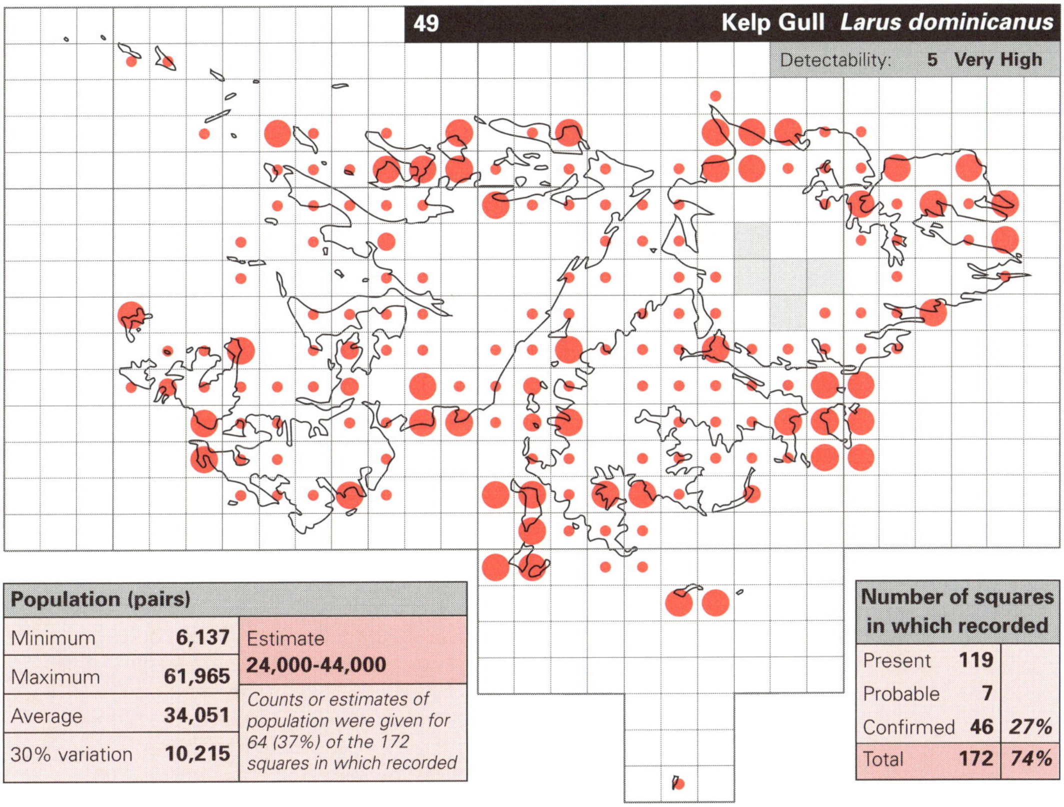

Population (pairs)		
Minimum	**6,137**	Estimate **24,000-44,000**
Maximum	**61,965**	
Average	**34,051**	*Counts or estimates of population were given for 64 (37%) of the 172 squares in which recorded*
30% variation	**10,215**	

Number of squares in which recorded		
Present	**119**	
Probable	**7**	
Confirmed	**46**	***27%***
Total	**172**	***74%***

Comparisons between species records and topography	Dry/Wet Number of wet tetrads					Inland/Coastal Amount of coast			Upland/Lowland Proportion of square > 500′				
	Nil	1-5	6-10	11-15	16-22	None	<10%	≥10%	≥1/2	≥1/4	<1/4	None	Total
Surveyed squares	38	104	52	24	16	17	14	203	7	25	73	129	234
% of surveyed squares	16	45	22	10	7	7	6	87	3	11	31	55	100
Grid squares in which recorded	19	82	40	19	12	4	6	162	2	15	54	101	172
% records in each category	11	48	23	11	7	2	4	94	1	9	31	59	100
Significant variation								*					

distribution of records and the topographical parameters shows a significant bias (p = 0·04) towards more records from coastal squares. Other comparisons show no significant results. Several observers commented on the large numbers (up to at least 1,000) which gathered at rubbish tips, including the comparatively new ones in the Mount Pleasant area.

World Distribution

The Kelp Gull has a circumpolar distribution in coastal regions of the southern hemisphere. In South America it occurs from 6°S in Peru and 23°S in Brazil to Tierra del Fuego. It also breeds at subantarctic islands and on the Antarctic Peninsula to about 68°S, in New Zealand, southern Australia and South Africa.

Implications for Conservation

This species is a successful opportunistic scavenger and conservation measures are unlikely to be necessary. The development of the fishing industry has broadened its available diet but it also takes any eggs that it can find and preys on ducklings and goslings. Its status as a pest under the Wild Animals and Birds Protection Ordinance of 1964 is understandable but not wholly justified because it performs a very useful function in disposing of refuse.

References

TBFI (1975) p164-9, GBFI (1988) p193-5

50 Brown-hooded Gull

Larus maculipennis

A. Gaviota cabecicafé

C. Gaviota cahuil

Local name: Pink-breasted Gull

History
Lichtenstein was the first to name this species scientifically as *L. maculipennis*, in 1823, from a specimen obtained at Montevideo, Uruguay. In 1934, Laubmann reconsidered its affinities and decided that it should be classified as a race of the European *L. ridibundus* (Hellmayr & Conover 1948). Hellmayr noted that Falkland birds were slightly smaller than specimens from the mainland but he considered the difference too slight to justify recognition of a separate race. However, later authors including Murphy (1936) found this South American gull sufficiently different to be considered as a separate species, the Brown-hooded Gull.

Garnot (1826) mentioned only one species of small gull, which he called *L. ridibundus*. Gould (1859), 'with a degree of hesitation', described a specimen sent by Abbott as a new species *Gavia roseiventris* but Abbott (1861) later decided that this was an immature in first year plumage. He thought that the species was migratory, 'arriving in East-Falkland Harbour about July 25th, almost to a day, though occasional stragglers occur all the year round'. Saunders (1877) understood that it was a 'well-known breeder' in the Falklands, yet Vallentin (1904) believed that it was 'very locally distributed' and had not seen it. Between 1909 and 1911, he saw it occasionally on West Falkland, reported that he had eggs from the Jason Islands and Dunnose Head, but made no other comments on distribution or numbers (Vallentin in Boyson, 1924). In 1926 Bennett stated that it was a resident but described it as 'not very common'. Cawkell & Hamilton (1961) agreed that it was not common; there were 'probably less than half-a-dozen breeding places, the sites of which tend to vary from year to year'.

Habitat and Breeding
Its favoured feeding habitats are offshore kelp beds where it takes small fish and shrimps. It also feeds at the edges of sandy ocean beaches when swarms of krill have been stranded. It does not normally come to chicken runs to take food but parties sometimes show the fearlessness for which they are known in South America, by gathering at a slaughter house at killing time. It breeds in colonies of up to about 200 pairs in a variety of coastal or inland sites, by fresh or salt water and often with a Dolphin Gull or South American Tern colony. Nests are usually close together, made of dry grasses and hold 2-3, occasionally 4, olive or buff eggs, heavily blotched with brown. It seems to lay later than other gulls, from the third week of December to early January. There is now no evidence that the Falkland population is migratory.

Breeding Birds Survey: analysis of data
Population estimates suggest that this gull is much less common than the other two breeding species in the family. Cawkell & Hamilton's statement about few breeding places is supported by the survey; breeding was 'probable' or 'confirmed' in only 20% of the grid squares in which it was recorded. Slightly more than half of all records came from East Falkland and there was a strong bias for more confirmed breeding records than from West Falkland. The analyses show only a significant bias ($p = 0{\cdot}02$) towards presence in squares described as coastal. During the period of the survey, Bourne & Curtis (1987) observed birds over kelp in inlets in summer, 'but flocks sometimes running into hundreds also fed with shags and penguins over tide-rips at the north end of Falkland Sound and in Port William from February to April, and 12 were feeding in a similar situation off Cape Dolphin on 27 April 1985'. Up to ten were seen around Stanley rubbish tip by John Cliffe on 10 July 1984. Phil Gregory observed a few along Stanley waterfront in September and October 1989 and 1990 and Nicholas Beswick saw up to 20 at Bertha's Beach in October 1992. In 1987, several reports also showed that noticeable flocks occur in the autumn; Tim Duggleby saw more than 200 in front of Stanley hospital in early March and later that month, Harro Müller noted up to 250 over the kelp north of Sea Lion Island.

World Distribution
The Brown-hooded Gull is widely distributed in southern South America, breeding on coasts and inland from about 40°S on the coast of Chile and

50 Brown-hooded Gull *Larus maculipennis*

Detectability: **3 Moderate**

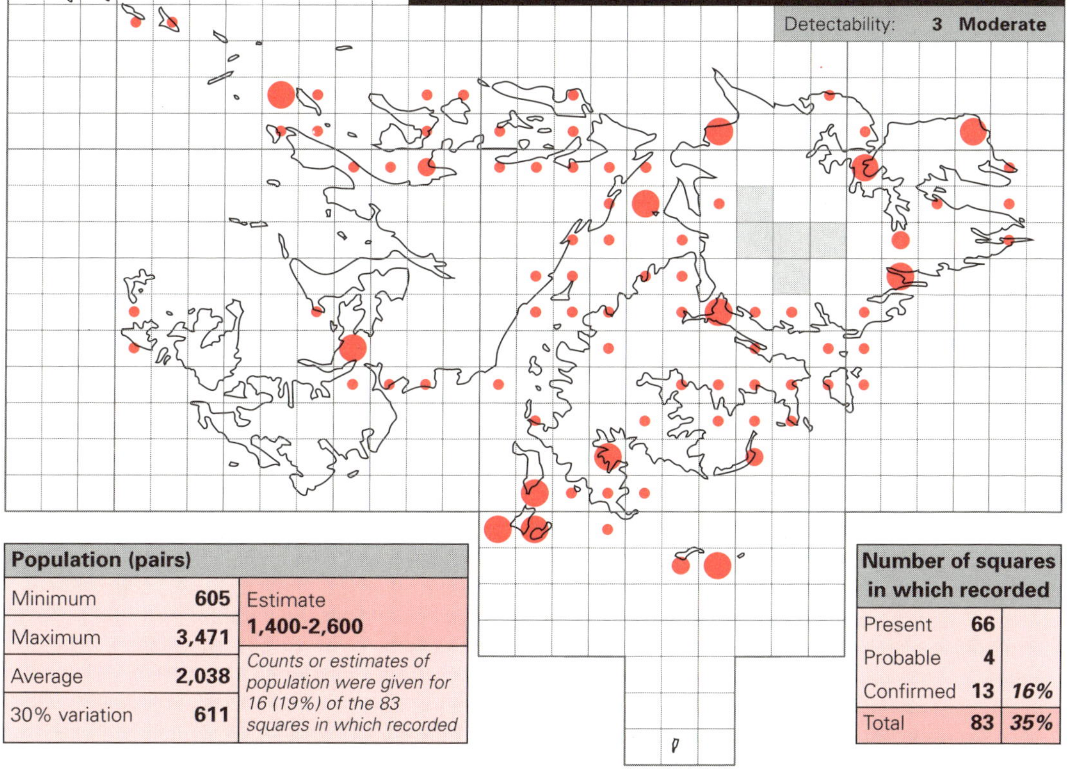

Population (pairs)			
Minimum	**605**	Estimate	**1,400-2,600**
Maximum	**3,471**		
Average	**2,038**	*Counts or estimates of population were given for 16 (19%) of the 83 squares in which recorded*	
30% variation	**611**		

Number of squares in which recorded		
Present	**66**	
Probable	**4**	
Confirmed	**13**	***16%***
Total	**83**	***35%***

Comparisons between species records and topography	Dry/Wet Number of wet tetrads					Inland/Coastal Amount of coast			Upland/Lowland Proportion of square > 500'				
	Nil	**1-5**	**6-10**	**11-15**	**16-22**	**None**	**<10%**	**≥10%**	**≥1/2**	**≥1/4**	**<1/4**	**None**	**Total**
Surveyed squares	38	104	52	24	16	17	14	203	7	25	73	129	234
% of surveyed squares	16	45	22	10	7	7	6	87	3	11	31	55	100
Grid squares in which recorded	6	45	18	10	4	0	3	80	1	4	25	53	83
% records in each category	7	54	22	12	5	0	4	96	1	5	30	64	100
Significant variation								*					

33°S in Uruguay, through lowland Argentina to southern Tierra del Fuego. The southernmost population leaves the Beagle Channel region in winter, when many migrate northwards to reach the coast of Brazil (10°S) and northern Chile (18°S), possibly occurring north to Peru.

Implications for Conservation

Although the population is small, there are no signs that it is endangered. Breeding colonies are thought to be susceptible to disturbance and have been known to suffer predation from rats. They are therefore more likely to be successful if they nest on islands without introduced predators. Their dependence on marine organisms means that any serious pollution of coastal waters would have a detrimental effect.

References

TBFI (1975) p169-73, GBFI (1988) p196-7.

51 South American Tern

Sterna hirundinacea

A. Gaviotín colilargo

C. Gaviotín sudamericano

Local names: Cassin's Tern, Sea Swallow, Split-tail, Swallow-tailed Gull

History
Garnot wrote about this species in 1822 at Port Louis, although he did not separate it from the similar Common Tern *S. hirundo* of the northern hemisphere. He found nests but made no comment on numbers of birds. Sclater (1860) remarked that Captains Pack and Abbott had both supplied specimens and he named the species *S. cassinii* in honour of John Cassin, one of America's foremost ornithologists of the nineteenth century (Mearns & Mearns 1992). However, this species had already been named as *S. hirundinacea* by Lesson from a Brazilian specimen in 1831 (Hellmayr & Conover 1948). Abbott (1861) stated that it arrived in East Falkland at the end of July, bred in communities on the sea beach, occasionally inland, and disappeared about the end of March. Vallentin (1904) said that it was very locally distributed in the Falklands and later (in Boyson 1924) he described it as migratory, leaving the islands in March and returning early in October. Bennett (1926) described it as a summer migrant, breeding in the Falklands and, mistakenly, at the South Shetlands, South Orkneys and South Georgia (where it is replaced by the Antarctic Tern *S. vittata*). Cawkell & Hamilton (1961) noted that it was common between October and April and that the sites of colonies, sometimes with hundreds of nests, tended to vary from year to year.

Habitat and Breeding
The South American Tern generally returns to inshore waters by late September and leaves by early April. It occurs around all Falkland coasts, over kelp beds and in the lower reaches of larger rivers. Breeding colonies, which are often near those of the three species of gulls, have been found in a variety of sites, mostly on shingle or rocky beaches and peninsulas or on isolated rock points sheltered by tussac grass or thrift. Colonies also occur at freshwater ponds inland. Nests are mere scrapes on beaches but are lined with grasses where available. Two eggs are laid, very variable in shape and colour, ranging from buff to pale blue or rich olive-brown, variably speckled and blotched with brown and smaller than those of the Brown-hooded Gull. Eggs have been found between mid-November and late December and fledged young may be seen from late January.

Breeding Birds Survey: analysis of data
Comparison between the distribution of records and the Inland/Coastal analysis shows a significant bias ($p = 0{\cdot}04$) towards occurrence in grid squares with considerable coastline. However, breeding was recorded in two inland squares containing the northern part of Lake Sulivan (TC85) and the Hawks Nest ponds (TC95) in central West Falkland. The population estimate suggests that it is about twice as numerous as the Dolphin Gull. Reports from observers have added valuable information. Simon Strickland saw small groups in San Carlos waters (UC59) from late August 1983 and Phil Gregory's observations of many in Stanley harbour on 21 October 1989 suggested an influx of migrants. The earliest record of a pair holding territory came from Nicholas Beswick at East Cove on 12 November 1992. These records suggest the possibility of a greater variation in arrival dates than was previously thought. The largest breeding colony reported was of 400 pairs observed by Norma and Roger Edwards in the 1986/87 season at Lake Sulivan South and reported by Donna O'Daniel. Autumn records included about 350 roosting on a sandy reef on Keppel Island, seen by Harro Müller in mid-March 1987. He also counted 312 migrating across the northwestern end of West Point Island between 1400 and 1530 hours on 24 March 1987.

World Distribution
The South American Tern is a common breeding species on both coasts of South America from about 14°S in Peru and 23°S in Brazil to Tierra del Fuego and Cape Horn. It ranges north in winter to about 5°S off Peru and 15°S off Brazil, though some are present in Falkland and Magellanic waters throughout the year.

51 South American Tern *Sterna hirundinacea*

Detectability: 5 **Very High**

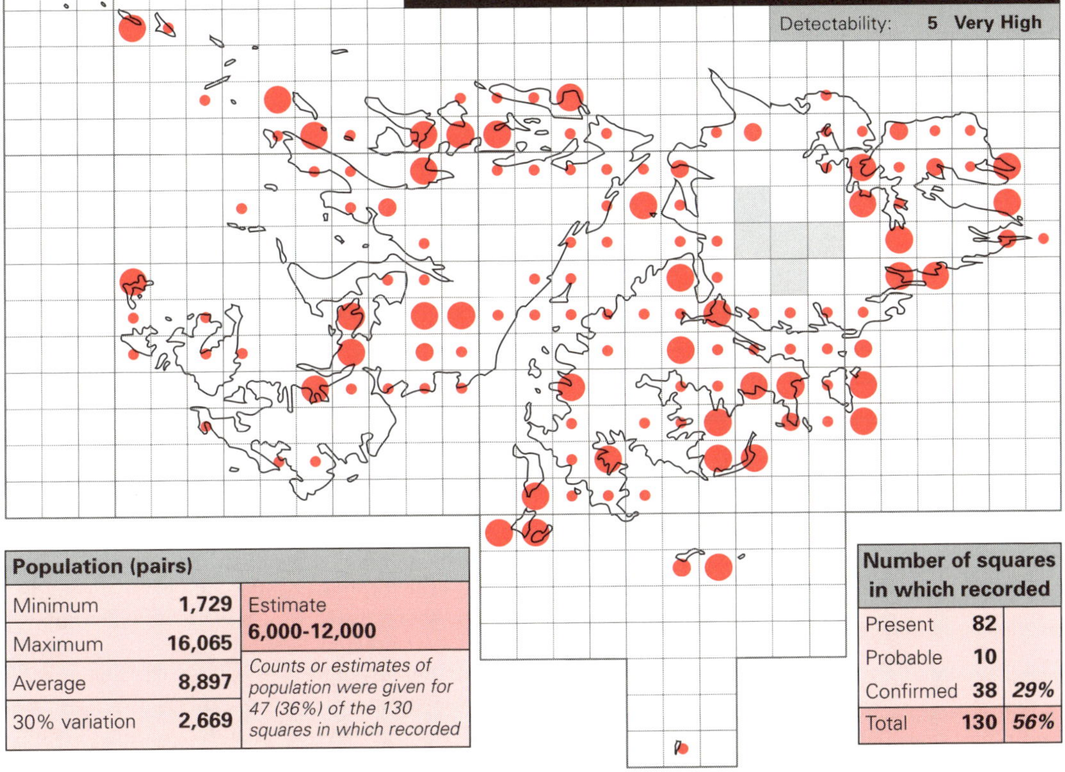

Population (pairs)		
Minimum	**1,729**	Estimate **6,000-12,000**
Maximum	**16,065**	
Average	**8,897**	*Counts or estimates of population were given for 47 (36%) of the 130 squares in which recorded*
30% variation	**2,669**	

Number of squares in which recorded		
Present	**82**	
Probable	**10**	
Confirmed	**38**	***29%***
Total	**130**	***56%***

Comparisons between species records and topography	Dry/Wet Number of wet tetrads					Inland/Coastal Amount of coast			Upland/Lowland Proportion of square > 500′				
	Nil	**1-5**	**6-10**	**11-15**	**16-22**	**None**	**<10%**	**≥10%**	**≥1/2**	**≥1/4**	**<1/4**	**None**	**Total**
Surveyed squares	38	104	52	24	16	17	14	203	7	25	73	129	234
% of surveyed squares	16	45	22	10	7	7	6	87	3	11	31	55	100
Grid squares in which recorded	14	58	29	18	11	2	5	123	1	11	46	72	130
% records in each category	11	45	22	14	8	1	4	95	<1	9	35	55	100
Significant variation								*					

Implications for Conservation

Although colonies are widely distributed around coasts and at inland lakes, the main items of food are not known in detail nor whether diet varies in different parts of the archipelago. The population seems large enough to be resilient to temporary adverse factors but commercial fishing and industrial pollution may prove harmful in the long-term. Terns are known to be susceptible to disturbance when nesting and visits to breeding sites by tourists should be carefully managed.

References
TBFI (1975) p173-4, GBFI (1988) p200.

52 Barn Owl

Tyto alba tuidara

A. Lechuza de campanarios

C. Lechuza

Local name: White Owl

History

Confusingly, Wace (1917) listed the 'Red-foot Owl *Stryx rufipes*' as breeding, noted specimens in the local collection (destroyed by fire in 1944) and gave it the local name 'Barn Owl'. Bennett (1926) notes under *Strix rufipes* Red-footed Owl, 'a rare bird which perhaps breeds in the Falklands. The feet of the local bird are not red but a pale brown'. On 14 September 1937, Bennett collected the corpse of a bird that hit Cape Pembroke Lighthouse. When he gave it to the Natural History Museum, London it was labelled "Rufous-legged Owl *Strix rufipes*" but this specimen was later found to be a Barn Owl. Bennett's unpublished notes show that he discovered the mistake in identification before 1938. There had been rumours of 'white owls' breeding in Port Howard camp at Shag Cove and near Port Purvis and several records of individuals seen, caught or found dead between the 1940s and the early 1980s. Birds were reported from Johnson's Harbour, George Island, Carcass Island, Port Stephens, Chartres, Moody Brook near Stanley and West Point Island and ten pellets containing the remains of mice were found at Limpet Creek, Salvador (Cawkell & Hamilton 1961, Woods 1988).

During and since the survey more earlier records have come to light. Ronnie Larsen reported that a Barn Owl was seen in the old shearing shed at North Arm for at least two months in 1949 and one caught in a garage at Port Louis on 9 October 1972, was reported by the children to David Carstairs (pers. comm.) and later released. In addition to the places named above, several records have come from the northeast corner of East Falkland. One was seen at Salvador by Robin Pitaluga, and at Rincon Grande, flash photos taken at a barbecue revealed a Barn Owl sitting on a fence post. Single birds, or more than one, seen at Limpet Creek house, Port Louis, Johnson's Harbour, Green Patch, Douglas, Estancia, and Riverview (The Malo) were reported by Denzil Clausen, Isobel McLeod, Sue and Mike Morrison and Clara and Bill MacKay.

The circumstantial evidence for resident and breeding Barn Owls was strong but before the survey, breeding in the Falklands had not been confirmed.

Habitat and Breeding

In Europe, the Barn Owl often breeds in old buildings with easy access and ample space around the nest site, but where there are no buildings, it will nest on cliffs or in caves. In the Falklands, introduced European gorse appears to be important and at most of the localities mentioned, gorse hedges, some of which were planted up to a century ago, are gradually increasing in extent and becoming the most dense and impenetrable, permanent shrubby cover. No nest is built, though a slight hollow may be made. Clutch size in the Falklands is unknown but may be small; in other areas most clutches are of 4-7 smooth, white elliptical eggs. More than one brood in a season can be reared, but this is unusual and depends on plentiful food or the first clutch being laid early in the season (Cramp 1985).

Breeding Birds Survey: analysis of data

It was recorded in eight coastal grid squares but there were insufficient records for comparison with the analyses of topographical parameters. The first record of breeding came from Seal Bay on the north coast of East Falkland, 12·5km (8 miles) from the nearest settlement at Johnson's Harbour in 1987/88. Mike Morrison photographed an adult roosting in an old stable and heard young calling from the nest in thick old gorse growing 3m (10 feet) high on a peat sod corral. A dead fledgling was seen there by Shane Wolsey in October 1988. In the 1988/89 season, Mike Riddy heard about an adult carrying food to a probable nest under dense gorse growing over an old sod wall at Port Edgar and of a bird seen flying at Fox Bay East settlement. Also in that season Mike Morrison reported a nesting pair at Loch Head House in the North Camp, not far from Seal Bay.

There have been several records of birds in the Moody Brook area, west of Stanley. Stuart Booth saw one in the spring of 1987/88 and later found a pair nesting in the concrete rafters of the generating hall at the old barracks. In late 1988, Phil Gregory reported a pair and two juveniles there and he saw four birds in May 1989. In June 1989, Kate and Robin Bayliss saw a family group of four around Stanley. The building was partly demolished in December as it was unsafe and the

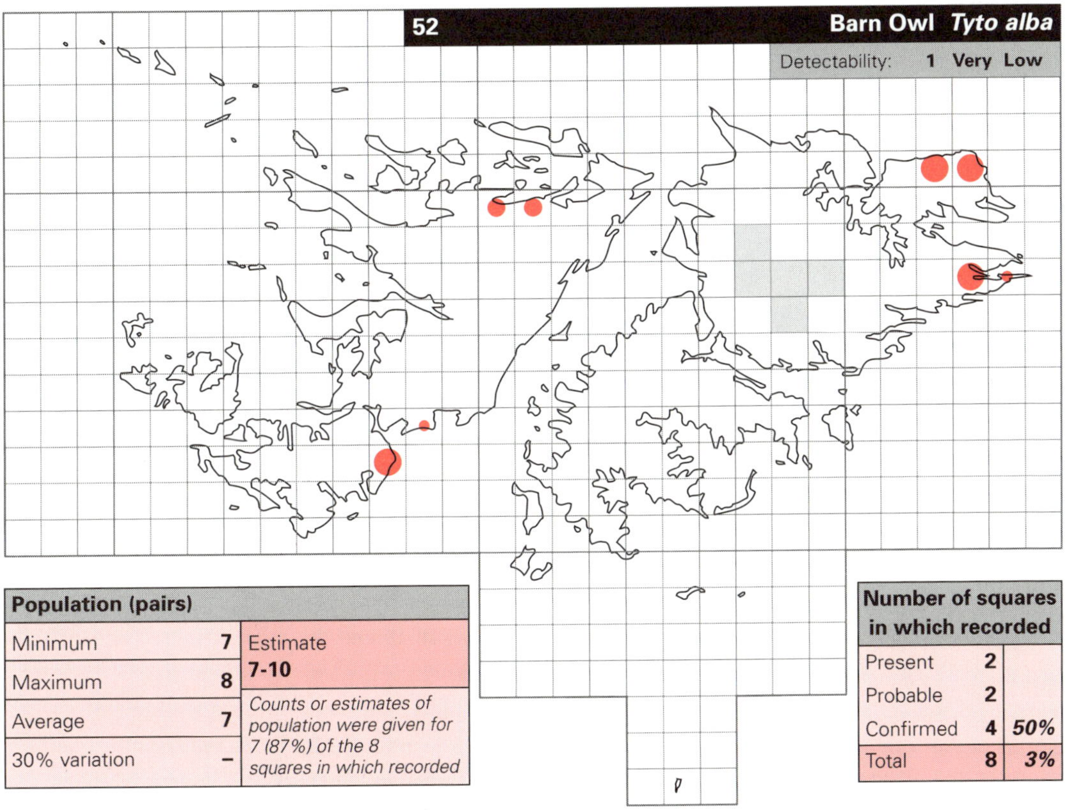

Population (pairs)		
Minimum	7	Estimate **7-10**
Maximum	8	
Average	7	*Counts or estimates of population were given for 7 (87%) of the 8 squares in which recorded*
30% variation	–	

Number of squares in which recorded		
Present	2	
Probable	2	
Confirmed	4	***50%***
Total	8	***3%***

work was completed in July 1990, depriving the Barn Owls of their nesting site. One bird was seen by Alan Henry near the market garden in August 1990 and one was found dead in the town in June 1991 (Abbott 1991). In 1989/90, Phil Gregory had an unconfirmed report of a pair at Port Stephens and a rumour of others in Port Howard camp. In the same season, Bobby Tulloch reported that Roy Dennis saw an adult at the western Turkey Rocks on 12 December, about 21km (13 miles) northwest of Port Howard (UC09) and in January, Bobby saw an adult sitting deep in a high crevice of the eastern Turkey Rocks (UC19), over one kilometre from the earlier sighting. He found many pellets and considered it likely that the birds were nesting there. Since the survey, Philip Hutton has reported seeing one or two birds at three of the shepherds' houses outside North Arm settlement before about 1990.

Sue Morrison suggested that the pair nesting in a gorse patch at Port Louis, probably in 1990/91, reared two broods in the season as young birds were calling from the nest site in September and the following February. Thirty-eight owl pellets collected from the Seal Bay site in December 1995 contained the remains of 132 mice, three rats and nine songbirds. Three long-dead Barn Owls were found in the 1995/96 season; one on the sand dunes behind Elephant Bay, Pebble Island, another beside the sod and gorse bank of the old corral at Salvador by Nick Pitaluga and a third in the old stable at Darwin by Sonia Felton. It is possible that all died from starvation because they were unable to find prey during the hardest Falkland winter in living memory.

World Distribution

About 36 races of the Barn Owl have been described. It is found throughout the world except in African deserts, parts of central Asia and northern Eurasia and North America. The race breeding in the Falklands appears to be *T. a. tuidara,* which inhabits South America from Guyana and Peru to Tierra del Fuego.

Implications for Conservation

Gorse thickets provide safe shelter for nesting owls, while introduced mice and rats are probably more numerous at settlements than in the surrounding camp. The very small population of Barn Owls is vulnerable to destruction of nest sites by fire or demolition and is likely to be intolerant of disturbance near nests. As one of the rarest breeding species in the Falklands, detailed information on the distribution and number of pairs and their preferred habitats and prey is necessary so that effective conservation measures can be taken.

References

TBFI (1975) p175-6, GBFI (1988) p204-5.

53 Short-eared Owl

Asio flammeus sanfordi

A. Lechuzón campestre

C. Nuco

Local name: Owl

History

Clayton (1774), on Saunders Island, remarked that there was only one sort of owl 'and that exactly like the European ones'. At Port Louis, Garnot (1826) noted an owl with a short crest and with plumage like 'the owl of Europe'. Darwin (Gould & Darwin 1838-41) obtained specimens, said that it 'harboured amongst low bushes' and flew readily in the day. The Antarctic Expedition of James Ross on HMS *Erebus* and *Terror* collected three specimens in the Berkeley Sound area in 1842 and Captains Kellett and Wood (of HMS *Herald* and *Pandora* respectively) collected another specimen in 1845 which was also given to the Natural History Museum in London. Abbott (1861) said that the owl came around the houses at Port Louis at night when searching for mice, though he had never seen one near Stanley. He was informed that it bred in the long grass but had not found a nest and he thought it was scarce in East Falkland. Bangs (1919) examined specimens collected in the Falklands in 1915 and found that they were smaller and generally darker than those from southern South America. He named the Falkland race after Leonard Sanford, one of the sponsors of the Brewster-Sanford Expedition (1912-1917) of the American Museum of Natural History, during which Rollo Beck collected around the coasts and islands from Peru to Cape Horn. Wace (1917) noted that it bred and Vallentin (in Boyson 1924) saw only one, which he shot after flushing it from some fachine bushes. Bennett (1926) described it as 'a common resident' while Cawkell and Hamilton (1961) stated that it was 'not uncommon' and that it preferred tussac islands for breeding. They also said that it was 'very partial to the smaller petrels and accessibility to a supply of them is probably a factor in its choice of residence', and that it had been seen in Stanley in the winter, possibly attracted by the large population of rats.

Habitat and Breeding

The Short-eared Owl is found on some of the larger islands and on tussac islands where it usually breeds. On Kidney Island, examination of pellets and corpses showed that it depended on Grey-backed Storm-Petrels, Common Diving Petrels and Camel-crickets, all of which were apparently caught at night, and Tussacbirds. Though there are few nest records, those available suggest that it nests on the ground in the shelter of thick grass or rushes and probably lays 2 rounded, white eggs. Nestlings have been heard calling in mid-November suggesting that eggs are laid from early October. Breeding adults can be very aggressive and may swoop on human intruders.

Breeding Birds Survey: analysis of data

The population estimate suggests that it is one of the least common landbirds and the map shows a scattered distribution with roughly twice as many records from the eastern half of the archipelago. The proportion of confirmed breeding records is low. The preference for offshore islands with tussac grass when nesting, and its tendency to hunt at night, make it a difficult species to locate and the population may be higher than shown. There was a highly significant bias ($p = 0{\cdot}0013$) towards grid squares with 6-15 wet tetrads and fewer records from squares with no standing water. This bias in distribution may be associated with the greater availability of ground cover for nest sites in wetter areas and the probable presence of more species of small birds as potential prey.

World Distribution

About eight races of the Short-eared Owl have been described, occurring mainly in the northern hemisphere from temperate to arctic regions, but also in the Hawaiian and Caroline Islands and South America. The race *A. f. suinda* is fairly common in open country from Venezuela to Peru and from southern Brazil through Argentina and Chile to northern Tierra del Fuego. Continental races mainly hunt at dawn and dusk (Fjeldså & Krabbe 1990).

53 * Short-eared Owl *Asio flammeus*

Detectability: **3 Moderate**

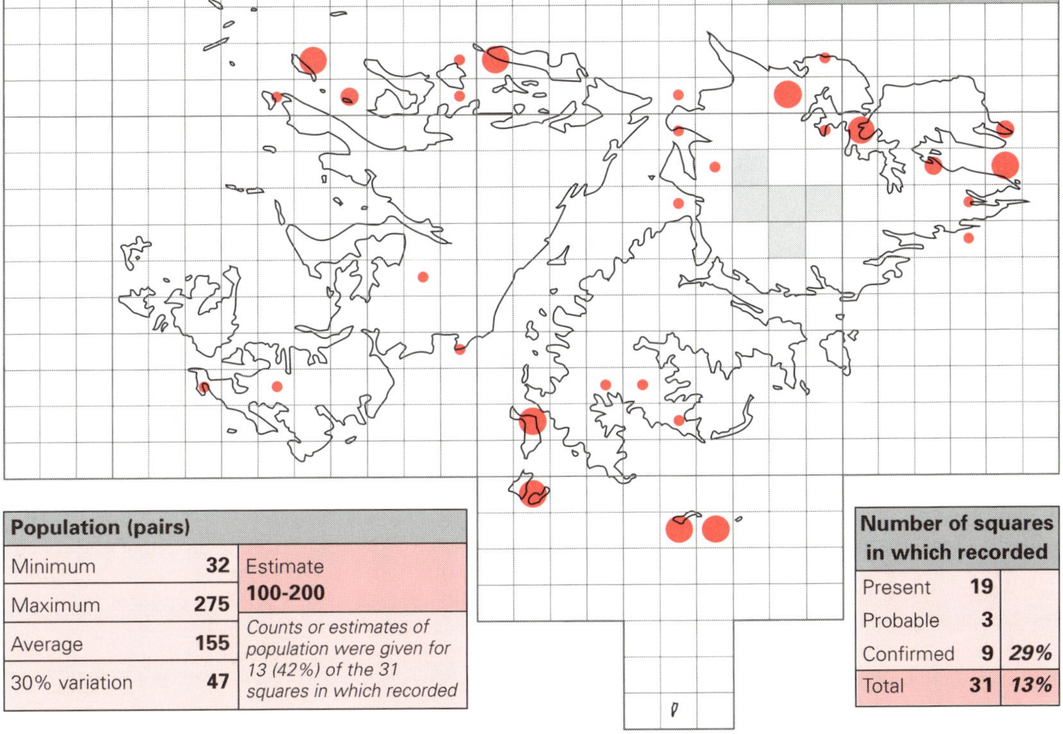

Population (pairs)		
Minimum	32	Estimate **100-200**
Maximum	275	
Average	155	*Counts or estimates of population were given for 13 (42%) of the 31 squares in which recorded*
30% variation	47	

Number of squares in which recorded		
Present	19	
Probable	3	
Confirmed	9	*29%*
Total	31	*13%*

Comparisons between species records and topography	Dry/Wet: Number of wet tetrads					Inland/Coastal: Amount of coast			Upland/Lowland: Proportion of square > 500′				
	Nil	1-5	6-10	11-15	16-22	None	<10%	≥10%	≥1/2	≥1/4	<1/4	None	Total
Surveyed squares	38	104	52	24	16	17	14	203	7	25	73	129	234
% of surveyed squares	16	45	22	10	7	7	6	87	3	11	31	55	100
Grid squares in which recorded	2	14	9	5	1	1	3	27	0	3	12	16	31
% records in each category	7	45	29	16	3	3	10	87	0	10	39	51	100
Significant variation			**	**									

Implications for Conservation

Tussac islands and paddocks with replanted or conserved tussac are very important for the Falkland race of the Short-eared Owl because they provide suitable dense vegetation for nesting and sufficient prey. It is not known whether introduced rats and mice are significant in its diet, but they may aid its survival during the winter when petrels have migrated and where native birds have declined through destruction of tussac, burning of grassland and erosion.

References

TBFI (1975) p177-9, GBFI (1988) p206-7.

54 Tussacbird

Cinclodes antarcticus antarcticus

A. Remolinera negra

C. Churrete austral

Local name: Black Bird

History

The first clear reference to the Tussacbird was made in 1764 at Port Louis by Pernety (1771). He remarked that there were few birds 'found upon land' but was impressed by the flock of birds upon the shore 'like small thrushes, of a brown grey, so tame that they came flying almost upon your finger' and they 'scratch among the seagrass which is thrown upon shore and eat the worms and small shrimps'. He said that he had killed ten in less than half an hour with a small switch, without changing his place. Clayton (1774) on Saunders Island described one or two 'kinds of little birds ... fond of coming in and about our houses like the Robins in England'. The Tussacbird was almost certainly one of these. Garnot (1826) was the first to name it scientifically, as *Certhia antarctica* (Antarctic Creeper) and he noted that it was difficult to distinguish among the dark rocks on the seashore.

Darwin remarked that it had 'long been noticed by voyagers to the Falkland Islands, for its extreme tameness' (Gould & Darwin 1838-41) and Sclater (1860) guessed that it was probably restricted to the Falklands. Abbott (1861) noted that it 'is not found except on the coast' and that he had only seen it on Kidney Island where, 'it is very abundant, and breeds among the Tussac, but I have never seen the eggs or obtained the nest'. Cobb (1910) said that it was very tame, friendly and inquisitive and that it flourished 'on islands where there are no cats or rats'. Brooks (1917) noted its tameness and found it abundant near the shore at Speedwell Island. He thought it was rare on the mainland of East and West Falkland and that it preferred outlying tussac islands. Bennett's comments (1926) confirmed Brooks' observations. Cobb (1933) added a few other notes on its inquisitive nature and modified his 1910 remark, saying that it did not usually 'live on islands where there are rats'. Cawkell & Hamilton (1961) stated that it was 'found exclusively on islands where there is a growth of tussac grass and is not to be seen on either of the main islands where this grass has, for all practical purposes, long been extinct'.

Habitat and Breeding

The Tussacbird is typical of boulder and sand beaches where dead kelp accumulates and dense tussac grass provides cover and soft ground where it can nest in holes, usually just below ground level. It is not restricted to tussac islands, however, and is able to take advantage of artificial cover such as sheds and walls. On Kidney Island between 1959 and 1963, a ringing project showed that pairs tended to remain in territory between October and mid-April. Of 41 birds colour ringed, four reached three and a half years and two survived at least five years. Field work in 1983 showed a population density in mixed tussac (40%) and grazed grass (60%) habitat on Carcass Island twice as high as that achieved in the coastal strip of Kidney Island, which was covered with dense mature tussac. It also occurred up to at least 500m (550 yards) inland and to a height of 200m (650 feet) on Carcass Island, where it was abundant (Woods 1984). It nests from early September to December with most starting to lay in October and often having second broods. The nest is a shallow cup of grasses lined with a few feathers, placed in a hole which may be under tussac grass, beneath loose rocks on a beach, under a shed or house, in an earth bank or in an old penguin or petrel burrow. It lays 1-3 slightly glossy white eggs, sometimes minutely spotted with red around the blunt end.

Breeding Birds Survey: analysis of data

The map shows concentrations of grid square records in groups of islands known to be free of rats, such as the Speedwell/George/Barren group, Lively, Sea Lion, Beauchêne, Kidney and Carcass Islands. The population estimate indicates that it is the most numerous passerine, nearly twice as common as the Long-tailed Meadowlark. The records are however, discontinuous around Falkland coasts, probably reflecting the presence of rats and cats on the two main islands. In the topographical analyses, there was a highly significant bias towards occurrence in squares with much coastline ($p = 0{\cdot}009$). There was also a strong, non-significant, bias towards more records from squares with no land above 500 feet ($p = 0{\cdot}05$).

54 #*Tussacbird *Cinclodes antarcticus*

Detectability: **5 Very High**

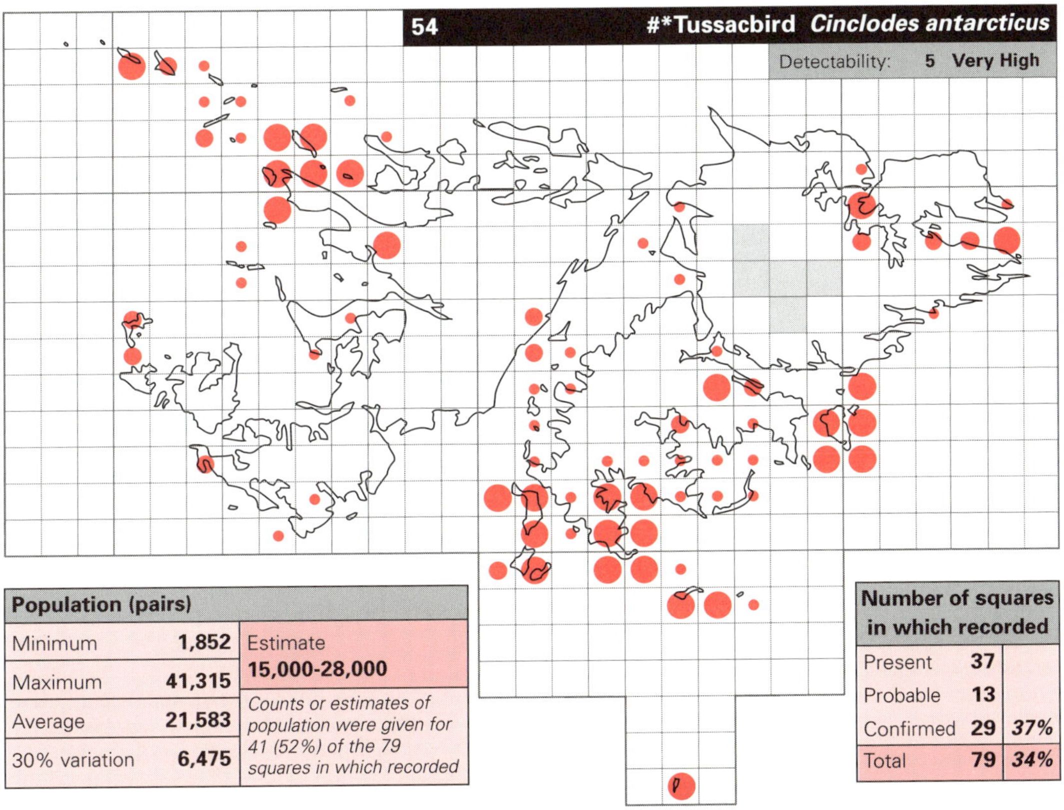

Population (pairs)		
Minimum	**1,852**	Estimate **15,000-28,000**
Maximum	**41,315**	
Average	**21,583**	*Counts or estimates of population were given for 41 (52%) of the 79 squares in which recorded*
30% variation	**6,475**	

Number of squares in which recorded		
Present	**37**	
Probable	**13**	
Confirmed	**29**	***37%***
Total	**79**	***34%***

Comparisons between species records and topography	Dry/Wet: Number of wet tetrads					Inland/Coastal: Amount of coast			Upland/Lowland: Proportion of square > 500′				
	Nil	**1-5**	**6-10**	**11-15**	**16-22**	**None**	**<10%**	**≥10%**	**≥1/2**	**≥1/4**	**<1/4**	**None**	**Total**
Surveyed squares	38	104	52	24	16	17	14	203	7	25	73	129	234
% of surveyed squares	16	45	22	10	7	7	6	87	3	11	31	55	100
Grid squares in which recorded	12	41	21	5	0	0	1	78	1	2	21	55	79
% records in each category	15	52	27	6	0	0	1	99	1	2	27	70	100
Significant variation								**				♦	

World Distribution

The Tussacbird or Blackish Cinclodes has a restricted distribution in extreme southern South America. Two races described are *C. a. antarcticus*, which is endemic to the Falklands and *C. a. maculirostris*, slightly larger with a heavier, shorter bill, which occurs only on coastlines of islands south of Tierra del Fuego, including Diego Ramirez (R Schlatter pers. comm.) and Cape Horn and on Staten Island (Ridgely & Tudor 1994).

Implications for Conservation

Formerly abundant around all coasts, the Tussacbird has become rare in the Falklands where cats and rats are present. Great care should be taken to guard against further introductions of potential mammalian competitors or predators. Its popularity with tourists is an important point in its favour. As Tussacbirds depend mostly on the invertebrate fauna of beached kelp for food, any commercial removal of kelp could be detrimental. They would also be at risk from any serious littoral pollution by oil or industrial waste.

References

TBFI (1975) p180-1, GBFI (1988) p210-1.

55 Dark-faced Ground-tyrant

Muscisaxicola macloviana macloviana

A. Dormilona cabeza parda

C. Dormilona tontita

Local names: News Bird, Blue Bird

History

Pernety (1771) mentioned a small bird, not often seen, 'like those that hover about flocks of sheep', which was probably the Ground-tyrant. Garnot (1826) was the first to describe it scientifically as *Sylvia macloviana* (Fauvette des Malouines = Falkland Warbler) which he said, 'perched on the shrubby trees'. Darwin noted that it was 'larger in all dimensions than *M. mentalis*' [of South America] and added that he could 'scarcely hesitate in thinking it a large-sized local variety, from some favourable condition in the Falkland Islands to its growth' (Gould & Darwin 1838-41). Sclater (1860) agreed with this opinion. Abbott (1861) said that it was 'not very common in East Falkland' and 'generally found near the shore'. In the breeding season 'it resorts to stone runs, or watercourses, where it breeds, no doubt, though I have never found its nest'. In contrast with Abbott, Vallentin (1904) felt that the 'Newsbird' was 'evenly distributed throughout the archipelago' and commented that he frequently saw 'numbers' ... 'along the Wickham Range of hills'. He noted that its food consisted 'chiefly of moths which it catches on the wing with great dexterity'. Brooks (1917) stated that it was common, with the majority found around stone runs, and he had found one nest with newly hatched young on 27 October, deep among the rocks. He thought that spiders were the 'chief food', though at Port Stephens a few were 'always feeding upon wingless flies on the beach'. Wace (1921) only reported that it bred in the Falklands and Bennett (1926), who called it Garnot's Ground-tyrant or Blue Bird, simply stated that it was 'a common breeding species'. Cory & Hellmayr (1927) noted that the seven Falkland specimens they had examined were so much larger than the mainland form 'its separation appears perfectly justified'.

Cawkell and Hamilton (1961) described it as a common resident, which was normally inconspicuous but called attention to itself by a faint callnote 'which carries remarkably far'. They said that the local name of News Bird came from the way the bird approaches people, giving the impression that it has come to hear or repeat news. On New Island, they observed that it 'falls an easy prey to the numerous cats and its remains are everywhere'. Pettingill (1974) found it more common about the settlement than elsewhere on New Island in 1953/54 and 1971/72. He considered that the vantage points for foraging provided by fence posts and the abundance of insects associated with the sheep, cattle, horses and poultry, made an ideal habitat. He believed it was resident and though many moved about the islands, there was no proof that any migrated to or from the Falklands.

Habitat and Breeding

It occurs inland at rock outcrops and stone runs and in all types of coastal areas throughout the islands but shows no dependence on the tussac habitat (Woods 1985). It nests between late October and late December, laying 2-3 slightly glossy white eggs closely spotted with red-brown and possibly has two broods in a season. The nest is of grass or root fibres well-lined with wool or feathers, placed up to 60cm (2 feet) below the surface of a stone run or in a deep rock crevice on a ridge or cliff.

Breeding Birds Survey: analysis of data

Records were obtained from nearly 75% of squares surveyed and there were no significant differences between the distribution of this species and the topographical parameters. It therefore appears to be well-adapted to most habitats in the Falklands. An important factor in its distribution may be the widespread occurrence of white grass which supports several moth species that are taken by the Ground-tyrant. In 1984 John Cliffe visited Saunders Island twice; in August, Ground-tyrants were seen mainly along the coast and in October they were common on inland heathland. Between September 1989 and January 1990, David Osborn also noted a decrease in numbers after September along the coast in the East Cove/Bertha's Beach area. In March 1987 on Beaver Island, Harro Müller was surprised to count 50 individuals because there is a substantial introduced population of Patagonian Grey Foxes on this island. There are two other interesting autumn records; Tony Godfrey noticed seven birds coming aboard at dawn in the northern approaches of Falkland Sound on 26 March 1985 (Bourne & Curtis 1987) and Brian Bewsher saw one bird on board a ship 71km (44 miles) south of Cape

55 * Dark-faced Ground-tyrant *Muscisaxicola macloviana*

Detectability: **5 Very High**

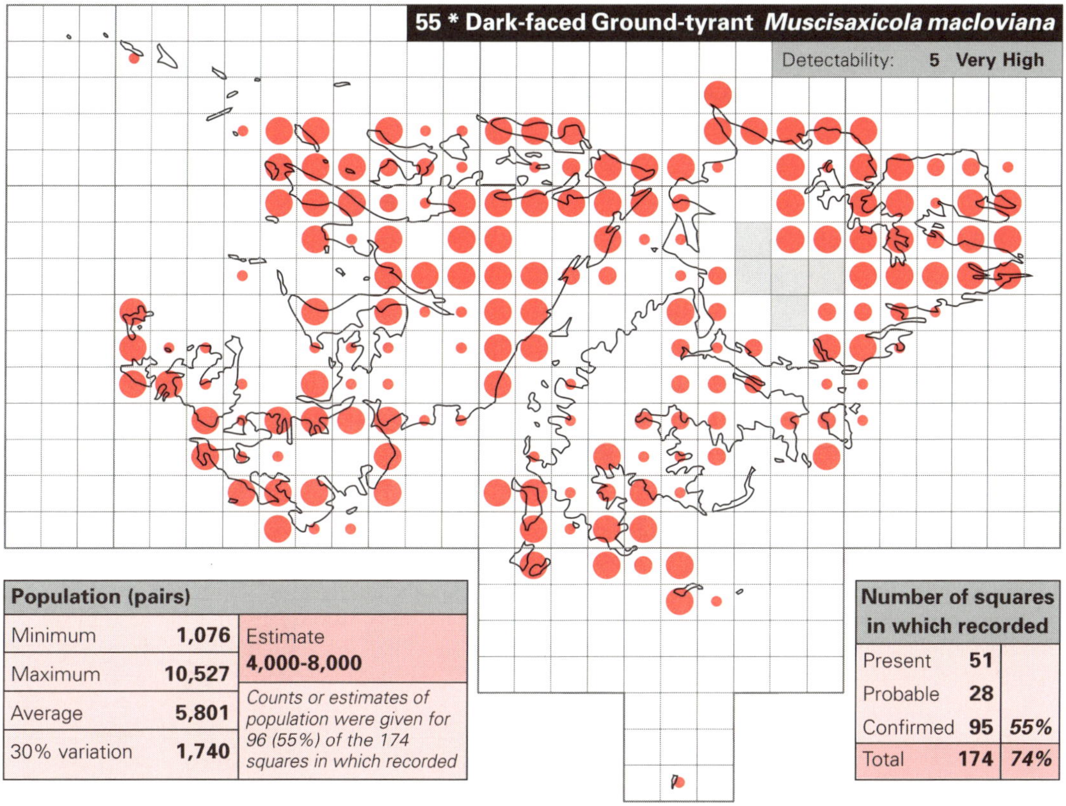

Population (pairs)		
Minimum	**1,076**	Estimate **4,000-8,000**
Maximum	**10,527**	
Average	**5,801**	*Counts or estimates of population were given for 96 (55%) of the 174 squares in which recorded*
30% variation	**1,740**	

Number of squares in which recorded		
Present	**51**	
Probable	**28**	
Confirmed	**95**	***55%***
Total	**174**	***74%***

Comparisons between species records and topography	Dry/Wet Number of wet tetrads					Inland/Coastal Amount of coast			Upland/Lowland Proportion of square > 500′				
	Nil	1-5	6-10	11-15	16-22	None	<10%	≥10%	≥1/2	≥1/4	<1/4	None	Total
Surveyed squares	38	104	52	24	16	17	14	203	7	25	73	129	234
% of surveyed squares	16	45	22	10	7	7	6	87	3	11	31	55	100
Grid squares in which recorded	23	80	43	19	9	12	7	155	5	21	62	86	174
% records in each category	13	46	25	11	5	7	4	89	3	12	36	49	100
Significant variation													

Meredith on 4 April 1984, which could have been wind drifted during migration from southern South America.

World Distribution

The endemic race of the Dark-faced Ground-tyrant *M. m. macloviana* is the only Falkland representative of the family Tyrannidae which has over 300 species in South America and about 120 in Argentina. The smaller race *M. m. mentalis* breeds in southern Chile from 42°S and Argentina from 30°S to Tierra del Fuego and Cape Horn. In winter some migrate north to 12°S in Peru and 22°S in northwestern Argentina, though part of the Magellanic population is resident (Venegas & Jory 1979). One bird, possibly of the migratory race *M. m. mentalis*, was seen and photographed at Bird Island, South Georgia on several days between 23 and 29 September 1994 (Prince & Croxall 1996).

Implications for Conservation

This species has taken advantage of increased insect populations around settlements, including introduced blowflies and it can co-exist with cats, rats and mice though cats apparently kill substantial numbers. With its adaptability to habitats from sea level to the higher rocky outcrops and crags it does not appear to be under any immediate threat.

References

TBFI (1975) p182-4, GBFI (1988) p215-6.

56 Chilean Swallow

Tachycineta leucopyga

A. Golondrina azul

C. Golondrina chilena

History

The first published record in the Falklands was of one shot from six or seven on 6 March 1937 at Cape Dolphin (Hamilton 1939). There is an earlier specimen in the Natural History Museum at Tring, caught on New Island by George Scott of the British Grahamland Expedition on 5 October 1936. No more were recorded until 1954 when one was seen on 24 February at Kidney Island, near Stanley (Pettingill 1960). Up to six occurred annually between 1958 and 1962 in Stanley between December and June, mostly from March. Don Davidson saw two at West Point Island in October 1976 and two survived there for over a month from March 1977.

Breeding Birds Survey: analysis of data

The first and only breeding record for the Falklands came from Port Stephens (TC32) where Peter Robertson found a pair nesting on the roof struts of a dutch barn about 6m (20 feet) above ground in the 1983/84 season. Eight other records between October and January were received, consisting of singles and parties of two to ten at New Island, Beaver Island, Beauchêne and Sea Lion Island, and at Bertha's Beach, Volunteer Point and Stanley. It now seems to occur almost annually in the Falklands, usually wind drifted on migration. From 1984 to 1991 there were over ten records between late February and April, including up to 30 birds at Sea Lion Island in the late summer of 1985. Brian Paul and Annabelle Spencer reported one on 24 February and three on the following day, looking thin and tired, hawking over a pond and above rotted kelp on a beach. Numbers increased to 20 or more by 8 March, when they looked much plumper. Only one was seen on 9 March, suggesting that after finding plentiful food on Sea Lion Island, they resumed migration. The map shows only records for the breeding season. There were insufficient records for any comparison with the analyses of topographical parameters though all reports came from squares with a considerable amount of coastline.

World Distribution

The Chilean Swallow is a common breeding species in southern Chile and Argentina from Tierra del Fuego to about 27°S in Chile and 37°S in Argentina where it nests in natural holes in trees, or under the eaves of buildings. Southern breeding populations are migratory, arriving in the Magellanic region between the last week of September and mid-October and departing northwards between mid-February and mid-March (Venegas & Jory 1979). Wintering birds have also been recorded in Peru and southeastern Brazil (Turner & Rose 1989) and there are two records from Antarctica; two birds were seen at South Georgia on 4 April 1982 and one specimen was collected at 67°17'S 50°30'W, 300km (160 nautical miles) southeast of Elephant Island, South Shetlands on 13 February 1985 (Prince & Croxall 1996).

Implications for Conservation

The flying insects which are its normal prey are generally scarce in the Falklands, except in settlements, around livestock and over dead kelp on beaches. No other breeding attempts have been reported but it is worth looking for signs of nesting at settlements in the spring if birds stay for several days. A larger number of suitable nest sites may become available as the trees planted in recent years come to maturity.

References

TBFI (1975) p184-5, GBFI (1988) p220-1.

56 Chilean Swallow *Tachycineta leucopyga*

Detectability: **3 Moderate**

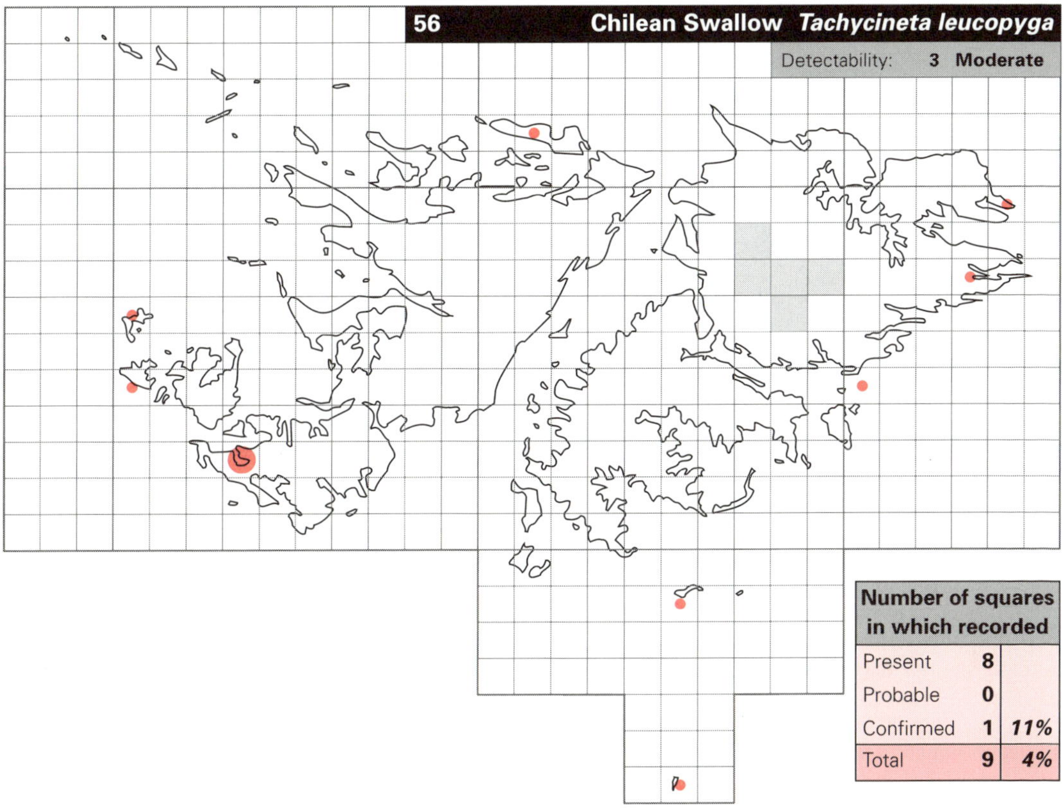

Number of squares in which recorded		
Present	**8**	
Probable	**0**	
Confirmed	**1**	***11%***
Total	**9**	***4%***

57 Falkland Pipit

Anthus correndera grayi

A. Cachirla común

C. Bailarín chico

Local names: Skylark, Lark

History

Captain Portlock noticed this pipit on Saunders Island in 1789 and called it the Cinereous Lark (Hellmayr 1935). Gould named it *A. correndera,* which was used by several later ornithologists (Gould & Darwin 1838-41). Darwin stated that it did not live in flocks but was very common. Abbott (1861) supposed that it left East Falkland about the end of April because he did not see any until the following September. Vallentin (in Boyson 1924) also considered it to be a migrant, 'arriving early in November and universally distributed over the islands' and Brooks (1917) recorded that it was abundant. Wace (1917) only mentioned it as breeding while Bennett (1926) said that it was a common breeding species and thought that some migrated for the winter. He also speculated that 'the resident form may differ from the migratory'. Cawkell & Hamilton (1961) stated that it rarely strayed away from the white grass and implied that it was common though 'less numerous in the winter'. Sewall Pettingill (1974), who studied passerines in 1953/54 and 1971/72 thought that it was much less common in mixed heathland where diddle-dee was extensive. In December 1980 Ron Lewis Smith found three dead pipits, believed to be of this species, that had apparently been sheltering beneath miniature tussac grass in the southeast of Beauchêne Island (Smith & Prince 1985).

Habitat and Breeding

During the breeding season the Falkland Pipit inhabits the very large areas of coarse white grass. It also uses wide sand beaches littered with dead kelp and is occasionally seen on rocky beaches but it does not utilise dense tussac grass (Woods 1985). It nests from late September to late December and has at least two broods in a season. The nest is of fine grasses sometimes lined with horsehair and is well-hidden in rough grass. Two to 4 eggs are laid, varying in colour from creamy-grey to dirty white and variably spotted and blotched with yellow-brown and purplish-brown with black hairlines near the blunt end. Young nestlings have dark brown-black down.

Breeding Birds Survey: analysis of data

Records indicate that it is widespread throughout the archipelago except for Beauchêne Island, which has no white grass. Comparisons with the analyses of topographical parameters show only a significant tendency ($p = 0{\cdot}049$) for more records from grid squares with more than 6 wet tetrads and fewer records from those without standing freshwater. It seems therefore that the Falkland race prefers wetter habitats, similar to those described for the continental races by Fjeldså & Krabbe (1990). Among the records were five estimates of between 100 and 1,000 pairs, from Danny Donnelly at Crooked Inlet (TC78), from Shane Wolsey at Warrah River plain (UC18), from Bill Bourne at Wreck Point & Ajax Bay (UC58) and Fanning Head & Race Point (UC59) and from Raymond Evans at Pebble Island (UD21). Four of these squares have 1-5 wet tetrads and one (UD21) has 10 wet tetrads. There were 67 estimates of 11 to 100 pairs. During the period of the survey, there were several records in different years of birds present in the winter and the possibility of regular migration to and from the Falklands seems unlikely.

World Distribution

The Correndera Pipit is found in South America from southern Peru and Bolivia in the andean region up to about 4000m (13,000 feet), through Chile to Tierra del Fuego and from 23°S in Brazil through Paraguay, Uruguay and most of Argentina. Five races have been described; *A. c. grayi* is endemic to the Falklands and is somewhat larger than *A. c. chilensis* found in southern Argentina, Chile and Tierra del Fuego. Some of the southern breeding populations migrate northwards in winter to about 20°S while others remain in northern Tierra del Fuego (Fjeldså & Krabbe 1990).

57 * Falkland Pipit *Anthus correndera*

Detectability: **4 High**

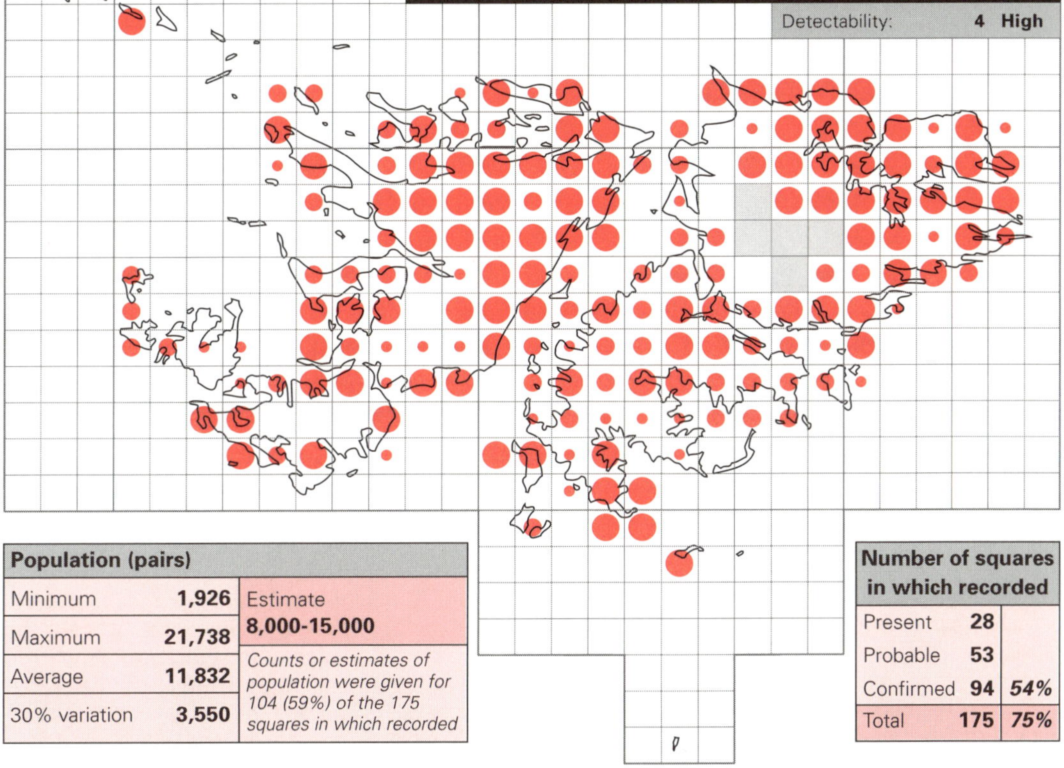

Population (pairs)		
Minimum	**1,926**	Estimate **8,000-15,000**
Maximum	**21,738**	
Average	**11,832**	*Counts or estimates of population were given for 104 (59%) of the 175 squares in which recorded*
30% variation	**3,550**	

Number of squares in which recorded		
Present	**28**	
Probable	**53**	
Confirmed	**94**	***54%***
Total	**175**	***75%***

Comparisons between species records and topography	Dry/Wet Number of wet tetrads					Inland/Coastal Amount of coast			Upland/Lowland Proportion of square > 500′				
	Nil	**1-5**	**6-10**	**11-15**	**16-22**	**None**	**<10%**	**≥10%**	**≥1/2**	**≥1/4**	**<1/4**	**None**	**Total**
Surveyed squares	38	104	52	24	16	17	14	203	7	25	73	129	234
% of surveyed squares	16	45	22	10	7	7	6	87	3	11	31	55	100
Grid squares in which recorded	12	77	48	22	16	15	12	148	6	19	60	90	175
% records in each category	7	44	27	13	9	9	7	84	>3	11	>34	>51	100
Significant variation			*	*	*								

Implications for Conservation

This species seems to prefer white grass which is the most widespread inland habitat. Now that the practice of grass burning, which often led to uncontrolled fires lasting several months (Bourne 1988), has lessened, feral cats and rats probably pose the greatest threat to its survival. More observations of its behaviour in autumn and winter could help to clarify its migratory status.

References

TBFI (1975) p195-6, GBFI (1988) p223-5.

58 Falkland Grass Wren

Cistothorus platensis falklandicus

A. Ratona aperdizada or R. sabanera
C. Chercán de las vegas
Local name: Tomtit

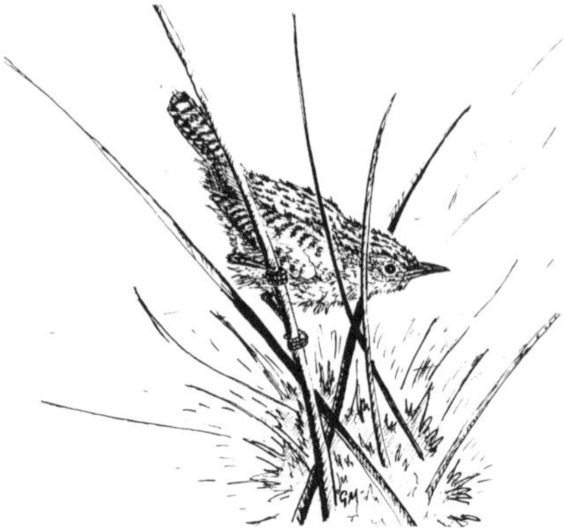

History
In 1822 Garnot noticed a little 'Fauvette' (warbler) which lived among the clumps of grass around Port Louis. He likened it to the 'Cisticole' or Fan-tailed Warbler of Sicily and Sardinia in Europe; the heavily streaked back and skulking behaviour of this warbler and the Grass Wren are very similar. The first published scientific record was by Darwin who commented that 'it is not uncommon' and 'lives almost exclusively, close to the ground in the coarse grass which springs from the peaty soil' (Gould & Darwin 1838-41). He did not think that he had ever seen a bird which, when it chose to remain concealed, was so difficult to disturb. Abbott (1861) thought it was 'singular' that 'this little bird should exist in such a place as the Falklands, where, if disturbed on a windy day, its power of flight is so weak that it is carried away by the wind!'. Vallentin (1904) did not see any wrens but was shown a nest from 'the neighbourhood of San Carlos' which, from the description, could have been of this species. In 1926, Bennett only stated that it was 'common on tussock-covered islands; breeding'. Frank Chapman (1934), of the American Museum of Natural History in New York, examined skins of several birds from southern South America. He compared continental Grass Wrens with four adult specimens collected by Rollo Beck in the Falklands in 1915/16 and found that the Falkland birds had heavier bills and less black in the plumage. He therefore named the Falkland race *C. p. falklandicus*. Cawkell & Hamilton (1961) considered that it was widely distributed 'and by no means confined to tussac islands as Bennett suggests'. It frequented places with plenty of cover, particularly beds of rushes. Pettingill (1974) described it as a 'common resident on tussac islands and in inland areas where there is a high, thick cover of rushes'. In field work in 1983, Grass Wrens were more than five times as numerous in dense tussac on the predator-free Kidney Island than in a re-established tussac and short grass paddock with introduced cats, rats and mice on West Point Island (Woods 1984).

Habitat and Breeding
The Falkland Grass Wren is found from sea level to at least 200m (650 feet), in damp areas with thick rush patches, long white grass and diddle-dee, gorse or marram grass. It also inhabits dense tussac grass, where higher population densities are possible and where most feeding takes place above ground level. The nest, usually built between 30 and 110cm (1 to 3½ feet) above ground, is an elongated ball of grasses 10 to 20cm (4 to 8 inches) in diameter, lined with wool or feathers, with a small round entrance, sometimes approached by a short grass tunnel. Five to 7 white eggs are laid between early October and mid-November; there is no evidence for more than one brood in a season.

Breeding Birds Survey: analysis of data
The records suggest a wide distribution on East and West Falkland and many offshore islands. The patchy appearance of the map may be due to difficulties in confirming the presence of this tiny wren. If birds are not singing when an observer is nearby, they are unlikely to be seen because they usually remain hidden in dense vegetation. In only seven of the 33 grid squares for which counts or estimates were given, the observers thought that 11-100 pairs could be present and this was the highest range category used. These were TC05 (Split Island, off Beaver Island, Faith Felton), UC44 (Drone Hill and Orqueta in Lafonia, Mike Riddy), UC58 (Wreck Point and Ajax Bay, Bill Bourne), UC89 (Douglas Station to Bombilla Hill, Clara and Bill MacKay), UC95 (Swan Inlet to Mount Pleasant House, Cumbria & Lancashire Expedition), VC28 (Long Island and Green Patch, Neil Watson) and VC48 (Kidney Island and Mount Low). Four of these were lowland squares with a large amount of coastline and 1-5 wet tetrads. Comparison of the distribution of records with the topographical parameters shows no significant differences.

58 *Falkland Grass Wren *Cistothorus platensis*

Detectability: **3 Moderate**

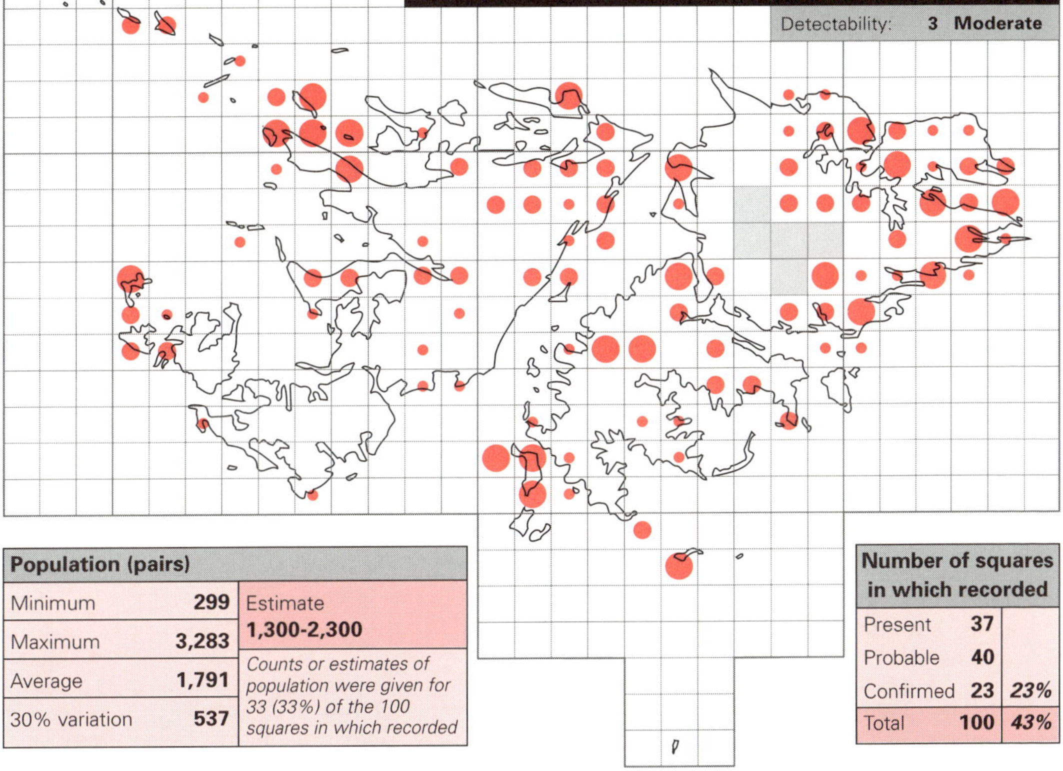

Population (pairs)		
Minimum	**299**	Estimate **1,300-2,300**
Maximum	**3,283**	
Average	**1,791**	*Counts or estimates of population were given for 33 (33%) of the 100 squares in which recorded*
30% variation	**537**	

Number of squares in which recorded		
Present	**37**	
Probable	**40**	
Confirmed	**23**	***23%***
Total	**100**	***43%***

Comparisons between species records and topography	**Dry/Wet** Number of wet tetrads					**Inland/Coastal** Amount of coast			**Upland/Lowland** Proportion of square > 500'				
	Nil	**1-5**	**6-10**	**11-15**	**16-22**	**None**	**<10%**	**≥10%**	**$\geq 1/2$**	**$\geq 1/4$**	**$< 1/4$**	**None**	**Total**
Surveyed squares	38	104	52	24	16	17	14	203	7	25	73	129	234
% of surveyed squares	16	45	22	10	7	7	6	87	3	11	31	55	100
Grid squares in which recorded	12	40	27	10	11	7	8	85	3	9	43	45	100
% records in each category	12	40	27	10	11	7	8	85	3	9	43	45	100
Significant variation													

World Distribution

Many races of the Grass Wren *Cistothorus platensis* have been described in the Americas. It is found from the northeastern United States and Canada through the Andes of Venezuela, Colombia, Peru, Argentina and Chile to Tierra del Fuego and Cape Horn and in the rest of Argentina and southeastern Brazil.

Implications for Conservation

Although this endemic race of the Grass Wren seems to be widespread, its total population may be small because it needs a particular combination of wet or flooded lowland with grasses or rushes tall and dense enough to provide shelter and nest sites. The much higher population density found on predator-free tussac islands suggests that eradication of these introduced mammals from offshore islands would be beneficial.

References

TBFI (1975) p186-7, GBFI (1988) p225-6.

59 Cobb's Wren

Troglodytes cobbi

A. Ratona común or Chochín común
C. Chercán común
Local name: Rock Wren

History
At Port Louis in 1764, Pernety (1771) mentioned 'great numbers of wrens like those in France', which suggests this species because it is very similar to *T. troglodytes*. However, none of the naturalists visiting mainland East Falkland in the 19th century reported a bird which is recognisable as Cobb's Wren. It is possible that this loud-voiced, very tame wren had disappeared from the mainland following the introduction of rats and the overgrazing of coastal tussac grass, before the naturalists arrived. There are no certain records of Cobb's Wren until it was named as a species from a Carcass Island specimen collected by Arthur Cobb in July 1908 (Chubb 1909). Hellmayr (1921) classified this wren as an island race of the continental House Wren (*T. musculus*, later merged with the North American *T. aedon*). Bennett (1926) stated that it was common and breeding on tussac islands while Cawkell & Hamilton (1961) remarked on its tameness and went further in saying that it was 'to be seen on tussac islands exclusively'. Following a study of its habitat preferences, behaviour and dimensions (Woods 1993), Cobb's Wren has once more been accepted as a full species, *T. cobbi* (Collar *et al.* 1994).

Habitat and Breeding
The optimum habitat for this wren appears to be dense tussac growing from high water mark behind boulder beaches with accumulated dead kelp in which invertebrates thrive. It also occurs in rushes and among rock outcrops up to 1·6km (1 mile) from coastal tussac on islands with no cats, rats and mice. These introduced predators seem to have a detrimental effect on populations of Cobb's Wren. In 1983, two similar open tussac paddocks were censused; on West Point Island which has all three mammals, there were no Cobb's Wrens but on Carcass Island only a few kms away, which has none of them, there was a high population density (Woods 1993). The nest is a domed ball with an entrance hole 6 to 8cm (2½ to 3") wide near the top, made of grasses and thickly lined with soft feathers of several species. It is usually well-hidden in a gap amongst tussac stems or a tussac pedestal or in a rock crevice, between ground level and about 60cm (2 feet) above ground. Three to 4 pinkish eggs, covered with many tiny spots of red or light brown, are laid between early October and December and two broods are probably reared in a season.

Breeding Birds Survey: analysis of data
The very scattered distribution of Cobb's Wren is clearly shown. Breeding was recorded as probable or confirmed at only 12 offshore islands, most of them in groups which are separated from each other by about 40km (64 miles) of ocean. Records, including counts or estimates of up to 100 or more pairs, came from South Jason Island (Nicholas Beswick and Carol O'Nians), Carcass Island (Sqn Ldr Craggs, Keith Reddick, Mike Reeves, Capt. Sherring, Maj. Towersley and Shane Wolsey), Speedwell, George and Barren Islands (Josie Larsen, Ian McLeod, Ken Maule, Kathy Patric and Paul FitzPatrick), Sea Lion Island (IME Lloyd, Alison Robson, Andrew Shaw, J Thwaite, Shane Wolsey), Beauchêne Island and Kidney Island (Andy Douse), Dunbar and Low Islands (Roddy Napier) and Lively Island (Ken Maule and Simon Bray). At Beaver Island settlement, Sally Poncet reported that single birds are seen annually in winter. Comparisons of the distribution of records with analyses of topographical parameters show no significant differences, only an unexplained tendency for more records in grid squares without standing freshwater. All records were from squares with much coastline and indicated that Cobb's Wren is closely associated with coastal habitats that have sufficient cover and invertebrate food. Since the end of the survey, it has been found in good numbers on Motley Island, southwest of Lively Island, (UC82 & 92) in January 1995 by an expedition from the Royal Naval College at Dartmouth. This island has a good mature tussac fringe and is apparently free of alien predators (Woods 1995).

World Distribution
Cobb's Wren *T. cobbi* is endemic to the Falkland Islands. Several races of the closely related House Wren *T. aedon* have been described and it is widespread from about 60°N in Canada through the United States and all of South America to Tierra del Fuego.

59 #** Cobb's Wren *Troglodytes cobbi*

Detectability: 5 **Very High**

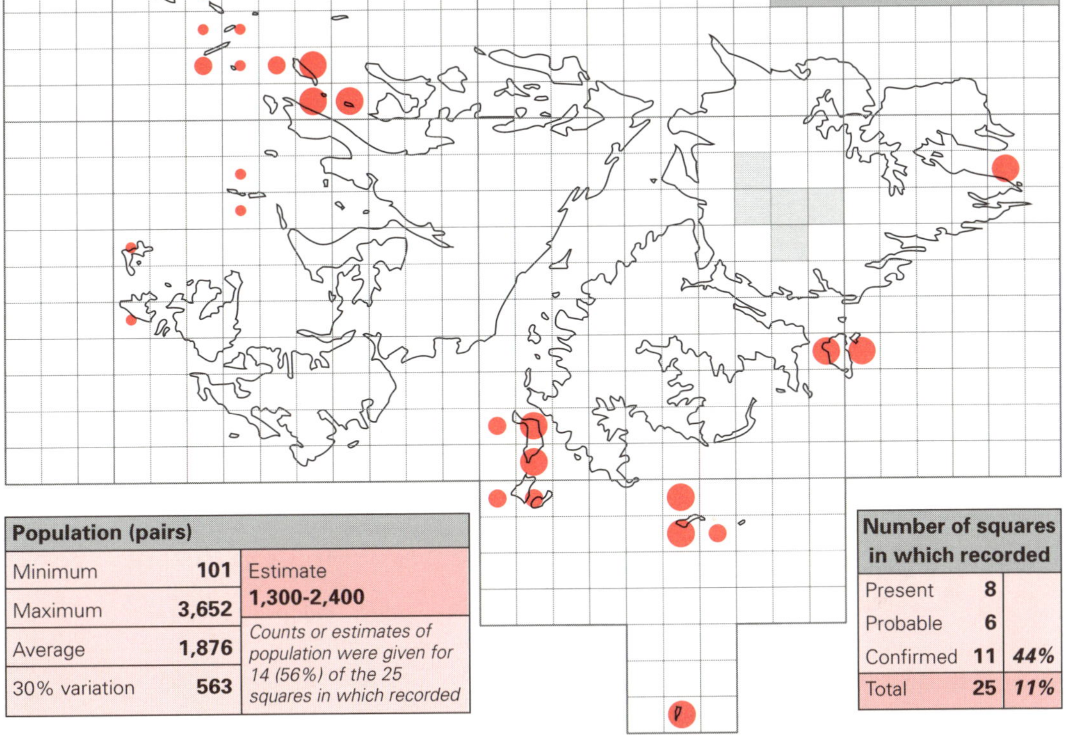

Population (pairs)		
Minimum	**101**	Estimate **1,300-2,400**
Maximum	**3,652**	
Average	**1,876**	*Counts or estimates of population were given for 14 (56%) of the 25 squares in which recorded*
30% variation	**563**	

Number of squares in which recorded		
Present	**8**	
Probable	**6**	
Confirmed	**11**	***44%***
Total	**25**	***11%***

Comparisons between species records and topography	Dry/Wet Number of wet tetrads					Inland/Coastal Amount of coast			Upland/Lowland Proportion of square > 500'				
	Nil	**1-5**	**6-10**	**11-15**	**16-22**	**None**	**<10%**	**≥10%**	**≥1/2**	**≥1/4**	**<1/4**	**None**	**Total**
Surveyed squares	38	104	52	24	16	17	14	203	7	25	73	129	234
% of surveyed squares	16	45	22	10	7	7	6	87	3	11	31	55	100
Grid squares in which recorded	9	12	3	1	0	0	0	25	0	0	8	17	25
% records in each category	36	48	12	4	0	0	0	100	0	0	32	68	100
Significant variation													

Implications for Conservation

This endemic species is classed by BirdLife International as '**Vulnerable**' due to the long-term destruction of tussac habitat, the introduction of mammalian predators and the small geographic range of its scattered population (Collar *et al.* 1994). Ecological studies of Cobb's Wren would be helpful in understanding the necessary conditions for its conservation. Records show that it can co-exist with a few domestic cats on a large island, but it is probable that rats and feral cats have destroyed whole populations of Cobb's Wren. Programmes to eradicate introduced predators, particularly from islands with tussac, would be beneficial.

References

TBFI (1975) p187-91, GBFI (1988) p226-7.

60 Falkland Thrush

Turdus falcklandii falcklandii

A. Zorzal patagónico

C. Zorzal

Local name: Thrush

History

Pernety (1771) mentioned a 'kind of thrush with a yellowish belly' that fed by scratching among the 'seagrass which is thrown upon shore'. Clayton (1774) remarked on a thrush-sized bird with 'a dun yellow breast and brown back' that was 'good food'! Quoy & Gaimard (in Freycinet 1824) were the first to name it 'Grive des Malouines' and *T. falcklandii*. Garnot (1826) said there were two species of thrush, this one and the 'Grive Guivrou' or 'Frosted Thrush' [perhaps the juvenile ?] found in various parts of the island. In 1833 and 1834 Darwin (Gould & Darwin 1838-41) observed the thrush in 'the more rocky and drier hills ... also the neighbourhood of the settlement [Port Louis] and very frequently ... within old sheds'. Abbott (1861) said that it was generally found among the rocks of the mountains, 'though sometimes frequenting the gardens in Stanley'.

Vallentin (1904) noted that it was 'universally distributed but nowhere very abundant'. Brooks (1917) stated that it was 'common, except near Stanley, where it was found on the hilltops among rocks'. Elsewhere, it occurred 'near the shore, especially around fern patches'. Bennett (1926) considered it a very common resident while Cawkell & Hamilton (1961) thought that it was formerly confined to rocky places but had in the previous 30 years spread all over the Falklands, occurring in bushes and introduced trees, and was therefore common in Stanley and at settlements. Pettingill (1974) described it as a common and widespread resident where rocks, ravines or shrubs provided shelter and he thought that it was particularly numerous on tussac islands and their beaches and around settlements.

Field work on Kidney Island between 1958 and 1963 showed that it was very successful in mature tussac (Woods 1970a) and census work in 1983 showed that it could achieve a high population density (four males per hectare) in dense tussac adjoining a boulder beach with dead kelp. Only about half this density was achieved on Carcass Island in a paddock of replanted tussac and short turf, but even this was about seven times higher than in a similar paddock on West Point Island where cats, rats and mice were present (Woods 1984). In a ringing programme at Stanley and on Kidney Island between 1957 and 1963, the greatest longevity was four and a half years and the longest movement recorded was 3·6km (2¼ miles) in six and a half months. Thrushes have a high mortality rate in Stanley and other settlements due to the many domestic and feral cats.

Habitat and Breeding

The Falkland Thrush occupies a variety of habitats, including tussac islands, settlements with bushes, trees and sheds, inland rocky outcrops and open slopes with ferns and diddle-dee. It usually nests between late August and December, occasionally from mid-July. The nest is a large deep cup of grass and root fibres, possibly including pieces of wool or string and it usually has an inner lining of mud or cattle dung, itself lined with fine grass and sometimes horsehair. Nests found on Kidney Island had no inner mud lining. Nests may be in almost any sheltered places, either natural or artificial and from ground level to 5 to 6m (15 to 20 feet) above ground, though most are at 1 to 2m (3 to 7 feet). It lays 2-3 blue-green eggs, closely marked with brown and purple, and may rear three or more broods in a season. Incubation lasts 14 to 16 days from laying the last egg and fledging takes 16 to 18 days; the interval to the next brood has been recorded as only 12 to 14 days.

Breeding Birds Survey: analysis of data

Records suggest that the Falkland Thrush is widely distributed in most parts of the Falklands, except on Beauchêne Island. A few interesting records of behaviour show how adaptable this species is to new circumstances. In late August 1983, Simon Strickland found a nest about 20m above ground, in a hanger at San Carlos. On 10 July 1984, John Cliffe saw 20+ thrushes at Moody Brook, west of Stanley, perching on derelict vehicles just before dusk, presumably about to roost. He noted that between June and October 1984, thrushes were seen occasionally scavenging in refuse around the coastels in Stanley harbour. John Cliffe also found about ten thrushes roosting in an outside shed at the shanty on Volunteer Point reserve on 20 July. 'Several of these birds

60 * Falkland Thrush *Turdus falcklandii*

Detectability: **5 Very High**

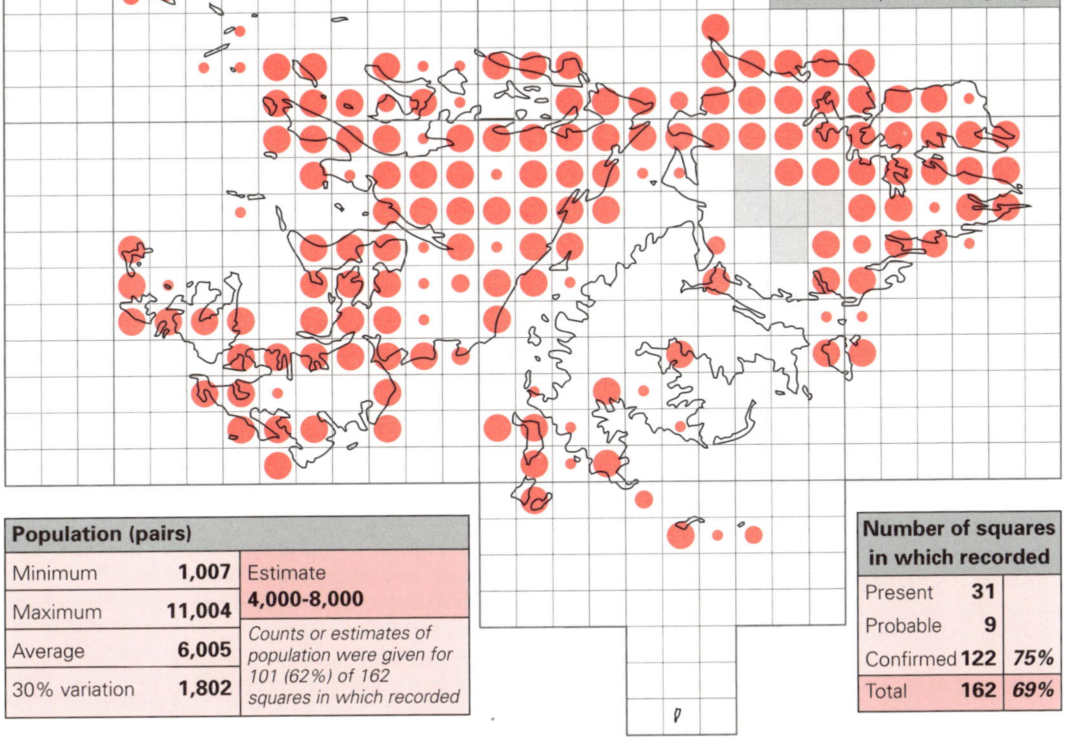

Population (pairs)		
Minimum	**1,007**	Estimate **4,000-8,000**
Maximum	**11,004**	
Average	**6,005**	*Counts or estimates of population were given for 101 (62%) of 162 squares in which recorded*
30% variation	**1,802**	

Number of squares in which recorded		
Present	**31**	
Probable	**9**	
Confirmed	**122**	***75%***
Total	**162**	***69%***

Comparisons between species records and topography	Dry/Wet Number of wet tetrads					Inland/Coastal Amount of coast			Upland/Lowland Proportion of square > 500'				
	Nil	**1-5**	**6-10**	**11-15**	**16-22**	**None**	**<10%**	**≥10%**	**≥1/2**	**≥1/4**	**<1/4**	**None**	**Total**
Surveyed squares	38	104	52	24	16	17	14	203	7	25	73	129	234
% of surveyed squares	16	45	22	10	7	7	6	87	3	11	31	55	100
Grid squares in which recorded	19	77	40	16	10	14	9	139	6	21	67	68	162
% records in each category	12	47	25	10	6	9	5	86	4	13	41	42	100
Significant variation											♦		

came and sat by the shanty door ... eagerly gobbling up any scraps thrown to them'.

Comparisons of the distribution of records with the topographical parameters show no large differences though there is a strong, non-significant tendency ($p = 0{\cdot}08$) for more records from grid squares with less than one quarter of the land above 500 feet. There are few records from Lafonia, particularly the northwestern area between North Arm settlement and Brenton Loch and none of these squares has any land above 500 feet. While it is very successful on tussac islands and around settlements, it is possible that the thrush still favours the higher ground on both East and West Falkland, as it did when Darwin visited 160 years ago.

World Distribution

The Falkland Thrush is an endemic race of the species known as the Austral Thrush in southern South America. The slightly smaller race *T. f. magellanicus* is resident from 27°S in Chile and 37°S in Argentina to Tierra del Fuego, Staten Island and islands south of the Beagle Channel. It occurs in various habitats, including woodlands and forest borders up to about 2,000m in central Chile and has adapted to gardens, parkland and agricultural regions (Ridgely & Tudor 1989).

Implications for Conservation

The Falkland Thrush has adapted to a wide variety of habitats though the highest densities have been found on tussac islands where there are no cats, rats or mice. With several broods in a season it can survive in the presence of introduced predators and it does not appear to be under any current threat.

References

TBFI (1975) p191-5, GBFI (1988) p229-31.

61 Rufous-collared Sparrow

Zonotrichia capensis

A. Chingolo

C. Chincol

History

This small sparrow was named by Latham in 1790 as the 'Rusty-collared Finch' from a specimen collected in Tierra del Fuego (Hellmayr 1938). It was first recorded in the Falklands in the summer of 1951 when three, one of which was singing, were seen at West Point Island (Cawkell & Hamilton 1961). Since then there have been about 20 records of up to four birds at several places, mainly on the western coast of the Falklands. Most have apparently been of the migratory southern race *Z. c. australis* which lacks black crown stripes. Two birds at West Point Island in 1973 survived from mid-April until at least mid-August. A more northern race (probably *Z. c. subtorquata*) with prominent black stripes either side of the crown, has been recorded twice, possibly three times. Each bird appeared in Stanley; one flew ashore 'a few years ago' [before 1958] from a ship that had come from Montevideo, Uruguay (Cawkell & Hamilton 1961) and the other two were trapped, examined and ringed in 1960 and 1962. These birds presumably reached the Falklands in the same way as the House Sparrow in 1919. The 1962 bird survived at least from May, through the winter and was last seen in December. The latest record before the survey was of one seen by J Pollock in the wood at Hill Cove, with Siskins and House Sparrows also present, in June/July 1982.

Breeding Birds Survey: analysis of data

The map shows only breeding season records since 1984. On Beaver Island, one bird, possibly two were seen between 13 October and mid-November 1986 by Tony Felton and up to four were recorded at Beaver and New Island by Sally Poncet and Tony Chater in May/June 1990. One photographed at Beaver Island by David Lee on 22 October 1991 was probably of the southern race *Z. c. australis*. Phil Gregory found four individuals on Carcass Island between 4 and 8 September 1990, one of which was singing. Two were reported from West Point Island before June 1991 by Roddy Napier, who informed Peter Abbott that individuals 'often survive the winter but tend to disappear in spring'. In 1991, Christophe Verheyden found singing males on Keppel Island on 5 September and at Beaver Island on 9 September and saw 'a pair' on Steeple Jason on 14 September. There were insufficient records to compare with the analyses of topographical parameters but all reports came from coastal grid squares. Since the survey there have been four further records during the winter. Two single birds were seen by Mike Morrison at Port Louis for several months, four were seen on West Point Island and three on Beaver Island. In mid-September 1995, James McGhie photographed a bird of a northern race, with head stripes, on Pebble Island and two birds, also with head stripes, travelled from Montevideo on the RRS *James Clark Ross* and left the ship as it approached the Falklands on 25 October (Steve Mee & Lesley Thomson pers. comm.).

World Distribution

The Rufous-collared Sparrow is widely distributed from southern Mexico through South America to Tierra del Fuego and the islands near Cape Horn. Of about 25 races described, the only one known to be migratory is *Z. c. australis*, which occurs from about 43°S to Cape Horn. Some of this race migrate north to about 20°S in winter while others remain in the Magellanic region, moving northeastwards, until the spring (Venegas & Jory 1979). It commonly occurs around human settlements and in most types of open or shrub-covered country from sea level to at least 5,000m (Fjeldså & Krabbe 1990).

Implications for Conservation

Though still considered an irregular vagrant from the Patagonian or Magellanic regions, this species is occurring more frequently and individuals have survived the winter and been heard singing. It is likely that the planting of trees and shrubs around houses at new settlements and the gradually increasing cover provided by established gorse hedges will enable the Rufous-collared Sparrow to breed in the Falklands.

References

TBFI (1975) p201-2, GBFI (1988) p237-9

61 Rufous-collared Sparrow *Zonotrichia capensis*

Detectability: **3 Moderate**

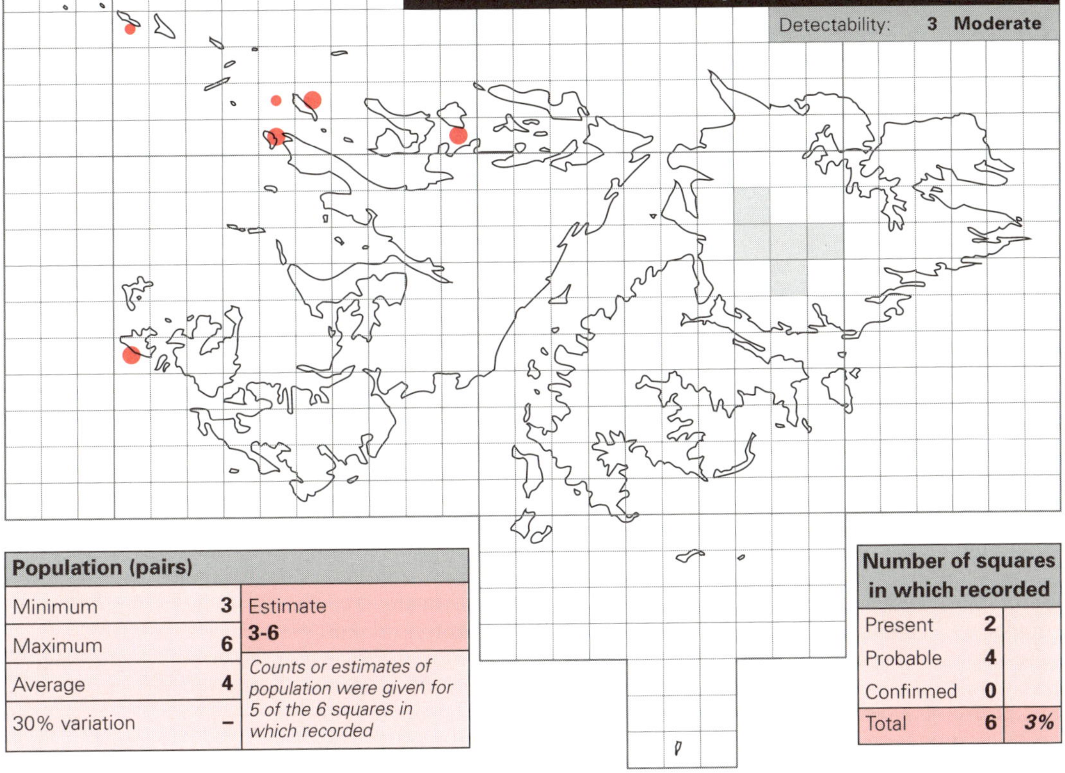

Population (pairs)		
Minimum	**3**	Estimate **3-6**
Maximum	**6**	
Average	**4**	*Counts or estimates of population were given for 5 of the 6 squares in which recorded*
30% variation	**–**	

Number of squares in which recorded		
Present	**2**	
Probable	**4**	
Confirmed	**0**	
Total	**6**	***3%***

62 Black-throated Finch

Melanodera melanodera melanodera

A. Yal grisiamarillo

C. Yal austral

Local name: Sparrow

History

Clayton (1774) on Saunders Island mentioned 'a kind of Sparrow' which could well have been the female of this species. Quoy and Gaimard (in Freycinet 1824) were the first to describe and name it scientifically and they remarked that it gathered in small flocks from February. Garnot (1826) noticed that these finches were usually to be found near the ruins of the old settlement at Port Louis. Darwin described it as 'extremely abundant in large scattered flocks' (Gould & Darwin 1838-41) and Abbott (1861) felt that it was 'plentiful everywhere, summer and winter'. Brooks (1917) named it 'Cock Sparrow' and described it as 'common' while Vallentin (1904) thought that it was the only indigenous species that nested close to human dwellings. Bennett (1926) said it was common and remarked that at least one of the males he collected had some yellow on the face, 'suggestive of hybridizing' with the Yellow-bridled Finch *M. xanthogramma*. Cawkell & Hamilton (1961) described it as common, handsome and widely distributed. They also remarked on possible competition from the House Sparrow at Teal Inlet. Pettingill (1974) considered that it was a common resident inland, nesting in mixed white grass and diddle-dee heathland and foraging on short turf. Field work in 1983 suggested that populations on the predator-free Carcass Island were about three times higher than on West Point Island where feral cats, rats and mice were present (Woods 1984).

Habitat and Breeding

The Black-throated Finch is found on heathland with dwarf shrubs, in dunes with marram grass and in open tussac grassland with short grazed greens near coasts. It occurs to at least 150m (500 feet) above sea level but does not inhabit dense tussac. Pairs are formed from at least mid-July and nesting occurs between mid-September and late December. Nests are usually on the ground, hidden in white grass or in a crevice between stones but have been found exposed at a height of 1·1m (3½ feet) on top of an old tussac clump. Fine grasses with a lining of horsehair, goose down or other feathers are used, the nest material being brought by the female accompanied by the male. The clutch is of 3-4 light blue-grey or grey-green eggs, with purple-brown markings concentrated around the blunt end, and there are probably at least two broods in a season. Young nestlings have silver-grey down. Flocks have been observed in December when other pairs are still incubating.

Breeding Birds Survey: analysis of data

Records give the impression that this species is evenly distributed in the Falklands. This is supported by comparison with the distribution of topographical parameters; there were no significant differences in distribution of records in any of the three analyses. It is likely that it was also present in most of the grid squares that were not surveyed. There were counts or estimates for more than half of the squares. Eleven were of less than ten pairs but in more than 60 squares, numbers were estimated as 11-100 pairs and on Sea Lion Island (UB58) observers estimated a population of more than 1,000 pairs.

World Distribution

The Black-throated Finch is restricted to extreme southern South America and the Falklands and two races have been described. The resident Falkland race *M. m. melanodera* is slightly larger than the continental *M. m. princetoniana*, which occurs from about 47°S in Chile and Argentina (Santa Cruz) through the Magellanic islands to the northern unforested regions of Tierra del Fuego. The population of this continental race has apparently fallen drastically due to heavy grazing by sheep and it may now be extinct in Tierra del Fuego (Fjeldså & Krabbe 1990).

62 #* Black-throated Finch *Melanodera melanodera*

Detectability: 4 High

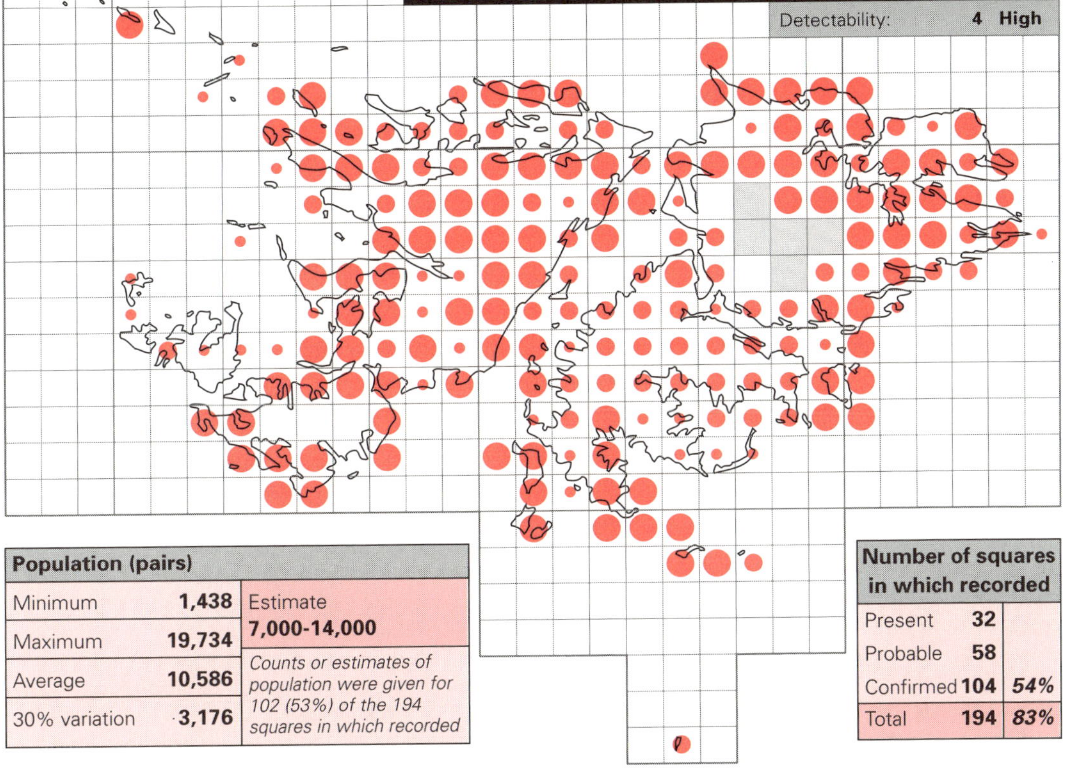

Population (pairs)		
Minimum	1,438	Estimate **7,000-14,000**
Maximum	19,734	
Average	10,586	*Counts or estimates of population were given for 102 (53%) of the 194 squares in which recorded*
30% variation	3,176	

Number of squares in which recorded		
Present	32	
Probable	58	
Confirmed	104	*54%*
Total	194	*83%*

Comparisons between species records and topography	Dry/Wet: Number of wet tetrads					Inland/Coastal: Amount of coast			Upland/Lowland: Proportion of square > 500′				
	Nil	1-5	6-10	11-15	16-22	None	<10%	≥10%	≥1/2	≥1/4	<1/4	None	Total
Surveyed squares	38	104	52	24	16	17	14	203	7	25	73	129	234
% of surveyed squares	16	45	22	10	7	7	6	87	3	11	31	55	100
Grid squares in which recorded	20	85	49	24	16	16	12	166	6	21	64	103	194
% records in each category	>10	44	>25	>12	>8	8	6	86	3	11	33	53	100
Significant variation													

Implications for Conservation

The Black-throated Finch has been classed as '**Near-threatened**' by BirdLife International (Collar *et al.* 1994). It is likely that the widespread practice of burning white grass in early spring destroyed many nests of this finch in the past and that introduced predators also reduce its numbers. However, in contrast to its reported decline on the continent, it does not appear to be adversely affected by overgrazing and seems to be thriving in the Falklands.

References

TBFI (1975) p199-201, GBFI (1988) p235-6.

63 Long-tailed Meadowlark

Sturnella loyca falklandica

A. Pecho-colorado grande

C. Loica

Local names: Robin, Military Starling

History

In the Falklands this brilliantly coloured species has been noticed by most visitors from the earliest days. In 1764, Pernety (1771) stated that he had 'killed a kind of Starling in the fields' and that the lower part of the neck and belly were 'of a very fine red, somewhat inclining to a flame colour'. In 1774, Clayton wrote that it was 'as large as a Magpye, with a fine Scarlett Breast, Speckled on the Back like a partridge with a long bill'. Garnot (1826) noted the differences in plumage between males, females and young birds, said that it was the only bird with brilliant colouring and that it lived in various parts of the island [East Falkland]. Darwin merely noted its presence (Gould & Darwin 1838-41) but Abbott (1861) called it the 'Red-breasted Starling' and said that it was very common in East Falkland. In 1889, Leverkühn described specimens from the Falklands as a distinct race *falklandicus* because they had much longer and heavier bills, more white on the outer tail feathers and were somewhat larger than birds from mainland South America (Hellmayr 1937).

Vallentin (1904) believed that it was locally distributed because, though it was common at Roy Cove, he had not seen one nearer to Stanley than in Berkeley Sound but other people had told him that it was seen in most parts of the Islands. Brooks (1917) said that it was common but local and was the least tame of the Falkland land birds. Bennett (1926) stated that it was very common and mentioned a semi-albino specimen in the old Stanley Museum. Cawkell & Hamilton (1961) remarked that it was a 'generally-distributed bird' and that the male was 'the most conspicuous of the smaller land birds on account of the brilliant scarlet breast and powerful voice'. They noted that it tended to flock in winter, a habit which Vallentin had also noticed. Pettingill (1974) said that it was widely distributed on inland heaths and common near settlements with shrub and tree plantations. Field work in 1983 suggested that population density in open tussac paddocks could be higher than on adjacent heathland (Woods 1984).

Habitat and Breeding

It occurs on white grass camp near the coast, on short turf around settlements, on open heathland with white grass and diddle-dee and in tussac grass paddocks with clearings of short turf. It also takes advantage of high perches such as planted trees or radio aerials. At settlements, its habit of digging for food has led to some conflict with people growing potatoes. It nests from late August to late November, laying 2-4 blue-white eggs blotched and streaked with purple and black, mainly around the blunt end. Two broods are probably reared in a season. The nest is constructed of dry grasses and may be well-hidden on the ground, in thick white grass sheltered by a clump, or above ground, up to 1m (3 feet) high, on a tussac grass pedestal. Nests on the ground have a noticeable pathway of up to 1m leading to the nest and those in tussac have a flattened side platform. The female apparently builds the nest and incubates alone. Recently-fledged young disperse and hide beneath grass clumps where they are fed by both parents. Flocks of up to 60 are formed from January and persist until about July.

Breeding Birds Survey: analysis of data

Records show that the Long-tailed Meadowlark is widespread on both East and West Falkland and that it occurs on most of the larger offshore islands. Analyses of records in comparison with the three topographical parameters show no significant differences, indicating that it finds suitable habitat in most regions of the Falklands. Over 50 of the population estimates were of 11-100 pairs. There were only three estimates of 100-1,000 pairs; from Danny Donnelly at Crooked Inlet (TC78), from Bill Bourne at Port San Carlos and Fanning Head (UC59) and from Robin Pitaluga in the Colorado Bay area south of Salvador (VC09).

World Distribution

Four races of the Long-tailed Meadowlark have been described from southern South America where it occurs from about 23°S in Argentina and 25°S on the coast of Chile to southern Tierra del Fuego and up to 3000m (10,000 feet) in northern

63 * Long-tailed Meadowlark *Sturnella loyca*

Detectability: **5 Very High**

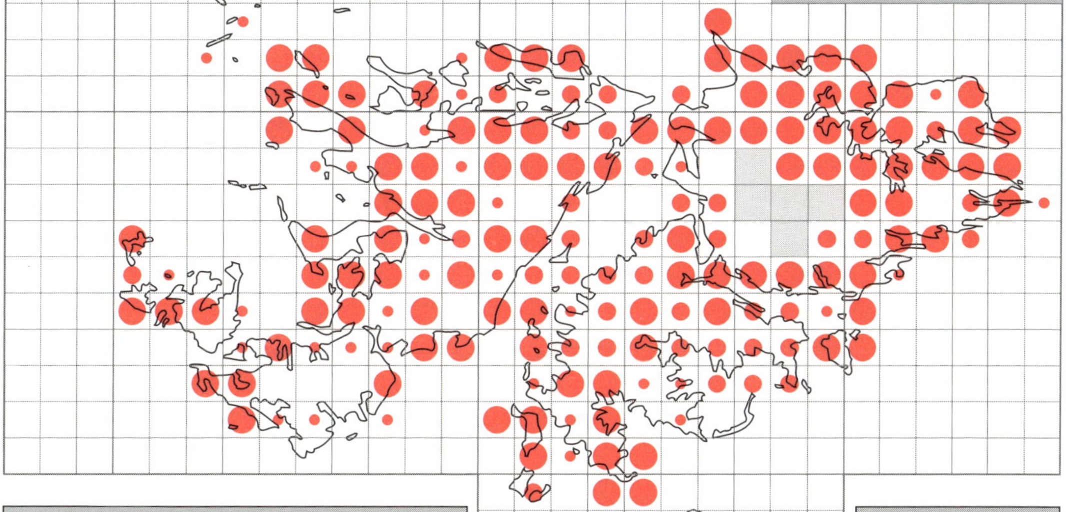

Population (pairs)		
Minimum	**1,378**	Estimate **6,000-10,000**
Maximum	**14,734**	
Average	**8,056**	*Counts or estimates of population were given for 94 (52%) of the 180 squares in which recorded*
30% variation	**2,417**	

Number of squares in which recorded		
Present	**36**	
Probable	**39**	
Confirmed	**105**	*58%*
Total	**180**	*77%*

Comparisons between species records and topography	Dry/Wet: Number of wet tetrads					Inland/Coastal: Amount of coast			Upland/Lowland: Proportion of square > 500′				
	Nil	**1-5**	**6-10**	**11-15**	**16-22**	**None**	**<10%**	**≥10%**	**≥1/2**	**≥1/4**	**<1/4**	**None**	**Total**
Surveyed squares	38	104	52	24	16	17	14	203	7	25	73	129	234
% of surveyed squares	16	45	22	10	7	7	6	87	3	11	31	55	100
Grid squares in which recorded	17	79	45	24	15	14	11	155	5	19	62	94	180
% records in each category	>9	44	25	>13	>8	8	6	86	3	11	34	52	100
Significant variation													

parts of its range. The endemic Falkland race *S. l. falklandicus* has a much longer, heavier bill than the mainland races (Hellmayr 1937). Some continental birds migrate northwards to about 23°S in winter though the southern race *S. l. loyca* is resident in the Magellanic region. A single male found at Prince Olaf Harbour on 9 April 1987 was the first record of this species at South Georgia (Curtis 1988).

Implications for Conservation

This species seems to be adaptable enough to take advantage of many changes in habitat since human settlement, though burning of white grass in early spring may have caused some destruction of nests in the past. It does not appear to be under any current threat in the Falklands.

References
TBFI (1975) p196-9, GBFI (1988) p239-40.

64 Black-chinned Siskin

Carduelis barbata

A. Sizerín de corbata

C. Jilguero

Local names: Siskin, Canary

History

At Port Louis in 1764, before shrubs such as fachine and christmas bush had been overgrazed, Pernety (1771) noted the presence of 'Siskins'. Clayton on Saunders Island in 1774, mentioned a 'beautyfull little bird like a goldfinch with its pretty notes, but very troublesome in the seed time', which was probably this species as its song resembles that of the European Goldfinch *Carduelis carduelis*. Later visitors to Port Louis from 1820 (Garnot, Quoy & Gaimard, Darwin, McCormick) did not record this species. Abbott (1861) reported one killed from a flock of five in a garden near Stanley in August 1860. This was the only East Falkland occurrence known to him, but he was told that it was very common on Keppel Island. After his third visit from 1909 to 1911, Vallentin described it (in Boyson 1924) as 'fairly plentiful round the settlement of Shallow Bay', 8 km (6 miles) south of Keppel Island, and noted that flocks of 10 to 15 were always present in winter and used dense gorse for shelter. He described a nest and said they were always placed some little distance from the ground, out of the reach of cats. Wace (1917) used the local name of 'West Falkland Canary' and Bennett (1926) stated that it was seldom seen on East Falkland except in winter, and breeds on West Falkland. Hellmayr (1938) mistakenly decided that this species was 'accidental on the Falkland Islands'. Since then there has been confusion because some authors have quoted Hellmayr while others have accepted that it is a resident breeding species.

Cawkell & Hamilton (1961) believed that it nested exclusively in bushes and that its distribution, before introduced trees and shrubs became established at settlements, was governed by the former natural distribution of the native shrub, box, which was said to be confined to West Falkland and adjacent islands. Since the 1920s it had been found breeding in Stanley, Port San Carlos and Teal Inlet, and was increasing in numbers. In 1952, Cawkell saw a pair of siskins on Kidney Island and speculated that this could indicate an extension of breeding range. Pettingill (1974) saw several flocks of ten to 12 birds on Kidney Island in November 1953. About ten birds were breeding there between 1960 and 1962, and one nest was found in tussac grass (Woods 1970a). Field work in 1983 on West Point Island and Carcass Island indicated that trees and bushes were favoured for nesting while the tussac planted in paddocks, up to 1·6km (1 mile) away, provided seeds for food during the breeding season. Nests examined at Carcass Island settlement averaged 2m (6½ feet) above ground, twice the height of the nests of other passerines in tussac (Woods 1984). Smith & Prince (1985) noted '16 individuals seen, mostly in pairs' at Beauchêne Island in December 1980, which confirms that siskins can live in a tussac-dominated habitat.

Habitat and Breeding

The Black-chinned Siskin occupies box thickets, tussac and planted trees and shrubs at settlements. Nests are neat cups of fine grass roots lined with hair, wool or grass root fibres, often placed in a fork of a large bush. One nest in a mature tussac plant was about 1·8m (6 feet) above ground in growing leaves about halfway in towards the centre from the leaf tips. It lays 3-5 pinkish-white eggs spotted or stippled with red-brown, between early September and December and up to three broods may be attempted in a season.

Breeding Birds Survey: analysis of data

This species is well-distributed and breeding around mainland coasts and on several offshore islands. West Falkland has 47 grid squares with siskins, compared with 34 more scattered squares on the East. There are few records from squares bordering Falkland Sound, perhaps because there are not many settlements. In comparison with the topographical parameters, there is a strong tendency for more records from coastal squares ($p = 0{\cdot}08$) and similarly ($p = 0{\cdot}06$) for more records from squares with less than one quarter of the land above 500 feet. Breeding season flocks of about 15 were seen at Hill Cove in September 1987 by MCA Clarke and two flocks of ten and 15+ were seen at Pebble Island in November 1990

64 Black-chinned Siskin *Carduelis barbata*

Detectability: 4 High

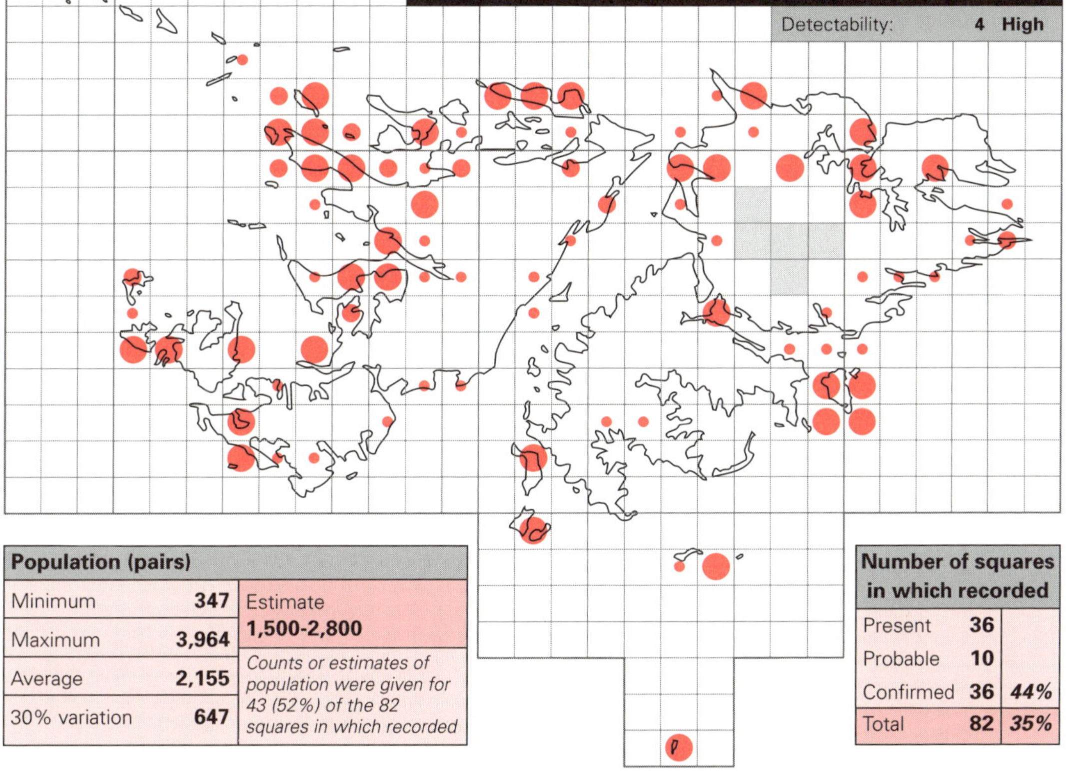

Population (pairs)		
Minimum	347	Estimate **1,500-2,800**
Maximum	3,964	
Average	2,155	*Counts or estimates of population were given for 43 (52%) of the 82 squares in which recorded*
30% variation	647	

Number of squares in which recorded		
Present	36	
Probable	10	
Confirmed	36	44%
Total	82	35%

Comparisons between species records and topography	Dry/Wet Number of wet tetrads					Inland/Coastal Amount of coast			Upland/Lowland Proportion of square > 500′				
	Nil	1-5	6-10	11-15	16-22	None	<10%	≥10%	≥1/2	≥1/4	<1/4	None	Total
Surveyed squares	38	104	52	24	16	17	14	203	7	25	73	129	234
% of surveyed squares	16	45	22	10	7	7	6	87	3	11	31	55	100
Grid squares in which recorded	10	46	17	8	1	1	3	78	2	10	38	32	82
% records in each category	12	56	21	10	1	1	4	95	>2	>12	>46	39	100
Significant variation								♦			♦		

by Andrew Clarke. Mike Riddy found them in remnant box along the north coast of West Falkland during the 1988/89 season and noted that they were more common than during his survey on East Falkland the previous year.

World Distribution

The Black-chinned Siskin is common and widespread from sea level to about 1500m in wooded or partly wooded country from Tierra del Fuego and the Magellanic islands to 27°S in the Andes of Chile and 37°S in Argentina (Fjeldså & Krabbe 1990).

Implications for Conservation

This species breeds successfully in introduced shrubs and trees and in mature tussac grass, which may have been a natural habitat before the mainland tussac fringe was destroyed. Replanting of tussac paddocks would be beneficial and efforts to eradicate introduced predators may also increase numbers of this most attractive songster. A taxonomic study may show that the Falkland population is a distinct geographic race.

References

TBFI (1975) p203-4, GBFI (1988) p232-4.

65 House Sparrow

Passer domesticus

A. & C. Gorrión

Local name: English Sparrow

History

Bennett (1926) stated that sparrows arrived on four different steamers from Montevideo, Uruguay, in November 1919 and he considered that by 1924 they had spread 'over the islands'. Hamilton (1944) saw some 20 sparrows aboard a whaling factory ship during the five day voyage to Stanley, 'about the end of October' and though the ship's cat caught a few, others survived and came ashore. A few pairs that reached Teal Inlet settlement about 48km (30 miles) northwest of Stanley (date not recorded) were killed because the manager thought they were driving away the native Black-throated Finches. Hamilton thought that there may have been other immigrants but apart from this small group, he believed that they had not spread outside Stanley.

By 1961, Cawkell & Hamilton noted that the species had spread westwards to Fitzroy (22km or 14 miles) and Darwin (97km or 60 miles). They stated that some damage was done to seedlings but House Sparrows had never increased much, 'probably due to cats'. It was considered common only in Stanley, though it was present at Green Patch (in Berkeley Sound) in January 1957. Kitty Bertrand reported that three pairs reached Carcass Island in late October 1959 and a similar number appeared at West Point Island in 1958/59, possibly from coasting vessels. Both colonies were still present in late 1965 though it had not been reported from mainland West Falkland. At West Point Island the colony increased to about 50 birds by early 1980. During the absence of four of the five human residents from May to August 1980, and all five during September, sparrow numbers declined sharply and they were thought to have died out. However, three birds and an occupied nest site under a roof were seen in October 1983.

Habitat and Breeding

House Sparrows breed colonially, making large untidy nests of grasses usually under roofs or in any available hole high on a building. On Carcass Island, the small colony was reported as nesting socially in a large cypress tree. Eggs are probably laid from September onwards. In England 3-5, occasionally 7 grey-white eggs are laid, finely spotted with brown and grey. The clutch may be smaller in the Falklands but no studies have been made. House Sparrows are seen as a nuisance in gardens at Stanley and the settlements because they eat seeds, buds and fruit, peck at flowers, dust-bathe in seed beds and take grain given to poultry. They usefully take scraps of kitchen waste.

Breeding Birds Survey: analysis of data

On and near East Falkland, it was recorded in Stanley, at Mount Pleasant Airport, 11 settlements and two inhabited islands. Around West Falkland, it was recorded at four mainland settlements, including Hill Cove, where it was noticeably less common than the Black-chinned Siskin, and on four offshore islands. Distribution is wide, from Stanley in the east to New Island in the west and from Pebble Island in the north to Sea Lion Island, the most southerly settlement. Work on building Mount Pleasant Airport in a formerly uninhabited area of East Falkland started in late 1983. Construction work was completed and the site occupied from early 1986, with most military personnel moving in by mid-1987. Mike Strickland noted that House Sparrows were present between February and June 1986 but Tim Duggleby remarked on their absence in April 1987. It is possible that the earlier colonists had not survived the winter of 1986 and that others have since become established. Phil Middleton and the Scouts found a House Sparrow egg at Paloma Beach huts, Port San Carlos camp (UD50), in January 1992, but no sparrows were seen. The analyses of records in relation to the topographical parameters show no significant differences, though almost all records are from settlements on lowland near the coast. Observations three years after the survey ended, suggest that the small populations at some settlements outside Stanley suffered from the extremely hard winter of 1995, when there was snow lying from early June to late September. Apart from one dead bird found on Carcass Island in mid-October by Robin McGill, no House Sparrows were seen at Pebble, Carcass, West Point or Sea Lion Islands in the 1995/96 season.

65 House Sparrow *Passer domesticus*

Detectability: **3 Moderate**

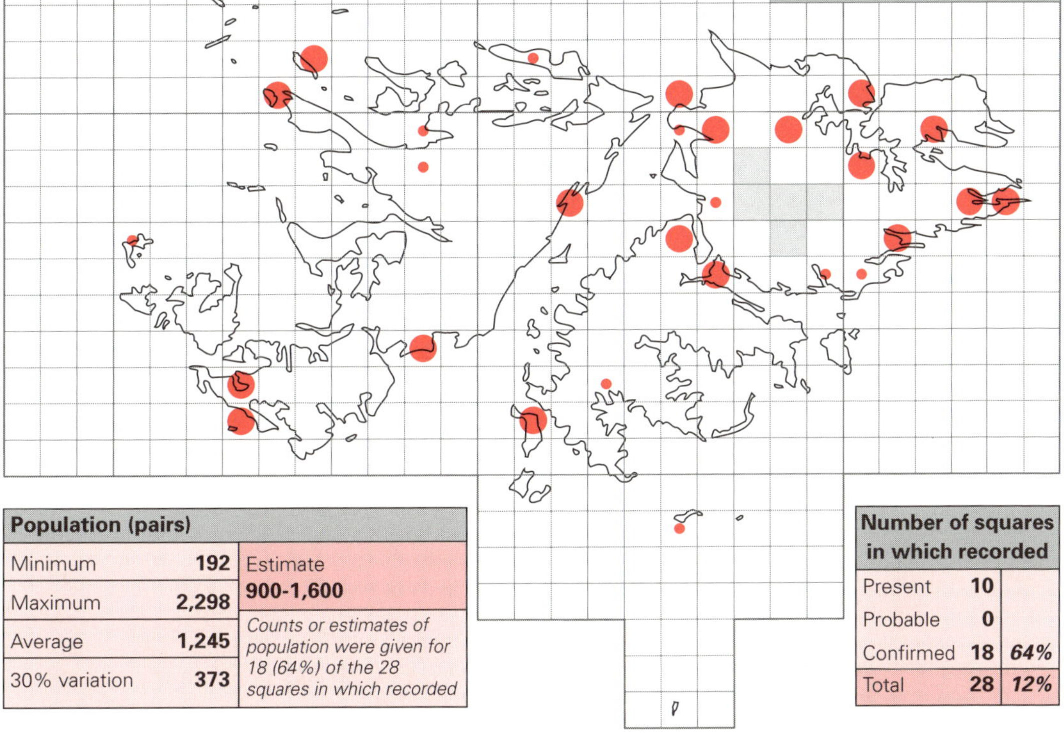

Population (pairs)		
Minimum	**192**	Estimate **900-1,600**
Maximum	**2,298**	
Average	**1,245**	*Counts or estimates of population were given for 18 (64%) of the 28 squares in which recorded*
30% variation	**373**	

Number of squares in which recorded		
Present	**10**	
Probable	**0**	
Confirmed	**18**	***64%***
Total	**28**	***12%***

Comparisons between species records and topography	Dry/Wet Number of wet tetrads					Inland/Coastal Amount of coast			Upland/Lowland Proportion of square > 500′				
	Nil	**1-5**	**6-10**	**11-15**	**16-22**	**None**	**<10%**	**≥10%**	**≥1/2**	**≥1/4**	**<1/4**	**None**	**Total**
Surveyed squares	38	104	52	24	16	17	14	203	7	25	73	129	234
% of surveyed squares	16	45	22	10	7	7	6	87	3	11	31	55	100
Grid squares in which recorded	2	12	6	7	1	0	1	27	1	4	13	10	28
% records in each category	7	43	21	25	4	0	4	96	4	14	46	36	100
Significant variation													

World Distribution

The House Sparrow was originally confined to Europe, North Africa and Asia but has successfully colonised most regions of the world. It remains closely associated with human habitation and very rarely migrates.

Implications for Conservation

This introduced species, confined to the settlements, does not appear to be in competition with any of the native Falkland birds. Little is known of its breeding biology in the Falklands and field work is necessary to determine the main items of its diet and whether it should continue to be classed as a pest.

References

TBFI (1975) p204-6, GBFI (1988) p231-2.

66 Great Grebe

Podiceps major

A. Macá grande
C. Zambullidor huala

History
It is likely that Abbott (1861) made the earliest reference to this species when he mentioned two specimens of a larger species of Grebe obtained in 1859, which he believed had been shot near Fitzroy river. AG Bennett notes that in August 1931 a very large grebe 'was seen by cuttermen at Johnson's Harbour in the sea' ... as 'big as a shag' with a 'longer crest' (Bennett unpublished). Cawkell & Hamilton (1961) published the first definite record of one male found long dead at San Carlos in the spring of 1941. A single bird spent several days in Carcass Island harbour during February 1966 and another single bird stayed in West Point Island harbour for almost two months from 16 January 1967. One was seen near Horse Point, Stanley Common between 24 and 27 October 1981 and two were seen close to the shore at Goose Green on 11 February 1983 after a period of strong winds.

Breeding Birds Survey: records and comments
Between 1984 and 1991, the following 14 records of Great Grebes were reported. All but two birds were seen in coastal waters and seven sightings were of two birds together. On 16 October 1984, John Cliffe described a probable immature bird near the coastels (floating hotels) moored in the Canache, east of Stanley. Mike Strickland saw two adults in Mare Harbour on 24 February 1986 and in late December 1986, Tim Duggleby saw one at New Island. One was first noticed in Chabot Creek, Johnson's Harbour by Christopher May on 21 December 1986. A second bird joined it on 19 January 1987 and they were photographed by Mike Morrison on 1 February. They fed amongst the kelp beds away from the shore and were heard calling. In late January 1987 Kate Layman saw two adults in Mare Harbour and Tim Duggleby reported one from the same place, that stayed for several days. On 24 March 1987, Harro Müller saw two in West Point Island harbour. In 1988, two were seen at Hill Cove on 16 and 17 December by Mike Riddy and two separate birds were watched on 'Beaver' Pond, Pebble Island on 4 and 18-19 December by Phil Gregory and John Reid. Although both birds were apparently adults, their colouration was different. Another was seen by John Croxall on 20 December 1988 at sea, east of the mouth of Mullet Creek, East Falkland. In 1989, there were again two adults in Mare Harbour, seen and photographed by Andrew Shaw on 13 January and seen again on 28 January. In the same month and into February one was present at Port Howard on Falkland Sound, reported by Diana Bradley. Mike Rendell saw one on the Graveyard pond at Port Louis on 3 January 1991 and Norma Edwards reported one on 27 March 1991 from Fox Bay West, where it had remained near the jetty for several days. Since the survey one was seen offshore at Sea Lion Island on 15 January 1994 (Henry 1994) and Mike Rendell reported one in Chabot Creek, Johnson's Harbour, from 19 December 1995. Some of these records could be of nonbreeding or immature birds wandering further than usual from the mainland but the apparent increase in records may be significant. Careful examination of the large ponds and lakes in the wetlands of Lafonia, in southern West Falkland and on Pebble Island may be necessary to establish whether this handsome grebe is breeding in the Falklands.

World Distribution
In southern South America the Great Grebe is common on the larger freshwater lakes with reeds and in estuarine marshes. It breeds from about 30°S in Chile and 25°S in southeastern Brazil through Paraguay, Uruguay and Argentina to Tierra del Fuego and on the coast of Peru between 5°S and 10°S. In winter it migrates into coastal waters and gathers in flocks, sometimes hundreds strong (Fjeldså & Krabbe 1990, del Hoyo *et al.* 1992).

References
TBFI (1975) p212, GBFI (1988) p100-1.

67 Grey-headed Albatross

Diomedea chrysostoma

A. & C. Albatros cabeza gris

History

The first Falkland record appears to be of three shot 150km (90 miles) to the south, over the Burdwood Bank, by the Scottish National Antarctic Expedition on 1 December 1903 (Ramsay 1913). Another report with a locality appears in *A List of the Birds of the Falkland Islands and Dependencies* (Bennett 1926), where he stated that this species 'has been obtained on West Point Island and breeds on South Georgia' and indicated that there were specimens of local origin in Stanley museum. Wace (1921) gave no clear references but he also mentioned examples in the local collection. They could both have been using 'local' to include South Georgia. Murphy (1936) referred to Wace and Bennett when reviewing the status of this species in South American waters. Other comments by Murphy are unclear when examined closely, but he had mistakenly accepted its breeding status in the Falklands. According to Roddy Napier, a single adult occupying a nest on West Point Island in a Black-browed Albatross colony was ringed in 1952 and returned annually until 1956 and Charles Gallimore reported another single bird on West Point Island in 1969. During an ecological study of Beauchêne Island in December 1980, Smith & Prince (1985) reported that the Grey-headed Albatross was not breeding and was 'rarely seen offshore'.

Breeding Birds Survey: records and comments

Several observers reported this species during the survey, particularly Bill Bourne and Bill Curtis. They noted that large numbers followed fishing boats south and east of the Falklands in winter but it was scarce inshore (Bourne & Curtis 1985). In February and March 1987, Bill Curtis saw a total of 46 birds (Wolsey 1987), probably with many other albatrosses and petrels off the southern entrance of Falkland Sound. Harro Müller saw between one and six birds off Pebble, Keppel, Beaver and West Point Islands in March 1987. Phil Gregory reported single birds off Kidney Island and in Berkeley Sound on 27 November 1988 and 25 January 1989 respectively. Nicholas Beswick saw one about 160km (100 miles) east-southeast of Cape Pembroke on 25 June 1992 and another only 16km (10 miles) east of the entrance to Choiseul Sound on 8 September 1992.

As far as is known, there have never been any substantiated breeding records in the Falklands.

World Distribution

Circumpolar in antarctic and subantarctic waters between about 65°S and 35°S, the Grey-headed Albatross reaches about 15°S in the Humboldt Current off western South America. The nearest known breeding colony, containing about 20,000 birds in 1981, is on the Diego Ramirez Islands south of Cape Horn, where Schlatter (1984) noted that both the Black-browed and Grey-headed Albatross were breeding. The colonies at South Georgia hold around 80,000 pairs, or about half the known world population (Croxall *et al.* 1984). It also breeds, often in association with the Black-browed Albatross, at Ildefonso Island, Prince Edward Island, and Kerguelen and Campbell Islands.

References

TBFI (1975) p78-9, GBFI (1988) p106-7.

68 Coscoroba Swan

Coscoroba coscoroba

A. Ganso blanco

C. Cisne coscoroba

History

Abbott (1861) only saw this species at Mare Harbour, where there was generally a flock of eight to ten birds. On 1 May 1860, he saw a pair with three small cygnets about a month old, which he felt, 'had been bred on some of the adjacent islands'. Sclater (1860) merely listed the species and noted that it was also reported by Captain Pack. Though Wace (1921) considered it a frequent visitor and mentioned that the 'Challenger' Expedition had noted it in the Falklands (between 23 January and 6 February 1876), Bennett (1926) classed it as 'a rare visitor only'. Cawkell & Hamilton (1961) quoted several records: an adult and immature spending three months in winter on floodwater near Stanley in 1951; one on Beaver Island in late October 1952; a pair near Champina Pond, Fitzroy all the summer of 1953, suggesting breeding; birds seen near Mount Pleasant in Fitzroy camp in 1954 and a very tame bird that spent several days along the shore of New Island in December 1954. They said that 'it has either become more common or the increased interest in ornithology in the Colony has led to more being reported'. A small flock of less than ten swans 'with no black on the neck' was seen flying towards Elephant Beach, Foul Bay in February 1959 and reported by Tony Nelson. On 15 February 1981, 14 birds photographed when swimming just off Northwest Point, Carcass Island were at first thought to be Coscoroba Swans but were later correctly identified as Feral Domestic Geese (Wolsey 1986). The Coscoroba Swan is larger, white with distinctive black wingtips, a slender, usually curved neck and pink-red bill which resembles a duck's bill. Some Domestic Geese are all white but the neck is much thicker with horizontal carriage of the head, the bill is orange, heavy-based and conical and when standing, a distinctive heavy paunch is visible.

Breeding Birds Survey: records and comments

A single bird reported from North Arm 'before 1986' was not documented with a description. There were no other reports of Coscoroba Swans during the survey, but there have been a few sightings since the 1992/93 season. James McGhie reported a group of seven, first sighted on 22 November 1993 on Ship Harbour Pond, Pebble Island, about five miles east of the settlement and last seen about 12 December. Identification was confirmed by Richard Coomber, a visiting ornithologist. About a week later, three swans appeared and stayed until late February. They were photographed by Ronnie Spafford in January 1994. Two birds (male and female) were seen at Bertha's Beach Pond, in Mare Harbour camp, by Bill Curtis on 19 and 24 October 1994, 15 November and 3 and 24 December. They were seen displaying on many occasions. Andrew Clarke and others also noted these two birds on 19 November 1994. Two were present on Big Pond, Pebble Island for at least three weeks in December 1995 and were photographed by Carol O'Nians.

There have been no records with any evidence of breeding since the family group was seen by Abbott in 1860.

World Distribution

In South America, the Coscoroba Swan frequents marshes and freshwater lakes with fringing vegetation (Madge & Burn 1988). It breeds from Tierra del Fuego to 45°S in Chile, through most of lowland Argentina, to Paraguay and southeastern Brazil and is partly migratory, with some birds reaching 25°S. There are also important winter quarters in northern Isla Grande, Tierra del Fuego. It is locally common and not globally threatened but in Chile numbers are decreasing and it may now be restricted to the extreme south (del Hoyo *et al.* 1992).

References

TBFI (1975) p108, GBFI (1988) p136.

69 Mallard

Anas platyrhynchos

A. & C. Not recorded

History
Cawkell & Hamilton (1961) were the first to mention this species in the archipelago when they reported that 'some were released in the East Falkland in the 1930s and are said now to be breeding in very small numbers'. The 1953 Bird Protection Ordinance listed it as a protected species, with several other rare or vagrant birds, presumably on the strength of a known introduction. There have been no reports to substantiate Cawkell & Hamilton's statement and no records of feral Mallard before the survey. However, many plumage features of the Mallard were seen in some domestic ducks in Stanley in 1983.

Breeding Birds Survey: records and comments
There were no records of any breeding behaviour by feral Mallard. The only report of a 'wild' duck resembling this species came from Andrew Shaw, who saw and photographed a male with a Domestic Goose, near the Sulivan House jetty in Stanley harbour on 30 March 1989. This duck had a blue sheen on the head, instead of green, and somewhat scaly flanks; otherwise it resembled a Mallard that had almost completed its post-breeding moult. Since the survey, there has been one report of a Mallard at Goose Green in August 1994 (Thompson 1994).

Several domestic ducks are descendants of wild Mallards. It is an aggressive and dominant species that has hybridised with at least 23 other forms in the family Anatidae, including some with a restricted range which are now in danger of extinction. A great variety of plumage forms has been produced, though most still show signs of their Mallard ancestry (Madge & Burn 1988, Marchant & Higgins 1990). It appears that efforts to introduce the Mallard to East Falkland in the 1930s were unsuccessful. A feral stock has not been established but some may have hybridised with other ducks. Any future suggestion that it could be reintroduced should be vigorously opposed.

World Distribution
The Mallard is common in the northern hemisphere and well-known because of its adaptability to almost any shallow aquatic environment with fringing vegetation and by its readiness to take artificially supplied food. It was introduced to Australia and New Zealand where populations in millions have become established. It is migratory in the northern parts of its range and sedentary in more temperate regions.

Reference GBFI (1988) p146.

---

70 Cinereous Harrier

Circus cinereus

A. Gavilán ceniciento
C. Vari común

History
Quoy and Gaimard (in Freycinet 1824) were the first to record this species in the Falklands in 1820. Darwin in 1833/4 said that it was 'very tame' and that it preyed on small quadrupeds, molluscs and even insects; one was seen feeding on the carrion of a dead cow (Gould & Darwin 1838-41). Abbott (1861) was certain that it was a 'far from plentiful' resident and he thought that it may have bred on islands off East Falkland

though he had never found a nest. Young birds had followed him when he was out hunting rabbits; it was bold for its size and 'very swift in the air'. Fitzroy provided the first of nine Falkland skins in the Natural History Museum, Tring; the most recent immature specimen was shot in April 1858. In 1921, Wace classed it as an accidental visitor, as did Bennett in 1926, though he stated that 'it may have been common in former days'. Cawkell & Hamilton (1961) had only one report suggesting breeding since 1940 and speculated that the annual grass burning may have adversely affected this ground nesting species. Don Davidson reported two sightings of individuals on West Point Island in February and August 1973 and John Peatfield (1981) had 'a number of scattered reports' but gave no details.

Breeding Birds Survey: records and comments
There were a few records during the survey but no evidence of breeding. At Philomel Pass, near Dunnose Head (TC65) on 8 January 1989, Chris Baxter, Robin Watt and Rosemary Wilkinson saw a probable female. At Pebble Island (UD21) on 12 November 1990, Andrew Clarke had distant views of a female or immature bird. Paul Hutchinson and Martin Davies had clear views of a male flying past in the Sussex Mountains (UC67) on 17 October 1992 and reported that two birds were seen in the area during the previous month by Toni and Richard Stevens of Port Sussex.

This former breeding species was probably extirpated by shooting and habitat destruction. If enough birds are drifted to the Falklands on migration and suitable habitat and prey are available, it could become re-established.

World Distribution
The Cinereous Harrier occurs in open grassland and marshy areas from Colombia and Peru through Chile and Argentina to the northern, unforested parts of Tierra del Fuego, and to 4500m from northwestern Argentina to Peru. Southern populations are partly migratory though birds breeding in southern Chile remain throughout the year (Venegas & Jory 1979).

References
TBFI (1975) p218-9, GBFI (1988) p156-7.

71 American Kestrel

Falco sparverius

A. Halconcito común

C. Cernícalo

History
Bennett (1926) published the first record of the American (Cinnamon) Kestrel in the Falklands, describing it as a rare bird that may breed. He mentioned one that was shot in 1909 and another seen flying about Stanley for most of the year 1924. However, it had occurred earlier as there are three nineteenth century Falkland specimens in the Natural History Museum at Tring, which unfortunately lack details on collectors or localities. There are four later specimens dated from 1930 to 1954. Cawkell & Hamilton (1961) give details of two sightings between 1951 and 1955 and state, 'the suspected breeding place is the north-east corner of the West Falkland and there cannot be more than two pairs'. Before 1983, there were more than ten records from different regions of the Falklands in the twentieth century. At least four of these birds spent periods of up to several months in the Stanley area, perhaps because House Sparrows and mice were available as prey.

Breeding Birds Survey: records and comments
Only four sightings of this species were made during the survey. All were in the late summer and winter and none of them included any evidence of breeding in the Falklands. Stuart Booth saw two at Rolon Cove on 14 February 1988. Mike Morrison reported a probable female at Port Louis on 25 February 1989 which stayed in the area until about 6 April. One, apparently the same individual, appeared on 17 May and was seen frequently, mainly in early mornings or evenings, until 30 August 1989. A female, possibly the same bird was seen at Long Island and Brookfield Farm during this period. An immature male was seen and photographed by

Phil Gregory at Moody Brook, west of Stanley on 14 and 27 May 1989, but was not seen after a spell of cold weather. A male, possibly the same individual, was seen and photographed west of Stanley by Robin Bayliss in June 1989. After the survey, a female was seen around the eastern end of Stanley from 22 May until at least 13 July 1993 (Stones 1993).

No adults with juveniles have been seen in the Falklands but individuals continue to appear and often survive for months. The possibility of residence and breeding in the Port Howard area should be investigated.

World Distribution
The American Kestrel is widespread and common in the Americas from the arctic tree line to Tierra del Fuego in open country and urban habitats, though absent from northern and eastern Brazil. About 16 races have been recognised, varying in size and colour. The birds recorded in the Falklands are presumed to be of the race *F. s. cinnamominus* which is found from southeastern Peru and Bolivia, Paraguay, Uruguay and extreme southeastern Brazil through Chile and Argentina to Tierra del Fuego.

References
TBFI (1975) p140-1, GBFI (1988) p162-3.

---

72 Least Seedsnipe

Thinocorus rumicivorus

A. Agachona chica
C. Perdicita

History
The first records for the Falklands were published by Bennett (1926) who stated that 'one was obtained on 15 May 1922, another in 1923 and two in 1924. One of the last, although it had an injured wing, reached the Zoological Society's Gardens, London, alive'. Bennett speculated that as the Least Seedsnipe has occurred during the summer, it may breed. Cawkell & Hamilton (1961) noted that Hamilton had seen three in the 1920s and that one was seen by Stan Bennett near Stanley in the summer of 1955. Between 25 August and 3 September 1972, David Carstairs (pers. comm.) saw up to three individuals on six occasions at Volunteer Green, at least one of which was recognised as male. On 8 September 1973, he flushed a male near the Big Pond, on Saunders Island. At the end of that breeding season, Stuart Booth reported a probable female near Surf Bay on 7 April 1974 and on 30 April 1974, Norman Drake reported a possible Least Seedsnipe at Port Purvis. Stuart Booth saw a male at Rolon Cove beach in January 1982.

Breeding Birds Survey: records and comments
There were only two records during the survey; a female was seen and photographed by Mike and Sue Morrison between 27 March and late May 1986 on the hard grass of a vehicle track in the Horse paddock at Port Louis and a very wary female was seen by Alan Henry on short turf at Gipsy Cove on 1 October 1989 (Gregory 1994). Since the survey there have been more records. Geoff McMullan saw a single male on 27 November 1993 near the track to Bertha's Beach. This bird was watched for about 15 minutes as it called from a mud mound. Mike and Sue Morrison photographed a bird that appeared to be a young male near Fish Creek, Port Louis on 23 June 1994.

This species is probably an occasional vagrant, though the possibility of breeding remains.

World Distribution
Four races of the Least Seedsnipe have been described from South America, where they inhabit dry plains, coastal dunes and cultivated open ground. *T. r. rumicivorus* is found from northern Tierra del Fuego to about 27°S in Chile and 40°S in Argentina. This race, which is commonest in the most southerly parts of its range, migrates to the plains of northeastern Argentina for the winter. The other races inhabit lower andean regions from northern Argentina and Chile through Bolivia to coastal Peru and southwestern Ecuador (Fjeldså & Krabbe 1990).

References
TBFI (1975) 158-9, GBFI (1988) 185.

73 Chilean Skua

Catharacta chilensis

A. Gaviota parda

C. Salteador chileno

History

Wace (1917) was the first to suggest that this species occurred in the Falklands. Describing it as a rare visitor, he said, 'it possibly frequents some of the more Western Islands'. Dabbene (in Wace 1921) misquoted this in translation as, 'posiblemente es más común en las [islas] del oeste del archipiélago' which Devillers (1978) accepted as 'perhaps commoner in the western part of the archipelago'. Dabbene and Devillers therefore both implied that the Chilean Skua occurred throughout the Falklands whereas Wace in his 1917 typescript was much more cautious. Bennett (1926) was sceptical when he stated 'I know of no reason to include here the Chilean Skua' and Murphy (1936) did not mention any occurrence in the Falklands. Hamilton (1937) shot two of three birds seen at Eagle Point, East Falkland in November and December 1930. He thought that the third bird was mated to a Falkland (Antarctic) Skua. In October 1936 he found an adult female dead in Stanley Harbour and noted (unpublished) that it was 'the fourth seen by me in the Falkland Islands (or anyone else)'. Surprisingly, Cawkell & Hamilton (1961) do not discuss this species, though Cawkell mentions that he saw a skua on 5 March 1951 which was 'cinnamon-coloured, especially on the head and neck', which suggests a Chilean Skua. In the summer of 1975/76, Pierre Devillers studied Falkland and Chilean Skuas in the Falklands and on the Patagonian coast. His only Falkland record of Chilean Skua was an adult seen at Carcass Island on 9 December 1975, which he was unable to find again during the next five days.

Breeding Birds Survey: records and comments

There was only one record of this skua during the survey; one was photographed at a nest near the Tussac Plantation paddock at Seal Bay in Port Louis camp by Mike Morrison, probably in 1984.

The present status of the Chilean Skua is uncertain. It may be a vagrant or an occasional breeding species. It would be worth examining skua colonies closely as Chilean Skuas may be nesting among the Falkland Skuas.

World Distribution

The Chilean Skua breeds on the western coast of South America from about 35°S to southern Tierra del Fuego and Cape Horn and on the coast of Argentina between Puerto Deseado and Rio Gallegos. It ranges north in winter to the coast of northern Peru and probably the Brazilian coast.

References

GBFI (1988) p188.

---

74 MagellanicTapaculo

Scytalopus magellanicus

A. Churrín común

C. Churrín del Sur

History

There has been only one definite record and one doubtful sighting of this species in the Falklands. It would be considered as a vagrant, except that Charles Darwin collected a specimen in autumn 1833 or 1834 and described its Falkland habitat in more detail than he did for most other Falkland species (Gould & Darwin 1838-41). He stated, 'It has found its way over to the Falkland Islands, where, instead of inhabiting forests, it frequents the coarse herbage and low bushes, which in most parts conceal the peaty surface of that island'. Darwin only visited East Falkland but his notes imply that he saw other tapaculos in the Falklands. He also remarked that it 'might at first be mistaken for a *Troglodytes* [wren] ... to obtain a good view is not always easy, and still less so to make it fly'. The specimen has been in the Natural History Museum since it was acquired from the Zoological Society of London in 1855. Crawshay (1907) later examined it and described it as 'immature'. He commented from his own experience in Tierra del Fuego on the 'remarkably powerful' voice and rapid, restless and mouselike behaviour of this tiny tapaculo.

Abbott (1861) remarked, 'nor have I ever seen or heard of any such bird as the *Scytalopus magellanicus* also mentioned by Mr. Darwin'. Cory & Hellmayr (1924) commented that Darwin's was the only record from the Falklands, 'where it has not been taken by any of the more recent travelers'. Vallentin (1904) refers to

'*Scytalopus magellanicus*, "Wren"' and says 'I have never seen this species', but his notes on a supposed nest appear to describe the nest of a Grass Wren *Cistothorus platensis*. Wace (1917) did not mention this species but Dabbene (in Wace 1921) noted Darwin's record. Bennett (1926) described it as a 'rare and accidental visitor' and merely added, 'I believe that I saw this bird in 1916 near Stanley'. Cawkell & Hamilton (1961) said they had 'no knowledge of this rare straggler'.

Breeding Birds Survey: records and comments
None was recorded during the survey, indicating that it could be a lost breeding species. From Darwin's comments it was resident in 1833 or 1834 inhabiting dense low scrub. Its skulking habits could partly account for the lack of records later in the nineteenth century. In addition, the sheep population had reached about 800,000 by the mid-1890s and much of the natural vegetation, such as fachine in valleys and the coastal fringes of tussac, had been eaten out.

World Distribution
Fjeldså & Krabbe (1990) list several races of the Magellanic Tapaculo in southern South America, some still unnamed. The nominate race, *S. m. magellanicus*, ranges from Isla Grande and the islands near Cape Horn, northwards in the Andes to Neuquén (38°S) in Argentina and Santiago (33°S) in Chile. Fjeldså & Krabbe describe all tapaculos as extremely difficult to observe, very reluctant to fly, very difficult to tell apart by plumage and often only distinguishable by voice. Ridgely & Tudor (1994) note that this species inhabits a variety of terrain including large areas of loose boulders in the north of its range and tall grass, scrub and stunted woodland in the Cape Horn region.

References
TBFI (1975) p226, GBFI (1988) p212.

75 Yellow-bridled Finch

Melanodera xanthogramma

A. Yal verde
C. Yal cordillerano austral

History
Charles Darwin was apparently the first to record the Yellow-bridled Finch in the Falklands, in 1833 or 1834 (Gould & Darwin 1838-41). He wrote: 'This species is common at the Falkland Islands and it often occurs mingled in the same flock with the last one [the Black-throated Finch *Melanodera melanodera*]. I suspect however, it more commonly frequents higher parts of the hills. These species have a very close general resemblance; but the marks about the head, which are white in *[M.] melanodera* are yellow in *[M.] xanthogramma* while the parts of tail-feathers which are white in the latter, are yellow in *[M.] melanodera*. This difference of colours does not hold in females, but they may be at once distinguished by the greater length of wing, when folded, of *[M.] xanthogramma*.' The earliest specimen at the Tring Natural History Museum is a female collected by the English Antarctic Expedition in 1842. Although registered as originating from Hermite Island, south of Tierra del Fuego, the original label states 'Falkland'. In the accessions register, the previous three specimens are listed as from 'Hermite Isl.' It is therefore impossible to know whether this specimen came from the Falklands or Hermite Island. The Museum also has what Hellmayr (1938) calls 'a wholly typical adult male' labelled as collected on 11 July 1850 by John MacGillivray, naturalist aboard HMS *Rattlesnake*, at Port Stanley. He also collected a Black-throated Finch on the same day at Stanley and he did not visit South America.

Wace (1917) classed the Yellow-bridled Finch as a doubtful accidental visitor but Dabbene in his revision (Wace 1921) mentioned the Antarctic Expedition and MacGillivray specimens. Bennett (1926) also listed it as an accidental visitor and according to Cawkell & Hamilton (1961), Hamilton thought he had seen one in a flock of Black-throated Finches. They also noted that on 14 August 1953 'a small flock of extremely yellow birds "wary and like Siskins but not them" was reported in Stanley. Neither of us saw them'. The immature finch collected by WS Brooks on 1 February 1916 at Port Stephens and named incorrectly as a new species *Phrygilus malvinarum* was considered by Hellmayr (1938) to be an example of the Black-throated Finch. In Bennett (1926) and Cawkell & Hamilton (1961) the scientific name was considered to be an alternative (synonym) for *Carduelis barbata*, a name used for the Black-chinned Siskin, while in Woods (1988) the Brooks' specimen was discussed as an example of the Yellow-bridled Finch. Comparison in 1993 of details in Brooks' paper (1916) with Black-throated Finch specimens at the Museum suggests, however, that his immature bird was indeed a Black-throated Finch.

Breeding Birds Survey: records and comments
There have been no further records since the unsubstantiated sighting in the 1950s. Hardly anything is known about the food preferences of the Yellow-bridled Finch. Uncontrolled grazing could have destroyed habitat and food sources occurring in the higher regions of the Falklands and led to the loss of this species from the avifauna. Its status is still uncertain, but it is probably best considered as a former breeding species that was only locally common.

World Distribution
Both Ridgely & Tudor (1989) and Fjeldså & Krabbe (1990) consider the Yellow-bridled Finch to be an uncommon and sometimes rare species throughout its range in continental andean South America. Two races have been described; *M. x. barrosi* inhabits andean regions to 3000m (10,000ft) from about 33°S in Chile and Argentina to the Magellan Straits (53°S) while *M. x. xanthogramma* occurs above the timberline in the mountainous southern region of Tierra del Fuego and in the islands of the Beagle Channel.

References
TBFI (1975) p230, GBFI (1988) p236-7.

Appendices

Appendix A
Introduced Mammals

Many observers recorded the presence or absence of introduced mammalian predators and/or herbivores. Cats, potentially the most directly destructive to birds in the Falklands (Moors & Atkinson 1984), were recorded in 73 10-km squares. Patagonian Grey/Silver foxes are confined to Weddell, Beaver, Chain, Tea and Staats Islands and Split Island south of West Point Island. Their damaging effect on ground-nesting birds is well-known and eradication is being attempted on some of these islands by Sally and Jérôme Poncet. Rats were recorded in 47 grid squares and mice in 37. Hares were recorded in 14 squares on East Falkland and rabbits in nine squares mostly on the West. Guanacos were present on Staats Island. The numbers of cattle, sheep and horses were recorded by many observers, but no overall assessment of their impact on the natural vegetation and habitats for birds was possible. Altogether, introduced mammals are widespread on East and West Falkland and occur on several offshore islands.

There has been no detailed assessment of the damage that these animals cause to bird populations in the Falklands but there are indications. In 1983, census work was carried out on the passerines (songbirds) breeding in two paddocks on West Point Island and adjacent Carcass Island, containing similar habitats of mature tussac grass with interspersed short turf. The West Point plot was about 60% tussac-covered with potentially more shelter and nest sites than the Carcass plot, which was only about 40% tussac-covered. On West Point Island, rats were seen within 400m, mice were abundant throughout the tussac and cats hunted in the plot which held seven passerine species at a density of about two territory-holding males per hectare. There were no cats, rats or mice on Carcass Island and the eight passerine species in the plot had a community density of about 15 territory-holding males per hectare.

The comparison suggests that cats, rats and mice could have reduced the community density of songbirds in the West Point plot by up to 85%. The individual impact of the three introduced animals on these songbirds could not be assessed, but for two bird species, the difference between the islands was very marked. Cobb's Wren was absent from the whole of West Point Island, but had a density of at least two males per hectare in the Carcass Island plot. Very few Tussacbirds are seen on West Point Island but in the Carcass Island plot it was by far the most numerous species, reaching a density of more than six males per hectare (Woods 1984).

The effects of alien predators on seabird populations in other regions of the world have been summarised (Croxall *et al.* 1984; Burger & Gochfeld 1994). By 1995, Falklands Conservation had become increasingly aware of the severe damage caused by these predators to Falkland wildlife. With support from the Royal Society for the Protection of Birds, they commissioned a preliminary study by Brian Bell, a conservationist from New Zealand, into the feasibility of eradicating introduced predators, particularly from offshore islands. Programmes for the eradication of rats from a few selected islands will be implemented when logistical support and funding have been obtained.

1 Domestic/Feral Cat
Felis catus

History
It is not known when domestic cats were introduced, but it is likely that they accompanied the British immigrants who re-established Port Louis in 1834 and Stanley in 1842. An early reference to domestic cats living ferally was made by Vallentin (in Boyson 1924) who noted that in recent years the Magellanic Snipe had 'greatly decreased in numbers owing to the abundance of domestic cats which have taken to a wild existence'. Vallentin was resident in Stanley for two periods between 1898 and 1901 and lived at Shallow Bay, northeast of Hill Cove between 1909 and 1911. Near Hill Cove settlement he found that snipe were always scarce along the shores of the pond 'owing to the presence of cats - the domestic cat turned wild'.

In the Falklands, the serious effects on the native birds of allowing cats to roam have long been known and publicised by some people. The owner of Carcass Island towards the end of the nineteenth century, Charles Hansen, started a 'no-cat' policy which has been followed by subsequent owners for many years. Eleanor Pettingill met Kitty Bertrand there in 1953 and was told that a large yellow cat that made no moves when a Tussacbird approached was the first cat on Carcass for many a year and was 'so old the birds are safe'. She commented that other farmers mentioned Carcass Island with pride because of its abundant, tame birds, 'but it never occurred to them they could use the same conservation measures on their farms'. Eleanor Pettingill (1960) also wrote in

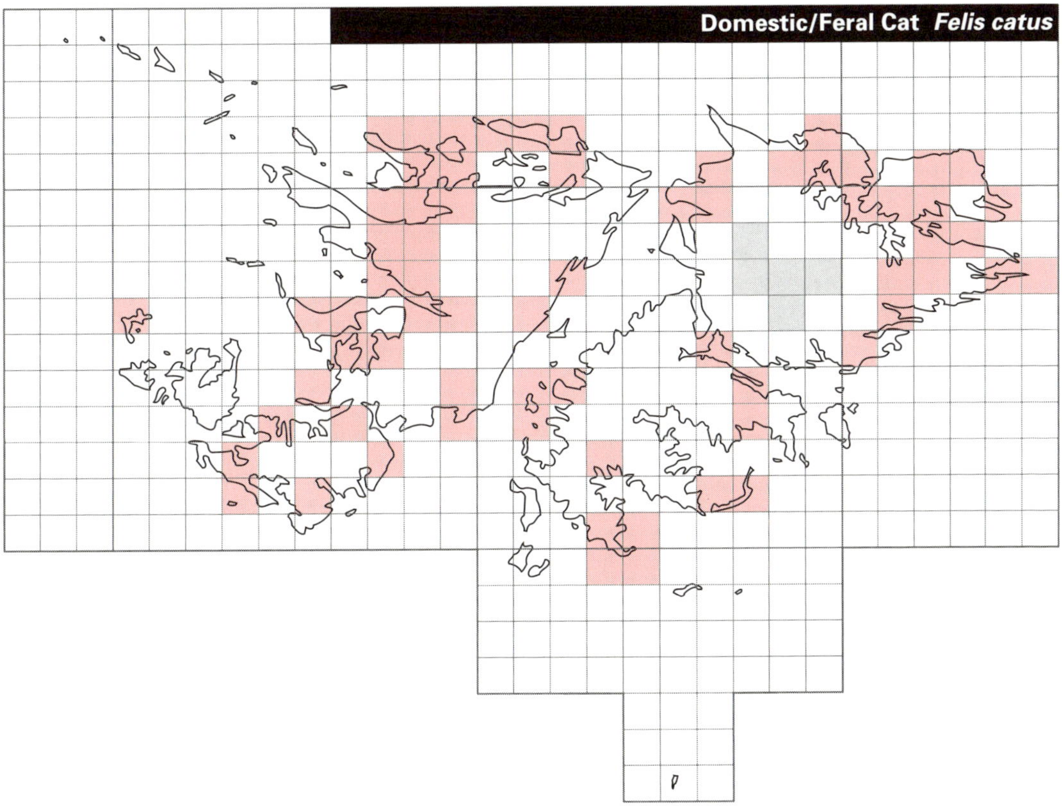

detail about the month that she and her husband Sewall spent on New Island from December 1953. They were told that there were about 75 cats on the island, to keep down the Thin-billed Prions and Magellanic Penguins or they would 'undermine the island with their burrows'.

EM Cawkell wrote about feral cats from his experience up to December 1955 (Cawkell, Maling & Cawkell 1960). He felt that they were 'widespread but not numerous' and noted that his dog which ran freely as he rode over the camp 'flushed them from time to time'. He considered that feral cats 'must take some toll of the bird life and on New Island they certainly do'. He believed that the owner had imported cats 'to check the spread of the prions'. He understood that there were about 300 cats on the island and that the 'home flock, which requires two buckets full of milk a day in the winter, numbers forty to seventy'. On New Island in April/May 1960, Fergus O'Gorman (1960) understood that 75 cats had been introduced and 35 of them were fed at the farmhouse. One was seen with a live prion and he saw the remains of several freshly killed prions. In the absence of prions during the winter, these cats would presumably prey on small birds, rabbits and rats.

Survey results and comments
Cats were reported from 73 10-km squares, 31% of those surveyed. Twenty-five of the mainland squares were on West Falkland and 32 on East Falkland. The wide distribution of records shows that feral cats could have access to all regions of both main islands. At the beginning of the survey, cats were also present at the following offshore islands: New, Beaver, Great, Speedwell, Bleaker, Pebble, Keppel, Saunders and West Point. Roddy Napier has since reported that he believed he had killed all feral cats on West Point Island in about 1984, Ronnie Larsen stated that he had eradicated feral cats from Speedwell Island in about 1986 and Sally Poncet reported the eradication of feral cats from Beaver Island in 1993. These eradications will have beneficial effects on the birdlife of the three islands but there must still be thousands of domestic and feral cats roaming freely in the Falklands.

Implications for Conservation of Falkland Islands Birds
The damaging impact of cats on native animals and birds has been publicised by the Australian Nature Conservation Agency, formerly the National Parks and Wildlife Service (1993b). Feral cats are found in every part of Australia. They inhabit remote central deserts with no permanent

water, eat an extremely wide range of native fauna and have two or even three litters a year which allow them to increase their population rapidly in favourable conditions. The ANCA states that although feral cats cannot be eradicated from mainland Australia at present, some islands have been cleared.

In another publication, specifically on cats, the ANCA (1993a) points out that it is not possible to define a cat as either domestic or feral because domestic cats stray and feral cats can become semi-domestic when food is scarce. Cats in Australia are known to hunt and kill over 100 species of native birds and ground nesting birds are believed to be particularly vulnerable. Data from Tasmania indicate that, over a three month period, 65,000 native animals, including birds, were killed by domestic and semi-domestic cats in the Hobart area alone. The ANCA states that several attempts to control rabbits, rats and mice by releasing cats have been unsuccessful.

The ANCA also notes that 'there are definitely millions of feral cats across Australia' and recommends 'Responsible Cat Ownership' as one way to reduce the cat problem. These are some of their suggestions which could be adapted for the Falkland environment: neutering domestic cats, limiting the number of cats kept, keeping cats home at night, either in a house or shed or in an outside run and identifying cats by collar and tag. Wearing a bell is ineffective because cats can stalk prey without the bell ringing. Being well-fed does not stop a cat hunting but stopping them from roaming is effective. The Shire of Sherbrooke in Victoria State was the first to introduce a local law to control cats, mainly to reduce their impact on local wildlife, particularly lyrebirds. Responsible cat ownership has been emphasised and the ANCA (1993a) reports that this initiative is working and that other local and State governments are considering this example.

2 Norway or Common Rat *Rattus norvegicus* Black or Ship Rat *Rattus rattus*

History
Rats were observed in the Falklands by Charles Darwin in 1833 and 1834. He stated that East Falkland was infested with the Common European Rat *Mus decumanus* and also said that he thought there were two species, one being of a brighter hue and less grey than the European variety. A rat was caught in a bay sometimes visited by ships, 30-40 miles (56km) from the settlement at Port Louis. He noted that rats had spread 'not only over the whole of East and West Falkland but even on some of the outlying islets' (Waterhouse & Darwin 1839). Moody (1842) however, said that there were not many rats, and these were mostly in the neighbourhood of the settlement at Port Louis. Bernsée (1855), whose ship *Courier* was wrecked in April 1854 on 'the S.W. point of East Falkland Island', [Bull Point, according to Southby-Tailyour (1985)] wrote to the Philadelphia Academy of Natural Sciences. His letter included the statement, 'I found a great many rats and mice on these Islands. Whether they are different from our own or not I am unable to say; their general habits are the same. The mice, however, I think are smaller than ours; and the legs of the rats are of a red-flesh color, and almost hairless above the knee joint'. Vallentin (1901), following his visit to Stanley from November 1898 to February 1899, remarked that the rat 'abounds in the settlement, and especially in the immediate vicinity of the two slaughter-houses which are located at either extremity of the town'. Rats and mice had 'been introduced by the many vessels which have from time to time visited this port, and in many instances remained, having been irreparably damaged off the Horn and elsewhere'.

Between 1908 and 1916 Messrs. Salvesens operated the only shore-based whaling station in the Falklands at New Island (Boyson 1924). Before that, the island had been used by sealers and whalers for more than a century. An anecdote from 'Dolly Penguin' (Duncan 1992) describes how, following the closure of this whaling station, 'without food, thousands of rats of all shapes, sizes and colours started to roam the island, so our cats soon had to earn their keep - we had about thirty on our farm !' Cobb (1933) commented that Tussacbirds 'do not as a rule live on islands where there are rats'. Cawkell (in Cawkell, Maling & Cawkell 1960) stated that 'where man has settled rats are abundant especially in Stanley near the slaughterhouse and in the rubbish dumps to the west of the town. They exist on some of the bird islands but it is hard to assess the damage they do'.

Cecil and Kitty Bertrand reported in 1983 that rats usually made themselves evident when they were present, for example on Flat Jason Island where Cecil had camped for a week. They reported that rats had been seen at the East Jetty in Stanley, walking down a pile and under water and then returning, having picked off a mussel. On the *Fennia*, a condemned hulk used as a floating warehouse in Stanley harbour from 1927 to 1967, heaps of mussel shells were seen by Bill Browning and Lou Sorensen in the peak where the anchor chain entered. Rats had been seen going down this anchor chain and returning with mussels.

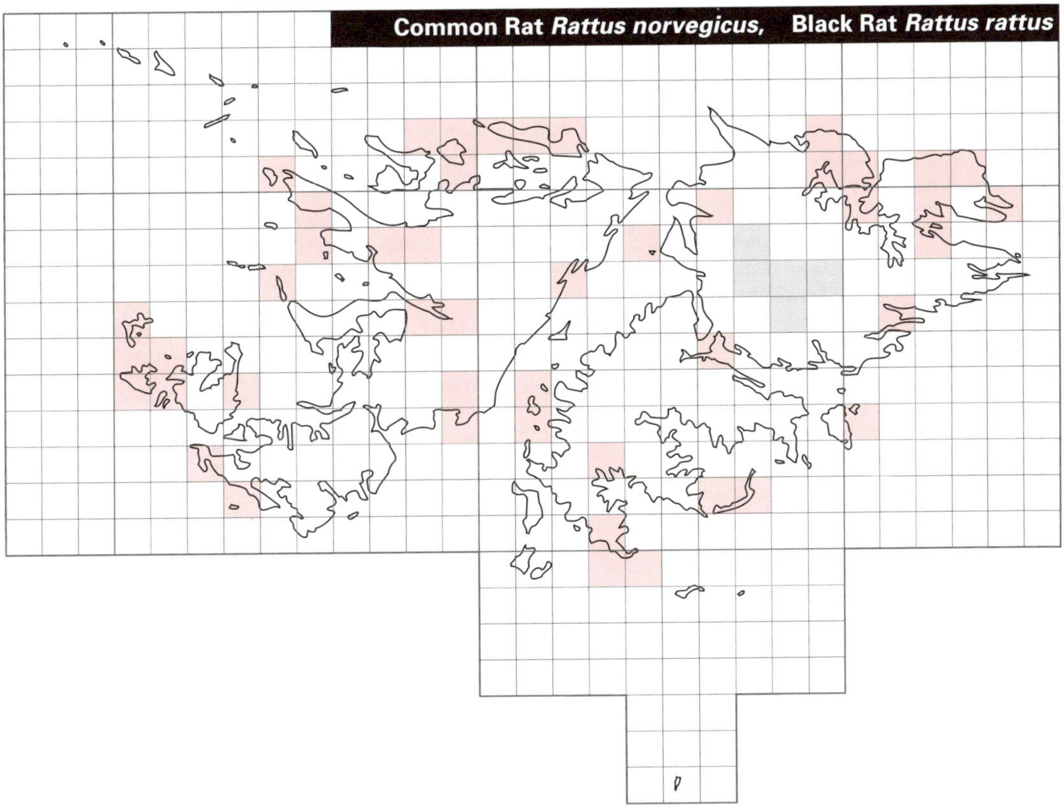

On the biggest island of the Arch Island group (in the season 1948/49) the Bertrands saw many rats of a grey-brown, russetty colour. One rat shot was 18 inches (46cm) from tip of nose to tip of tail. 'Dozens used to sit on tops of tussac bogs in early frosty mornings'. They ate the best part of a 30lb (14kg) tin of putty and were known to walk out and eat the snouts of Elephant Seals trapped in mud wallows. They also chewed through a bullock skin hung on a high wire in a shed. The skin fell down and the rats ate it all except for a pile of hairs. A successful rat trap was made from an upright 40 gallon drum with some water in the bottom and a tight paper fixed across the top. A cross was cut in the paper and a bone suspended over the centre. The Bertrands did not identify Black Rats but remarked that on the western Porpoise Island (off Bull Point, East Falkland) there were rough-haired rats, which may have been of either species. On New Island, they understood from a former owner, Jack Davis, that the rats were pied or black with smooth or rough fur.

Survey results and comments

Rats were recorded as present in 47 10-km squares, 20% of those surveyed. On Outer Northwest Island, in Falkland Sound, Simon Strickland camped for three days in October 1983. He reported 'brown rats' near their camp site on the beach at night. The survey records do not include the town of Stanley nor the 10-km squares in which the Arch Islands are situated and where the Bertrands saw many rats nearly 50 years ago. Rats were definitely present at the Stanley rubbish tip in 1983 where, according to official notices, they were being poisoned. The wide spread of records illustrates their presence at settlements on East and West Falkland, at some coastal regions (Bull Point and Volunteer Point) and on several of the offshore islands.

The Black Rat was identified in the Falklands, apparently for the first time, on New Island in 1986/87 by Kate Thompson (pers. comm.). It is not known whether this species occurs elsewhere in the Falkland Islands but it is likely to be present because, since 1850, it has been more commonly found aboard ships than the Norway Rat (Atkinson 1985).

Implications for Conservation of Falkland Islands Birds

Evidence of the detrimental effects of rats on Falkland birds is not easily obtained, partly because islands with rats are often those with populations of mice and cats and partly because most predation by Norway Rats on ground nesting birds occurs at night. There is however, some evidence from the 1983 field work on songbirds suggesting that their numbers are greatly reduced if rats are present (Woods 1984). Evidence from other countries is that rats can cause considerable damage. Black Rats, which are agile climbers weighing up to 180g, are known to have been responsible for the extermination of five species of forest bird within three years of colonising Big South Cape Island, New Zealand in 1962 (Moors 1985). Norway Rats weigh up to 450g and are much more likely to prey on birds nesting in burrows, such as the smaller petrels (Atkinson 1985). Some interest in eradicating introduced predators has developed recently among Falkland land owners. Sally Poncet reported in 1995 that a Tussac Islands Restoration Group of owners and farmers hoped to gain support in the Falklands and from abroad for the restoration of tussac habitat and the eradication of some island populations of rats (Poncet 1995).

3 House Mouse *Mus musculus*

History

Mice were probably introduced to the Falklands by early visitors, who inadvertently imported them within cargoes of grain or other foodstuffs. Darwin recorded that mice were present (Waterhouse & Darwin 1839) and one specimen was 'procured on a grassy cliff on East Falkland Island at the distance of a mile from any habitation'. He thought it 'singular that so delicate an animal should be able to subsist under the cold and extremely humid climate of the Falkland Islands and on its unproductive soil'. Darwin (1845) also stated that he had 'reason to suspect ... there is a field-mouse'. Vallentin (1901) stated that 'the common house-mouse *Mus musculus* is also very abundant in the wooden houses in Stanley' and he also wrote, 'the indigenous field-mouse, whose presence was suspected by Darwin, eluded my most careful search; neither did I ever hear of one ever been [sic] seen by the inhabitants'.

Cawkell (1960) noted that he had found mice on New Island 'feeding on dead sheep' and in Cawkell & Hamilton (1961) it is stated that 'the mice have taken to a completely wild life. In a remote place, six miles from an intermittently inhabited house, they were a nuisance in Hamilton's encampment and a specimen collected was identified at the British Museum as *Mus domesticus [= musculus*]. The mice are at least partly carnivorous. They have been known to eat parts of an Upland Goose hanging for the table and to eat beetles' and it 'is likely that the eggs and nestlings of smaller ground nesting birds are destroyed by these animals'. Fergus O'Gorman (1960) collected a specimen in the settlement at New Island on 3 May 1960 which was also identified at the British Museum as a House Mouse. It was a female of unknown age, with the following dimensions: body length 86mm; tail 73·3mm; left and right forefoot 20·1mm; left hindfoot 18·7mm; right hindfoot 18·2mm; ear lengths 13·6mm. The measurements were made as follows: forefoot measured with dividers, paw stretched flat; hindfoot from heel measured with calipers, paw stretched flat; ear measured from notch to tip, with dividers. He was informed that they had another House Mouse from the Falklands in the museum, collected in 1924 by AG Bennett. Fergus O'Gorman and I caught a mouse in diddle-dee on Stanley Common in 1957 which was also identified as a House Mouse.

Survey results and comments

The House Mouse was recorded as 'present' in 37 grid squares on both East and West Falkland and some of the offshore islands. On West Point Island in September 1983, there were heaps of chewed flower heads of tussac grass with mouse excreta intermixed, in large tussac plants, on top of the pedestal about 1m above ground. Roddy Napier reported that he had seen them far from the settlement under sheep carcasses and around the house in winter. Most observers merely stated that it was present but there were a few comments on numbers. These varied from '1,000' at Long Island Coast from Neil Watson in 1986/7, 'many' at Port Howard in 1984/5 from Andy Douse, 'abundant' at Steeple Jason in 1987/8 from Kate Thompson, and reports from Reg Davis at Salvador of about '200' in the 10-km squares VC09 and VD00 up to 1986. North Arm schoolchildren reported 'a lot' in the settlement in 1985/6. The only counts were of three at Black Shanty House, Fox Mountain in 1987/8 from Donna O'Daniel and of two found dead 'in a nest under the pillows' in the cabin at The Neck on Saunders Island by RG Thompson in 1994. It was also useful to have reports from several observers that the House Mouse was 'absent' from some islands, notably Beaver, George, Speedwell, Sea Lion, Lively, Outer Northwest, Dunbar, Low, Carcass, and South and Grand Jason. Kidney Island is considered to be free of mice. However, a House Mouse was found dead there, lodged in a shelf of a mist net set to trap Grey-backed Storm-Petrels, on 16 September 1983. The dimensions of the mouse, measured with dividers and ruler were: body and head 68mm; tail 80mm; hindfoot 17mm; ear 12mm. This mouse had grey-brown underparts. Possibly it had been caught on the mainland opposite by a Short-eared Owl, carried back to Kidney Island and dropped when the owl bounced against the

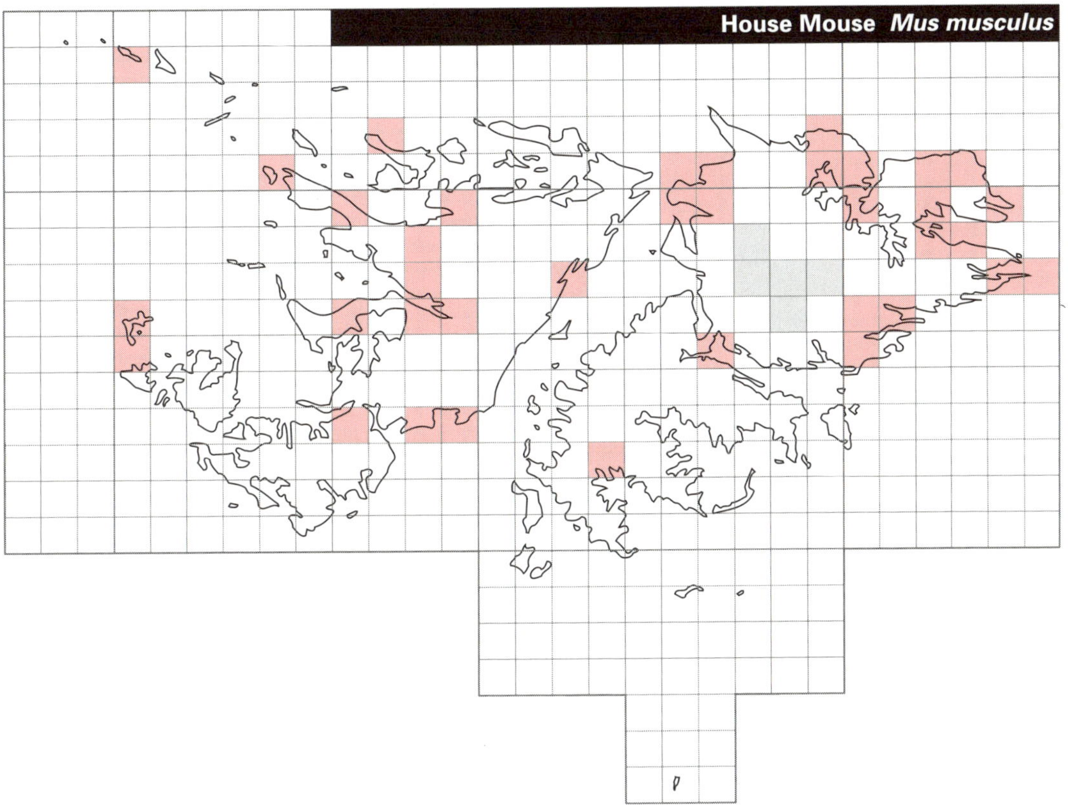

mist net. Short-eared Owls were calling that evening and had been caught in mist nets on other occasions.

Kitty Bertrand writing in *Warrah* (Wolsey 1986), posed the question 'Is there a Falkland Mouse?' Dr Hamilton had been told that such a mouse existed and it 'was said to have a white tummy'. He had asked her 40 years previously to look out for some! She described how in June 1986, after catching two 'ordinary' mice, she caught a small mouse with white underparts which she thought might possibly be an example of the 'Falkland Mouse'. However, this was in Stanley and Kitty decided that it was probably descended from a 'foreign mouse'. Soldiers had told her that, when the containers stacked along the grass verge outside her house were lifted, 'there were mice going in all directions'. There is still no evidence of an endemic Falkland Mouse.

Implications for Conservation of Falkland Islands Birds

There is no direct evidence that mice prey on small birds in the Falklands but, from the songbird census work on West Point and Carcass Islands in 1983, there is circumstantial evidence that mice have a detrimental effect. However, recent field work (Woods in press) in the Port Louis camp of East Falkland has shown that mice are a very important food source for the tiny resident population of Barn Owls. Mice are now so widespread that eradication would be almost impossible and every effort should be made to guard against their introduction to other islands.

Hares and Rabbits

During the survey, hares were found only on East Falkland and rabbits were present on three offshore island groups and in one mainland area of West Falkland. Rabbits were also reported from one square on East Falkland. One small island (30ha) in Choiseul Sound is called Hare Island and 11 islands, varying in size from a few hectares to 178ha, are named Rabbit Island on the DOS Maps. Few survey forms which included these islands were received and it would be interesting to know if the name was related to the presence of these herbivores. They were introduced to the Falklands for the purpose of providing sport and improving the diet and it is not known if they have caused significant damage to the habitat for birds. While the hares and rabbits seem to be thriving, there is no evidence that they have become pest species by competing with birds for food or burrows or causing soil erosion.

4 Brown Hare

Lepus capensis

History

On 26 December 1881 Sir William Wiseman wrote to his wife from the Falklands while commanding HMS *Dwarf* on an official visit to dissuade illicit sealing. He had visited Sparrow Cove (VC47), near Stanley as one of a shooting party 'to get hares and geese'. Wiseman said that they 'started a good many hares, but ... they were very wild and bolted a long way out of shot'. On 30 December they visited Lively Island and then landed on nearby Philimore Island, where five officers shot thirty hares though it was 'blowing and raining all the time' (Layman & Cameron 1995). When Sally Poncet and others visited Philimore Island in March 1994, no hares were seen, suggesting that they had been extirpated by hunters or had died out. Several were seen singly at different times in the sand dunes behind Yorke Bay, east of Canopus Hill, Cape Pembroke in the early 1960s. Between 1976 and 1984, Steve Whitley reported the presence of hares only in Johnson's Harbour camp and the Cape Pembroke area.

Survey results and comments

Hares were noted as present in 14 grid squares and possibly present in two others, all on the main island of East Falkland. Ten of these squares were adjacent to other squares from which hares were reported. None was recorded from West Falkland or the adjacent islands. The largest numbers reported were 'approximately 100' in the North Arm settlement area (UC32) by North Arm schoolchildren in 1985/86 and '30+' around Port San Carlos (UC69) by Gavin Hardcastle in 1986/87. Most other records were of 'a few hares'. The main centres for this species seem to be the southwestern section (North Arm camp), northwestern section (Port San Carlos camp) and the Mount Pleasant to Bertha's Beach area. Since the survey, further records have come from the Salvador and Port Louis camps of northern East Falkland.

5 European Rabbit

Oryctolagus cuniculus

History

In 1770 Erasmus Gower (1803) noted that, 'Pigs and rabbits were brought to Falkland Islands by ourselves, and multiplied exceedingly'. This was almost certainly on Saunders Island, which Gower reached on 16 April 1770, having previously visited either in January 1765 or in 1766, to establish the first British settlement. Clayton (1774) stated that there were 'some' rabbits on both Saunders and Marble [Pebble] Islands. In 1797 Edmund Fanning was at New Island. He commented that, 'several of the islands are also overrun with the white rabbits' (Fanning 1833). In 1836 Admiral George Grey visited New Island and found that it 'abounds in rabbits' and he remarked that 'rabbits run about Saunders Island in thousands' (Grey 1836-37). In 1833 and 1834 Darwin (1845) found that the introduced rabbit 'has succeeded very well; so that they abound over large parts of the island. Yet, like the horses, they are confined within certain limits; for they have not crossed the central chain of hills, nor would they have extended even so far as its base, if, as the Gauchos informed me, small colonies had not been carried there'. He saw grey, black and piebald individuals. Governor Moody (1842) reported from Port Louis that rabbits, 'which together with the rats and mice have evidently been imported, are literally in myriads in the vallies near the settlement, and are easily caught by dogs. Many skins were formerly exported, and it appears to have paid well'. In 1856 a passenger from the *Great Britain* rode from Stanley to Port Louis where he met a shepherd and his family who lived mainly on wildfowl and rabbits (Walker 1856-57).

Sir William Wiseman wrote on 19 January 1882 at Keppel Island, 'After luncheon I landed with the Doctor and Seymour to shoot. The island is alive with rabbits, but they are very wild and have such a lot of holes that the only chance of getting them is early in the morning or late at night. We spread out and walked up a valley that leads to some ponds in hopes of getting a few unwary bunnies, but the result was only one to Seymour's gun'. In the evening 'we spread out for bunnies again and managed to get ten between us by the time we got down to the boat' (Layman & Cameron 1995). A crew member of HMS *Dwarf* also wrote an account of their experiences, particularly noting hunting successes. They encountered rabbits apparently only at Keppel Island, where there were plenty but were 'very hard to get near' (Anon. 1882). Mowat (1896) visited all of East and most of West Falkland in January, surveying the sheep farms. He wrote: 'There are no rabbits to contend with here. You occasionally meet with a few reserved on some of the small islands for the man-o-war's men to shoot'.

Fergus O'Gorman (1960) visited New Island to study Fur Seals and collected two rabbit specimens there on 28 April 1960. He caught one behind the settlement and shot the other at the southern end of the island. Both were immature females, weighing about 1·45 kg (3lb 3ozs). They were identified as the European Rabbit by Mr Hill of the Mammal Department at the British Museum who informed Fergus that they were the first

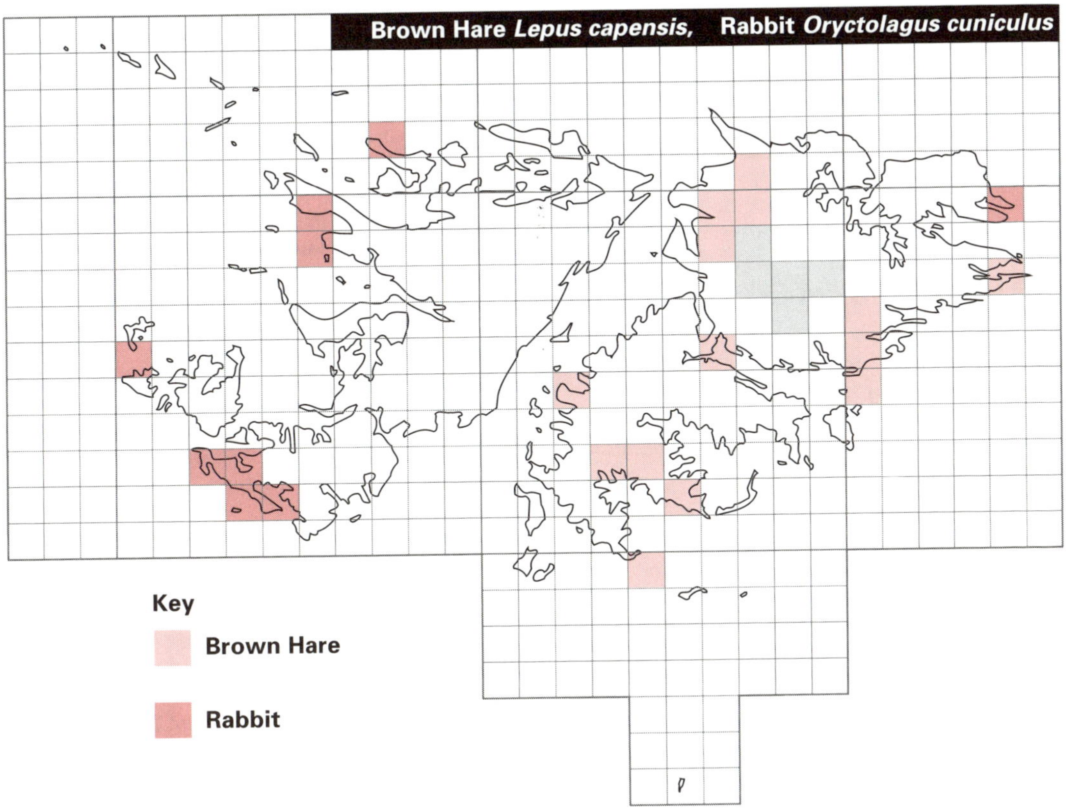

rabbit specimens from the Falkland Islands in the national collection. Between 1976 and 1984 Steve Whitley reported that there were rabbits on Flat Tyssen Island (UC14) in Falkland Sound, on New Island, at Inside Volunteer camp and northwest of Johnson's Harbour settlement (VC29). He also recorded their presence in two squares to the west and southeast of Port Stephens settlement and at Fanning Head, near Port San Carlos, East Falkland.

Survey results and comments
Rabbits were reported from only three island groups and two mainland areas. Their presence was confirmed at the northwestern point of Saunders Island, on the two Bense Islands northwest of Roy Cove settlement and on New Island where Bill Bourne and CPO Pringle reported hundreds in 1983/84. The largest area occupied was the southwestern part of West Falkland, between Ten Shilling Bay, south of Port Stephens and Cape Orford, covering four grid squares. At Port Stephens settlement, Brian Bewsher reported in a letter of 20 July 1985 that there was good cover for numerous rabbits in the gorse hedges. They were also reported from around the shanty in the Volunteer Point area of East Falkland, which is part of Inside Volunteer camp. Since the survey, rabbits have been confirmed as present at Port Louis settlement.

Implications for Conservation of Falkland Islands Birds
Both species are maintaining populations in different areas, though the population density of hares seems to be low, apart from the North Arm camp, and the only reports of large numbers of rabbits came from New Island. There is no evidence that grazing by hares or rabbits reduces the availability of food or vegetational cover for any Falkland birds while both contribute to the diet of some birds of prey. Studies in Australia have shown that rabbits are highly selective grazers, eating the most nutritious plants to below ground level, and wool production has decreased where rabbits are abundant (ANCA 1993b). There has been no systematic study of their food or breeding requirements in the Falklands and it would be useful to investigate the factors behind their successful naturalisation. Rabbits may compete for burrowing space with Magellanic Penguins in a few places, but the interaction between these species has not been studied. Hares and rabbits were introduced to the islands to provide sport and food for seamen and residents. They are excluded from any protection in the 1964 Wild Animals and Birds Protection Ordinance and there are no apparent grounds for altering this status.

Appendix B

Grid squares for which records are lacking

The following tables (10, 11 and 12), list those regions of the Falkland archipelago where survey work was either lacking or only partly covered the land area or the species likely to be present. The '■' mark indicates the full 100 sq km grid squares within the military practice area to which the public has no access and the abbreviations 'W' or 'E' show whether the region is located on or near either West or East Falkland.

Table 10. Grid squares with no records by 1992/93

Square	West or East Falkland	Area in sqkm		Area in sqkm	Offshore islands
TC13	W			<12	Pillar Bluff, Weddell; Sea Dog I
TC46	W	<24	Dunnose Head	<8	S end 1st Passage I
TC57	W			<4	Rabbit Rocks
TC62	W	<90	Gibraltar Hill to Saucepan Hill, NW of Port Edgar		
TC67	W			<8	Hummock & Middle Is
TD04	W			<4	NW tip of Steeple Jason I
TD52	W			<4	W tip of Sedge I
TD92	W			<4	Government Islet off Pebble I
UB29	E			<4	Tea Pt E end of Barren I
UC36	E	<4	NW tip Hope Cottage Rincon		
UC50	E	<4	Bay N of Driftwood Point	<4 <4	Islet 2.5 km W of Driftwood Pt, Great I & Button I
UC60	E	<8	Driftwood Point	<4	Cassie's & Driftwood Is, Cassard Pt, Bleaker I
■UC77	E	100	Mt Usborne/Rabbit & Rodeo Mts		
■UC78	E	100	3rd Corral (E) to 3rd Corral House		
UC81	E			<4	Shag Rock (S of Motley I)
■UC86	E	100	Black Rock House to Blue Mt and Mustard Mountain		
■UC87	E	100	Goat, Jacks & White Mts		
■UC97	E	100	Starvation Peak to Mt Simon		
UD31	W			<4	E end Pebble I, Tamar Pass (N)
VC25	E			<4	Outer Knob
VC46	E	<12	Seal Pt (E), Lake & Horse Points		

Table 11. Grid squares from which 1 to 5 species were recorded by 1992/93

Seventeen squares which were not fully surveyed are listed. Only two contain 100 sq km of land, two have substantial amounts of land and the others have only very small islands or peninsulas.

Square	West or East Falkland	Area in sqkm		Area in sqkm	Offshore islands
SD84	W			<4	Jason West Cay
SD94	W			<4	Jason East Cay
TC47	W			<12	1st Passage (N); 2nd Passage (E)
TC52	W	100	Tiger Bay Shanty/Mt Young/ Carew Harbour House		

TD22	W			<4	S Jason I (centre N) Elephant Jason (W tip)
TD23	W			<8	Flat Jason (most); The Fridays
TD30	W			<4	Gibraltar Rock (The Slipper)
TD31	W			<4	South Fur I
TD33	W			<4	Seal Rocks (most); NE tip Flat Jason
TD62	W			<8	Sedge I (most)
TD72	W			<4	Wreck Is
TD82	W			<4	Port Egmont Cays
UC02	E			<4	Elephant Cays; Stinker I
UC76	E	100	High Hill House/Ceritos House/ Table Rock (SE Mt Usborne)		
UD02	W			<4	White I (N of Pebble I)
UD10	W			<50	Broken/Golding (E)/East/ Middle/River Islands
UD52	E			<4	Eddystone Rock

Table 12. Grid squares from which 6 to 10 species were recorded by 1992/93

In 11 squares, between six and ten species were recorded. Six are inland squares and two contain substantial amounts of land. These six squares probably have more species present, though the three on West Falkland, with at least half of the land above 500 feet, are unlikely to have coastal species. Of the three on East Falkland, only one is mountainous. One is lowlying with large ponds and the other adjoins the San Carlos river and settlement and both squares could support many more species. The two squares with 20-50 sq km of land, one off West and one on East Falkland, probably support 20 or more species.

Square	**West or East Falkland**	**Area in sqkm**		**Area in sqkm**	**Offshore islands**
TC35	W			<20	Loop Ridge to Sealer Cove (Weddell I)
TC44	W	<4	Penguin Pt (W end Queen Pt)		
TC60	W			<4	Clump/Tussac/Natural Arch/ S end of Big Arch Is
TC98	W	100	Mt Robinson camp to French Flats		
TD61	W			<4	The Holy City & Elephant Pt (Saunders I)
UC07	W	100	Saddle House/Muffler Jack Mt		
UC17	W	100	Stewart's Rock/Green Mt/Tent Mt/ Mt Maria		
UC33	E	100	Wreck Gate/White Hill Gate/ Pidgeons Park Ponds		
UC41	E	<40	Cow Pt to Cattle Pt	<4	Cattle Point I
UC68	E	100	Inner & Outer Verde camp		
VC27	E	100	Challenger/Vernet/Kent/Estancia Mts		

Appendix C
Scientific names of plants and animals

Local name	**Scientific name**
Green algae	*Ulva* spp
Red algae	*Porphyra* spp
Giant Kelp	*Macrocystis* and *Lessonia* spp
Small Fern	*Blechnum penna-marina*
Tall Fern	*Blechnum magellanicum*
Brown or Short rush	*Rostkovia magellanica*
Diddle-dee	*Empetrum rubrum*
European Gorse	*Ulex europaeus*
Fachine	*Chiliotrichum diffusum*
Marram or Sand grass	*Ammophila arenaria*
Monterey Cypress	*Cupressus macrocarpa*
Native Box	*Hebe elliptica*
Sea cabbage	*Senecio candidans*
Thrift	*Armeria macloviana*
Tussac grass	*Parodiochloa flabellata*
White grass	*Cortaderia pilosa*
Lobster-krill	*Munida gregaria*
Camel-crickets	*Parudenus* spp.
Blowfly	*Calliphora* spp.
Mussels	*Mytilus* spp.
Argentine Shortfin Squid	*Illex argentinus*
Patagonian Squid	*Loligo gahi*
Blue Whiting	*Micromesistius australis*
Rock Cod	*Notothenia* spp.
Guanaco	*Lama guanicoe*
Patagonian Grey/Silver Fox	*Dusicyon griseus*
Southern Sea Lion	*Otaria flavescens*
South American Fur Seal	*Arctocephalus australis*
Elephant Seal	*Mirounga leonina*

Bibliography

Abbott, CC (1860) The Penguins of the Falkland Islands. *Ibis* 2:336-8.

(1861) Notes on the birds of the Falkland Islands. *Ibis* 1:149-167.

Anon. (1882) The cruise of HMS 'Dwarf' amongst the Falkland Islands. *The Falkland Islands Journal* 1971, eds. JA & JICB Jones, pp36-40.

Ansible Information (1996) *AnsibleIndex Plus* v4.10, indexing programme, Reading, RG1 5AU.

Ashford, JR, Croxall, JP, Rubilar, PS & Moreno, CA. (1995) Seabird interactions with longlining operations for *Dissostichus eleginoides* around South Georgia April-May 1994. *CCAMLR Science.*

Ashton-Tate (1984) dBASE II v2.41, database programme licensed through First Software Ltd., Pangbourne, Berkshire.

Atkinson, IAE (1985) The spread of commensal species of *Rattus* to oceanic islands and their effects on Island Avifaunas, pp35-81, in *Conservation of Island Birds* ICBP Tech. Pub. No.3, Cambridge.

Australian Nature Conservation Agency (1993a) *Cats in Australia*, ANCA, [formerly Australian National Parks and Wildlife Service] Canberra, April 1993.

(1993b) *Introduced wild animals in Australia*, ANCA, Canberra, May 1993.

Bangs, O (1919) Notes on South American Short-eared Owls. *Proc. New England Zool. Club* Vol.VI:95-98.

Beck, JR & Brown, DW (1971) The breeding biology of the Black-bellied Storm-Petrel *Fregetta tropica*. *Ibis* 113:73-90.

Beck, RH (1917) Bird Photographing on the Falkland Islands. *Amer. Mus. Journal* Vol.XVII:428-460.

(1918) Photographs of Falkland Island bird life. *Bird Lore* 2:1-8.

Bell, BD & Robertson, CJR (1994) Seabirds of the Chatham Islands, pp219-228 in *Seabirds on Islands*, BirdLife International, BirdLife Conservation Series No.1, Cambridge.

Bennett, AG (1922) Notas sobre aves sub-antarticas: I, Problemas que presenta la migracion en algunas especies de aves sudamericanos. *El Hornero* Vol.II(4):255-258.

(1926) A List of the Birds of the Falkland Islands and Dependencies. *Ibis* 12th series, No.2:306-333.

(1930) Nesting of the Grey-backed Storm Petrel *Garrodia nereis* Chubb. *Oologists Record* 10(4):79.

(1931a) Additional notes on the birds of the Falkland Islands and Dependencies. *Ibis* 13th series, No.1:12-13.

(1933) *The Penguin Population of the Falkland Islands in 1932/3*. Unpublished report to the Governor, dated 29 March 1933.

(1935) Two records from the Falkland Islands. *Ibis* 13th series, 5(2):436. [Cobb's Wren (?) nest on Kidney I.]

Unpublished ms. notes to 1938 on a copy of his 1926 List.

Bernsée, F (1855) Letter from Captain F Bernsée of ship *Courier* wrecked on SW point of East Falkland Island in April 1854. *Proc. Acad. Nat. Sci's Philadelphia* 7:287-8.

Beswick, NW (1993) Bird Watching around Mount Pleasant. *The Adjutant*, 22:56-65.

Bingham, M (1994) Falkland Islands Seabird Monitoring Program, *Summary Report SMP/4*. Falklands Conservation, Stanley.

(1995) Falkland Islands Seabird Monitoring Program, *Summary Report SMP/5*. Falklands Conservation, Stanley.

Blake, ER (1977) *Manual of Neotropical Birds* Vol.1. The University of Chicago Press, Chicago and London.

Bourne, WRP (1988) The effect of burning and grazing on the grassland birds of north-west Britain, the Falklands, and other oceanic islands, pp97-103 in *Ecology and Conservation of Grassland Birds*, ICBP Tech. Pub. No.7, Cambridge.

& Curtis, WF (1985) South Atlantic Seabirds. *Sea Swallow* 34:18-28.

(1986) South Atlantic Bird Islands. *Sea Swallow* 35:24-34.

(1987) Birds seen in Falkland waters 1982-87. (in litt.).

& Warham, J (1966) Geographical variation in the Giant Petrels of the genus *Macronectes*. *Ardea* 54:45-67.

Boyson, VF (1924) *The Falkland Islands*. Clarendon Press, Oxford.

Brooks, WS (1916) Two undescribed birds from the Falkland Islands. *Proc. New England Zool. Club* Vol.VI:25-27.

(1917) Notes on some Falkland Islands birds. *Bull. Mus. Comp. Zool. Harvard College* 61(7):135-160.

Brown, RGB, Barker, SP, Gaskin, DE and Sandeman, MR (1981) The foods of Great and Sooty Shearwaters in eastern Canadian waters. *Ibis* 123:19-30.

Burger, J & Gochfeld, M (1994) Predation and effects of humans on island-nesting seabirds, pp39-67 in *Seabirds on Islands*, BirdLife International, BirdLife Conservation Series No.1, Cambridge.

Carins, M (1974) The Blue Petrel *Halobaena caerulea* in the Falkland Islands. *Ardea* 62:239-241.

Carstairs, DN (1989) The status of the Rufous-chested Dotterel *Z. modestus* in the Falkland Islands. *Bull. B.O.C.* 109:166-170.

Cawkell, EM & Hamilton, JE (1961) The Birds of the Falkland Islands. *Ibis* 103a:1-27.

Cawkell, MBR, Maling, DH & Cawkell, EM (1960) *The Falkland Islands.* Macmillan, London.

Chapman, FM (1934) Descriptions of new birds from Mocha Island, Chile and the Falkland Islands. *Am. Mus. Novit.* No.762:8pp.

Chubb, C (1909) [No title: first published note on Cobb's Wren *Troglodytes cobbi*] *Bull. B.O.C.* 25:15-16.

Clark, G (1985) *S.R.V. 'Totorora' Expedition to Chile 1983* see Winser N & S, 1985.

(1988) *The Totorore Voyage; An Antarctic Adventure.* Century Hutchinson, New Zealand and London.

Clark, R (1986) *Aves de Tierra del Fuego y Cabo de Hornos; guia de campo.* L.O.L.A., Buenos Aires.

Clayton, SW (1774) A Short Description of Falklands Islands... Letter to P Stephens Esq. Public Record Office Adm 7/704.

(1776) An Account of Falkland Islands. *Phil. Trans. Roy. Soc. London* pp99-108, in *The Falkland Islands Journal* 1981, eds. S Miller, J Abbott & RN Spafford, pp4-7.

Cliffe, J (1988) *Bird Sightings in Falkland Islands* 23 June to 23 October 1984, Unpublished report.

Cobb, AF (1910) *Wild Life in the Falkland Islands.* Gowans's Nature Books No.26, Gowans & Gray, London & Glasgow.

(1933) *Birds of the Falkland Islands.* Witherby, London.

(1906-1923) Unpublished manuscript notes, held by the Falkland Islands Company, Stanley.

Coleman, GJ & Coleman, SC (1986) AMSTAT1 v1·0 and AMSTAT2 v1·2; Computer software statistics packages for hypothesis testing, Ashby-de-la-Zouch.

Collar, NJ, Crosby, MJ & Stattersfield, AJ (1994) *Birds to Watch 2: The World List of Threatened Birds.* BirdLife International, Cambridge.

Cory, CB (1918-1919) Catalogue of birds of the Americas and the adjacent islands. *Zool. Series,* Vol.II, parts 1 and 2. Field Museum of Natural History, Chicago.

& Hellmayr, CE (1924-1927) Catalogue of birds of the Americas and the adjacent islands. *Zool. Series,* Vols.III, IV & V. Field Museum of Natural History, Chicago.

Cott, HB (1953, 1954) Egg exploitation of Wild Birds. *Ibis* 95:409-449 & 643-675; 96:129-149.

Cramp, S (1985) *Handbook of the Birds of Europe, the Middle East and North Africa,* Vol.IV, Oxford University Press, Oxford.

Crawshay, R (1907) *The Birds of Tierra del Fuego.* Quaritch, London.

Croxall, JP (1992) Status and trends of Antarctic and Subantarctic seabirds, in *Report of the Eleventh meeting of the Scientific Committee for the Conservation of Antarctic Marine Living Resources (SCCAMLR)*

(ed) (1994) Penguin Conservation Assessment; Antarctic and Subantarctic Species. *Scientific Committee on Antarctic Research, Bird Biology Subcommittee.*

, Hall, AJ, Hill, HJ, North, AW & Rodhouse, PG (1995) The food and feeding ecology of the white-chinned petrel *Procellaria aequinoctialis* at South Georgia. *Journal of Zoology, London* 237:133-150.

, McInnes, SJ & Prince, PA (1984) The status and conservation of seabirds at the Falkland Islands, pp271-291 in *Status and Conservation of the World's Seabirds,* ICBP Tech. Pub. No.2, Cambridge.

& Prince, PA (1980) The food of Gentoo Penguins *Pygoscelis papua* and Macaroni Penguins *Eudyptes chrysolophus* at South Georgia. *Ibis* 122:245-253.

, Prince, PA & Baird, P (1985) The diet of the Southern Rockhopper Penguin *Eudyptes chrysocome chrysocome* at Beauchêne Island, Falkland Islands. *J. Zool. Soc. London* (A)(1985) 206:485-496.

Curtis, WF (1988) Highlights of a South Atlantic Tour 1986/87. *Sea Swallow* Vol.37:3-8.

Daciuk, J (1975) La fauna silvestre de las Islas Malvinas, in Campaña Cientifica en las islas Malvinas, 1974. *An. de la Soc. Cient. Arg.* Vol.199, (4-6):153-176.

Darwin, C (1845) *Journal of Researches into the Natural History and Geology of countries visited during the voyage of HMS 'Beagle' round the world.* John Murray, London.

Davies TH, Dickson, IA, McCrea, CT, Mead, H & Williams, WW (1971) *The sheep and cattle industries of the Falkland Islands.* Foreign & Commonwealth Office, Overseas Development Administration, London.

del Hoyo, J, Elliott, A & Sargatal, J (eds) (1992 & 1994) *Handbook of the Birds of the World.* Vols.1 & 2, Lynx Edicions, Barcelona.

Devillers, P (1978) Distribution & Relationships of South American Skuas. *Le Gerfaut*, 68:374-417.

& Terschuren, JA (1978) Relationships between the Blue-eyed Shags of South America. *Le Gerfaut* 68:53-86.

& Terschuren, JA (1980) Les pétrels géants *Macronectes* des îles Falkland et du sud de l'amérique du sud. *Le Gerfaut* 70:447-454.

Duggleby, TR (1987) *Falkland Islands Bird Observations, September 1986 -April 1987.* Unpublished report to Army Bird Watching Society.

Duncan, D (1992) *Penguins and Seashore Friends*, by 'Dolly Penguin', Duncan, Oban.

Fanning, E (1833) *Voyages and discoveries in the South Seas 1792-1832.* Salem Marine Research Society 1924 (Reprint).

Fjeldså, J & Krabbe, N (1990) *Birds of the High Andes.* Zoological Museum University of Copenhagen and Apollo Books, Svendborg.

Freycinet, LC (1824) *Voyage autour du monde. . . .1817-1820. . . . Uranie et Physicienne.* Zoologie, Vol.1, Paris.

Garnot, P (1826) Remarques sur la zoologie des îles Malouines... *Annales de Sciences Naturelles* 7:39-59.

Goodall, JD, Johnson, AW & Philippi, RA (1951 & 1957) *Las Aves de Chile.* Vol.1 & Supplement, Vol.2. Platt Establecimientos Gráficos, Buenos Aires.

Gould, J (1859) List of birds from the Falkland Islands. *Proc. Zool. Soc. London* 27:93-99.

& Darwin, CR (1838-41) *The Zoology of the voyage of HMS 'Beagle'.* Part 3 Birds:8-145.

Gower, E (1803) The Loss of the 'Swift'. *The Falkland Islands Journal* 1970, ed. WH Thompson, pp24-29.

Gregory, P (1991) *Penguin News* Vol.3 No.1, 25 January 1991.

(1994) Notes on new and scarce birds in the Falkland Islands 1988-1990. *Bull. B.O.C.* 114(1):12-20.

Grey, G (1836-7) *The diary of Admiral the Hon. George Grey.* Extracts in *The Falkland Islands Journal* 1969, ed. WH Thompson, pp54-68.

Hamilton, JE (1937) The Chilean Skua in the Falkland Islands. *Ibis* 14th series, 1:177-178.

(1939) Additions to the Falkland Islands List. *Ibis* 14th series, 3:139-140.

(1944) The House Sparrow in the Falkland Islands. *Ibis* 86:553-554.

(1951) The breeding place of *Pachyptila belcheri* Mathews. *Ibis* 93:139-140.

(Unpublished) MS notes on copy of AG Bennett's 1926 List.

Harradine, J (1976) Geese in the Falkland Islands. *The Falkland Islands Journal* 1976, eds. S Miller, J Smith & J Abbott, pp9-20.

Harrison, P (1983) *Seabirds: an identification guide.* Helm, London.

Hellmayr, CE (1921) Review of the Birds collected by Alcide D'Orbigny in South America. *Novitates Zoologicae* XXVIII:171-213, 230-276.

(1929-1938) Catalogue of birds of the Americas and the adjacent islands. *Zool. Series*, Vols. VI to XI. Field Museum of Natural History, Chicago.

& Conover, B (1942-1949) Catalogue of birds of the Americas and the adjacent islands. *Zool. Series*, Vol. I, parts 1 to 4. Field Museum of Natural History, Chicago.

Henry, A (1994) Recent Bird Records from the Islands. *The Warrah*, Newsletter of Falklands Conservation, No.5:12.

Humphrey, PS, Bridge, D, Reynolds, PW & Peterson, RT (1970) *Birds of Isla Grande (Tierra del Fuego).* Smithsonian Institution, Washington, D.C.

& Livezey, BC (1983) Giant Petrel *Macronectes giganteus* nesting in Chubut, Argentina. *Le Gerfaut* 73:3-8.

, Livezey, BC & Siegel-Causey, D (1987) Tameness of Birds of the Falkland Islands: An Index and Preliminary Results. *Bird Behaviour* 7:67-72.

Jouventin, P, Stahl, JC, Weimerskirch, H & Mougin, JL (1984) The seabirds of the French Subantarctic Islands and Adelie Land, their status and conservation, pp609-625 in *Status and Conservation of the World's Seabirds*, ICBP Tech. Pub. No.2, Cambridge.

Kerr, JA (ed.) (1994) Proceedings of the First Farmers' Forum on Tussac Grass and Sand Grass planting in the Falkland Islands. *Farmers' Association & Department of Agriculture*, Stanley.

Keymer, IF (1988) An investigation of Rockhopper Penguin mortality in the Falklands during the 1985-1986 breeding season. *Falkland Islands Foundation Project* report.

Layman, CH & Cameron, J (1995) *The Falklands and The Dwarf*. Picton Publishing (Chippenham) Ltd.

Madge, S & Burn, H (1988) *Wildfowl: an identification guide.* Helm, London.

Marchant, J, Prater, AJ & Hayman, P (1986) *Shorebirds: An identification guide to the waders of the world.* Croom Helm, London.

Marchant, S & Higgins, PJ (co-ordinators) (1990) *Handbook of Australian, New Zealand and Antarctic Birds*, Vol.1 Ratites to Ducks. Oxford University Press, Melbourne.

McAdam, JH & Walton, DWH (1990) *Ecology and Agronomy of Tussac Grass*, Department of Agricultural Botany, The Queen's University of Belfast.

Mearns, B & Mearns, R (1992) *Audubon to Xántus, The Lives of Those Commemorated in North American Bird Names*, Academic Press, London.

Moody, RC (1842) Despatch from Lieutenant Governor Moody to Lord Stanley, in *The Falkland Islands Journal* 1969, ed. WH Thompson pp4-33.

Moore, DM (1968) The Vascular Flora of the Falkland Islands. *British Antarctic Survey Scientific Reports* No.60. NERC.

Moors, PJ (1985) Eradication campaigns against *Rattus norvegicus* on the Noises Islands, New Zealand, using Brodifacoum and 1080, pp143-155 in *Conservation of Island Birds*, ICBP Tech. Pub. No.3, Cambridge.

(1986) Decline in numbers of Rockhopper Penguins at Campbell Island. *Polar Record* 23:69-73.

& Atkinson, IAE (1984) Predation on seabirds by introduced animals and factors affecting its severity, pp667-690 in *Status and Conservation of the World's Seabirds*, ICBP Tech. Pub. No.2, Cambridge.

Mowat, JT (1896) Letter from the 'Marlborough Express', in *The Falkland Islands Journal* 1970, ed. WH Thompson, pp33-35.

Müller, HH (1990) Ornithologische Notizen von den Falkland Islands. *Seevôgel* 11(3):41-46.

Munns, RPD (1989) Shot Striated Caracaras and Slaughtered Geese: the failure of Conservation. *Falkland Islands Birding News*. Summer 1989, Vol.1(3):18-19.

Murphy, RC (1936) *Oceanic Birds of South America.* Two vols. The Macmillan Company & American Museum of Natural History, New York.

Napier, RB (1968) Erect-crested and Rockhopper Penguins interbreeding in the Falkland Islands. *Bulletin British Antarctic Survey* 16:71-72.

O'Gorman, F (1960) The Fur Seal *Arctocephalus australis* on New Island, West Falkland Islands. (Unpublished report to Colonial Secretary, FI).

Orgeira, JL & Fogliatto, ON (1991) The Blacknecked Swan *Cygnus melanocoryphus* in Antarctica. *Marine Ornithology* 19:140-143.

Osborn, D (1990) *Falkland Island Birds.* Unpublished report, records from September 1989 to January 1990.

(1991) *Falkland Island Birds.* Unpublished report, records from January to December 1990.

(1992) Falkland Islands Birds, Commemorative Report. *Newsletter of the Royal Air Force Ornithological Society* No.55.

Oustalet, JFE (1891) *Mission scientifique du Cap Horn 1882-1883*, Vol.VI, Zoologie; Oiseaux.

Payne, MR & Prince, PA (1979) Identification and breeding biology of the diving petrels *Pelecanoides georgicus* and *P. urinatrix exsul* at South Georgia. *NZ Journal of Zoology* 6:299-318.

Peatfield, J (1981) Birds and Sea Mammals Report, *The Warrah*, Journal of The Falkland Islands Trust, ed. D. Davidson, No.1:7-19.

Pernety, AJ (1771) *The history of a voyage to the Malouine (or Falkland) Islands...* Translated for T Jeffreys, London.

Pettingill, ER (1960) *Penguin Summer*. Cassell, London.

Pettingill, OS jr (1960) The effects of climate and weather on the birds of the Falkland Islands. *Proc. XIIth Int. Ornith. Congress, Helsinki* 1958:604-614.

(1964) Penguins ashore at the Falkland Islands. *The Living Bird* 3:45-64, Cornell Laboratory of Ornithology.

(1965) Kelp Geese and Flightless Steamer Ducks in the Falkland Islands. *The Living Bird* 4:65-79.

(1974) Passerine birds of the Falkland Islands: their behaviour & ecology; *The Living Bird* 12:95-136.

(1975) *Another Penguin Summer*. Charles Scribner's Sons, New York.

Phillips, JA (1916) A new form of *Chloephaga hybrida*. *Auk* 33:423-424.

Prince, PA (1982) The Black-browed Albatross *Diomedea melanophris* population at Beauchêne Island, Falkland Islands. *Comité Nat. Franc. de Recherches Antarctiques* 51:111-117.

& Copestake, PG (1990) Diet and aspects of fairy prions breeding at South Georgia. *Notornis* 37:59-69.

& Croxall, JP (1996) The birds of South Georgia. *Bull. B.O.C.* 116(2):81-104

& Payne, MR (1979) Current status of birds at South Georgia. *British Antarctic Survey Bulletin* 48:103-118.

, Rothery, P, Croxall, JP and Wood, AG (1994) Population dynamics of Black-browed and Grey-headed Albatrosses *D. melanophris* and *D. chrysostoma* at Bird Island, South Georgia. *Ibis* 136:50-71.

Punta, GE, Saravia, JRC & Yorio, PM (1993) The diet and foraging behaviour of two patagonian cormorants. *Marine Ornithology* 21:27-36.

Ramsay, LNG (1913) Ornithology of the Scottish National Antarctic Expedition in, *Sci. Res. Scot. Nat. Ant. Exped.* 4:Sect.II:207-209.

Rasmussen, PC (1991) Relationships between coastal South American King and Blue-eyed Shags. *The Condor* 93:825-839.

Ridgely, RS & Tudor, G (1989, 1994) *The Birds of South America*. Vols.1 & 2. OUP, Oxford.

Salvadori, T (1895) *Catalogue of Birds in the British Museum*. Vol.27.

Saunders, HE (1877) On the *Laridae* collected during the Expedition, in Reports on the Collections of Birds made during the voyage of HMS *'Challenger'*, *Proc. Zool. Soc.* Nov.20, pp794-800.

Schiavini, A & Yorio, P (in press) Distribution and Abundance of seabird colonies in the Argentine sector of the Beagle Channel, Tierra del Fuego. *Marine Ornithology*.

Schlatter, RP (1984) The status and conservation of seabirds in Chile, pp261-269 in *Status and Conservation of the World's Seabirds*, ICBP Tech. Pub. No.2, Cambridge.

Sclater, PL (1860) Catalogue of the birds of the Falkland Islands. *Proc. Zool. Soc. London*, Part 28:382-391.

(1861) Additions and corrections to the List of Birds of the Falkland Islands. *Proc. Zool. Soc. London*, February 1861:45-47.

(1868) Extract from: The Secretary on additions to the menagerie, *Proc. Zool. Soc. London*, November 12, 1868:527-529.

(1881) List of Birds' Eggs obtained during the Challenger Expedition, *Appendix I to Reports on collections . . . 'Challenger'*: Birds.

& Salvin, O (1876) Revision of the Neotropical Anatidae. *Proc. Zool. Soc. London*, April 4 1876.

Sharrock, JTR (1976) *The Atlas of Breeding Birds in Britain and Ireland*. Poyser, Berkhamsted.

Siegel, S (1956) *Nonparametric Statistics: for the behavioral sciences*. McGraw-Hill, New York.

Sitters, HP (1988) *Tetrad Atlas of the Breeding Birds of Devon*. Devon Bird Watching and Preservation Society, Yelverton.

Smith, RIL & Prince, PA (1985) The natural history of Beauchêne Island. *Biol. J. of the Linnean Society* 24:233-283.

Snow, WP (1857) *A two years' cruise off Tierra del Fuego, the Falkland Islands, Patagonia and the River Plate*. 2 vols. South American Missionary Society, London.

Soames, B (1980) *Keeping Domestic Geese*. Blandford Press, Poole.

Southby-Tailyour, E (1985) *Falkland Islands Shores*. Conway Maritime, London.

Stanford, W (1989) *Observations of Birds in the Falkland Islands 1982-1987*. Army Bird Watching Society.

Stattersfield, AJ (1995) How important are the Falkland Islands for Birds? *The Warrah*, Newsletter of Falklands Conservation, No.7:4.

Stones, T (1993) Recent Bird Records from the Islands, *The Warrah*, Newsletter of Falklands Conservation, No.4:3

Strange, IJ (1968) A breeding colony of *Pachyptila turtur* in the Falkland Islands. *Ibis* 110:358-359.

Summers, RW (1985) The size and composition of Sheld-geese populations and their food consumption on different vegetation types in the Falkland Islands. *J. Appl. Ecol.* 22:1-17.

& Dunnet, GM (1984) Sheld-Geese and Man in the Falkland Islands. *Biol. Conserv.* 30:319-340.

& Grieve, A (1982) Diet, feeding behaviour and food intake of the Upland Goose and the Ruddy-headed Goose in the Falkland Islands. *J. Appl. Ecol.* 19:783-804.

& McAdam, J (1993) *The Upland Goose.* Bluntisham Books, Huntingdon.

Swales, MK (1965) The seabirds of Gough Island. *Ibis* 107:17-42 & 215-219.

Swann, HK (1924-30) *Monograph of the Birds of Prey.* British Museum (Natural History), London.

Thompson, D & Duck, CD (1995) Southern Sea Lions *Otaria flavescens* in the Falkland Islands: population size, foraging behaviour and diet. *Sea Mammal Research Unit*, NERC, Cambridge.

Thompson, KR (1989) An assessment of the potential for competition between seabirds and fisheries in the Falkland Islands. *Falkland Islands Foundation Project* report, FIF, Brighton.

(1990) Report on Seabird Monitoring in the Falkland Islands 1989/1990. *Falkland Islands Foundation Project* report, FIF, Auchterarder.

(1992) Quantitative analysis of the use of discards from squid trawlers by Black-browed Albatrosses *Diomedea melanophris* in the vicinity of the Falkland Islands. *Ibis* 134:11-21.

& Rothery, P (1991) A census of the Black-browed Albatross *Diomedea melanophris* population on Steeple Jason Island, Falkland Islands. *Biol. Conserv.* 45:39-48.

Thompson, RG (1994) Falkland Islands 3rd-26th August 1994. *Bulletin of the Army Ornithological Society* 2/94.

Tickell, WLN (1967) Movements of Black-browed and Grey-headed Albatrosses in the South Atlantic. *Emu* Vol.66(4): 357-367.

& Woods, RW (1972) Ornithological observations at sea in the South Atlantic Ocean, 1954-64. *British Antarctic Survey Bulletin* No.31:63-84.

Turner, A & Rose, C (1989) *A Handbook to the Swallows and Martins of the World.* Christopher Helm, London.

Vallentin, R (1901) Notes by a Naturalist on his Voyage to the Falklands and back, with remarks on the Fauna and Flora of those Islands. *Journal of the Royal Institute of Cornwall* Vol.14, No.47:339-364.

(1904) Notes on the Falkland Islands. *Manchester Memoirs* Vol.XLVIII, No.23:1-51.

(1924) Birds, pp 285-335 in *The Falkland Islands*, VF Boyson, Clarendon Press, Oxford.

Venegas, C & Jory, J (1979) *Guia de campo para las aves de Magallanes.* Instituto de la Patagonia, Punta Arenas.

Verheyden, C (1992) *List of the birds observed at the Falklands from 14 August to 15 September 1991.* Unpublished report.

Wace, RH (1917) *A List of the Wild Fowl of the Falklands.* Unpublished typescript at Natural History Museum Library, Tring.

(1921) Lista de Aves de las Islas Falkland. *El Hornero* II:194-204.

Walker, JS (1856-57) Three days in the Falkland Islands. *Naturalist (Morris)* 6:209-212 & 7:74-77.

Waterhouse, GR & Darwin, CR (1839) Zoology of the voyage of HMS '*Beagle*', Part II, Mammalia.

Weller, MW (1972) Ecological studies of Falkland Islands' waterfowl. *Wildfowl* 23:25-44.

(1975) Ecology and behaviour of the South Georgia Pintail *Anas g. georgica. Ibis* 117:217-231.

Wetmore, A (1926) *Observations on the Birds of Argentina, Paraguay, Uruguay and Chile.* Bulletin 133, Smithsonian Institution, Washington.

Winser, N & Winser, S (1985) *Expedition Yearbook 1983.* Expedition Advisory Centre, London.

Woehler, EJ (ed) (1993) The distribution and abundance of Antarctic and subantarctic penguins. *Scientific Committee on Antarctic Research*, Cambridge.

Wolsey, S (1986) *Warrah* 1985; Annual Report of Falkland Islands Trust, Stanley.

(1987) Recent reports in *Falkland Islands Trust News* No.5:7.

Woods, RW (1962) Unpublished report on banding of Black-browed Albatrosses at West Point Island, to U.S. Antarctic Research Program, Baltimore.

(1966) Letter to editor *Animals*, 8:224-225.

(1970a) The avian ecology of a tussock island in the Falkland Islands. *Ibis* 112:15-24.

(1970b) Great Shearwater *Puffinus gravis* breeding in the Falkland Islands. *Ibis* 112:259-260.

(1975) *The Birds of the Falkland Islands.* Anthony Nelson, Oswestry.

(1982) *Falkland Islands Birds.* Anthony Nelson, Oswestry.

(1984) A census of breeding Falkland Islands passerine birds in tussac grass. (Unpublished).

(1985a) Falkland Islands avian ecology expedition 1983, abstract in *Expedition Yearbook 1983*, Expedition Advisory Centre, London.

(1985b) Towards an Atlas of Falkland Islands Birds. *Falkland Islands Foundation Newsletter* No.4:10-11.

(1986) How many islands in the Falkland Islands ? *Falkland Islands Foundation Newsletter* No.5:8-9.

(1988) *Guide to Birds of the Falkland Islands.* Anthony Nelson, Oswestry.

(1989) Falkland Islands Trust & Foundation Breeding Birds Survey, Species review to 1987/88. *Falkland Islands Foundation Newsletter* No.8: 6-9.

(1990) Falkland Islands Trust & Foundation Breeding Birds Survey, Progress report after six seasons. *Falkland Islands Foundation Newsletter* No.10:10-11.

(1993) Cobb's Wren *Troglodytes (aedon) cobbi* of the Falkland Islands. *Bull. B.O.C.* 113:195-207.

(1995) Motley Island: on the Falklands Ecology Map. *The Warrah,* Newsletter of Falklands Conservation No.7:8-9.

Index

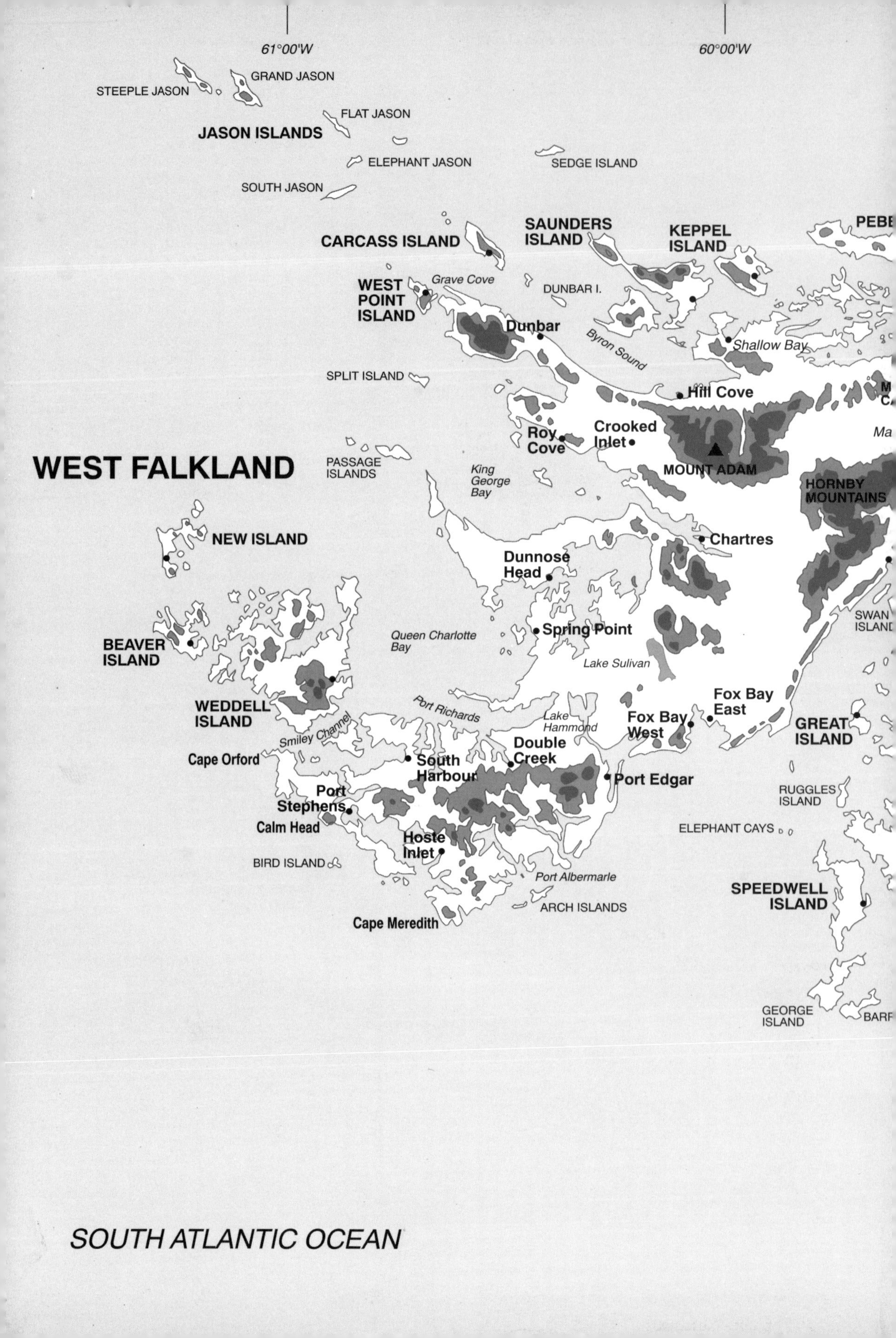

61°00'W
60°00'W
STEEPLE JASON
GRAND JASON
FLAT JASON
JASON ISLANDS
ELEPHANT JASON
SEDGE ISLAND
SOUTH JASON
CARCASS ISLAND
SAUNDERS ISLAND
KEPPEL ISLAND
WEST POINT ISLAND
Grave Cove
DUNBAR I.
Dunbar
Byron Sound
Shallow Bay
SPLIT ISLAND
Hill Cove
Roy Cove
Crooked Inlet
MOUNT ADAM
WEST FALKLAND
PASSAGE ISLANDS
King George Bay
HORNBY MOUNTAINS
NEW ISLAND
Chartres
Dunnose Head
Spring Point
Queen Charlotte Bay
BEAVER ISLAND
Lake Sulivan
SWAN
WEDDELL ISLAND
Port Richards
Fox Bay East
Fox Bay West
Lake Hammond
GREAT ISLAND
Smiley Channel
Cape Orford
Double Creek
South Harbour
Port Edgar
Port Stephens
RUGGLES ISLAND
Calm Head
ELEPHANT CAYS
Hoste Inlet
BIRD ISLAND
Port Albermarle
ARCH ISLANDS
SPEEDWELL ISLAND
Cape Meredith
GEORGE ISLAND
SOUTH ATLANTIC OCEAN